Archaeologies of Touch

Archaeologies of Touch

Interfacing with Haptics from
Electricity to Computing

David Parisi

University of Minnesota Press

Minneapolis

London

Portions of chapter 1 were previously published as "Shocking Grasps: An Archaeology of Electrotactile Game Mechanics," *Game Studies* 13, no. 2 (2013). Portions of chapter 2 were previously published as "Tactile Modernity: On the Scientific Rationalization of Touch in the Nineteenth Century," in *Media, Technology, and Literature in the Nineteenth Century: Image, Sound, and Touch,* ed. Collette Colligan and Margaret Linley (Farnham, UK: Ashgate, 2011), 189–214; copyright 2011. Portions of chapter 5 were previously published as "Fingerbombing, or 'Touching Is Good': The Cultural Construction of Technologized Touch," *The Senses & Society* 3, no. 3 (Berg Press, 2008), 302–27; copyright 2008.

Published by the University of Minnesota Press
111 Third Avenue South, Suite 290
Minneapolis, MN 55401-2520
http://www.upress.umn.edu

ISBN 978-1-5179-0058-8 (hc)
ISBN 978-1-5179-0059-5 (pb)

A Cataloging-in-Publication record for this book is available from the Library of Congress.

Printed in the United States of America on acid-free paper

The University of Minnesota is an equal-opportunity educator and employer.

22 21 20 19 18   10 9 8 7 6 5 4 3 2 1

For Kristen

What men make, men may also unmake; but what nature makes no man may dispute. To identify the role of human agency in the making of an item of knowledge is to identify the possibility of its being otherwise. To shift the agency onto natural reality is to stipulate the grounds for universal and irrevocable assent.

—Stephen Shapin and Simon Schaffer,
The Leviathan and the Air-Pump

Contents

Preface

Interrupting the Networked Body

The idea animating *Archaeologies of Touch* emerged at the intersection of two collisions. The first collision took place on December 27, 1990, when my sister—five years old at the time—was involved in a car accident that fractured her spine. The resulting swelling compacted a portion of her spinal column, leaving her paralyzed from the waist down (these generalizations are always insufficient: we would later learn that it was a level T 10–12 injury, meaning that she retained some sensation in her lower abdomen, but limited use of the muscles in that region). Bruising the spinal cord is different than severing it. In a severing, there is a complete break in the spinal cord—a total cutting off of one portion of the body from the command center of the brain. With bruising, the spinal cord compresses to the point of becoming functionally useless—the connection is still there, but the data channel is effectively compromised.

Though this crucial distinction between bruising and severing had serious implications for her therapy and potential recovery, it was unclear to me at the time. The important fact was the doctors' growing certitude in their diagnosis that her condition would be a permanent rather than a temporary one. Any chance at recovery would be evidenced by her responsiveness to the sensitivity tests they administered when they visited her hospital room; with each passing day that she failed to respond to their cutaneous probes, the prospects of recovery dimmed. My knowledge about the medical aspects of her condition (it was 1990, there was no WebMD, and I was a science-averse teenager) came in jumbled packets, assembled from whatever hazy conversations with doctors I happened to be present for, or relayed imperfectly by my parents. Regardless, I gradually came to understand it as a problem of information transmission: the vital connection between the brain and body had been damaged, impeding the successful reception of sensory data from the extremities and making the transmission of commands impossible.

The second collision had taken place a year earlier in the pages of the comic book *Iron Man,* when a bullet fired by a jilted ex-lover collided with Tony Stark's spine.[1] As with my sister's injury, Stark's spinal cord was not severed—the bullet's hit had not been a direct one—but it did leave Stark, like my sister, without the use of his legs. Prior to the injury, when not wearing his Iron Man armor, Stark had inhabited the stereotype of a billionaire playboy, image-obsessed and a tabloid celebrity. But after the injury, he struggled with his new identity as a paraplegic, quickly falling into a reclusive depression. Stark's adjustments to his chair over the subsequent months prefigured those my sister faced when she returned home from her three-month stay in the hospital. The reactions she received mirrored Stark's. In addition to having to navigate the constraints of a body with an imperfect sensorimotor system, she also had to confront the infrastructural impediments and social stigmatizations that accompanied the wheelchair. Stark, however, quickly rigged the Iron Man armor with "microcircuits" that allowed him the use of his legs. Though the writer never explained how the mechanism worked, Iron Man was back on his feet and in action, telling his friend "I never realized how much I'd miss walking, even after just a few weeks."[2] The solution, however, further bifurcated his identity: while in the armor, he had full use of his legs, but upon exiting it, he returned to his chair, where he once again faced both his physical limits and the steady stream of tabloid stories that presented the wheelchair-bound Stark as a pitiable and tragic figure.

With Stark's paralysis as a vehicle, the plot allowed writer David Michelinie to soberly confront some of the realities that those in wheelchairs face every day (though Stark's use of a $1.2 million *hover*chair exempted him from some of these struggles). But, wary of losing action-hungry readers, Stark's time in the chair was confined to a few panels each issue. After six short issues, the situation had become untenable, and Stark doggedly set himself to work leveraging his vast wealth to ameliorate his injury. The solution came in the form of an experimental chip implanted in his spine—"an organic computer"—that regrew the damaged nerve tissue, effectively restoring the connection between his brain and his lower extremities. The healing process, though miraculous, had not been instantaneous.[3] It required Stark to undergo an extensive rehabilitation routine that mirrored my sister's: both relearned to walk with the assistance of parallel bars. Kristen's rehabilitation required the further help of a custom-fitted brace that allowed her to be locked into a standing position—but this cruel exoskeleton, forced on her at a doctor's insistence, served more as a reminder of what had been lost than a promise of what could be regained.

In Stark's case, an impossible machine had solved an intractable problem: the chip overcame the data transfer interruption by regrowing the nerves, by hacking the body so that it would heal itself.[4] Given the rapid forward progress in computing and medicine, what his doctors did on the page seemed like it should have at least been plausible to replicate in the real-life lab. Deluded and desperate for hope, the ease of the solution ate at me as I watched my sister acclimate to her body's new parameters. If only we could find some way to restore the damaged connection between her brain and her limbs, all these struggles would melt away. If only someone at the Spinal Cord Injury Research Lab had been reading *Iron Man,* they could have set themselves to work on the project, treating paralysis, as Stark's doctors had, through an alchemic mixture of electrical engineering, computer science, and molecular biology.

The plotline that played out in *Iron Man*—comic book time proceeds according to the commercial logic of a monthly release cycle—mapped onto my embodied relationship with Kristen's new disability. When she returned home from the hospital after months of treatment and rehab, our rural home was unprepared to accommodate her limited mobility. A short flight of steep stairs loomed between the driveway and the doorway to the house. To get to her shared bedroom on the second floor of the house required ascending yet another flight of stairs. My role in all of this was to carry her up the stairs to the bedroom at night and then down the stairs in the morning, playing the part of Hodor to her Bran. My encounter with her disability, punctuated each month by a new issue of *Iron Man,* was very much an embodied one, as we struggled to adjust our living situation to her body's limits. Problems invisible to those who enjoy what we now call "ableist privilege" became impossible to ignore. The height of drinking fountains, the width of doors, and the pitch of ramps were suddenly reordered as unfair obstacles. And while my parents, buoyed by Congress's passing of the Americans with Disabilities Act the same year as my sister's injury, fought valiantly for these infrastructures to be made accommodating, Iron Man's chip lingered in my imagination, promising a way to circumvent the lifetime of hardships that lay ahead for Kristen.

Nearly ten years later, I learned of McLuhan's formulation of media as imperfect extensions of the human nervous system, functioning as fragmented, selective, and necessarily imperfect externalizations of the senses. Languages, as "stuttering extensions of our five senses" that vary in "ratios and wavelengths," had a particularly disruptive effect on the body.[5] McLuhan's metaphor seemed to be literalized in my sister's condition. The body's fragile internal communication network—the nervous system McLuhan thought to be "outered" by media technologies—frequently stutters. And sometimes it does not just stutter. Sometimes it breaks altogether. But as much as I found McLuhan's work to be revelatory, his discussions of touch were often maddeningly immaterial and metaphorical. Touch—as an ongoing feedback loop of action and reaction, of sensing and movement—was not a *mental* process of synesthetically translating between sense modalities, as McLuhan frequently claimed. Instead it was a fundamentally embodied and mechanistic process, involving the stimulation of dense and variegated mechanoreceptors, the exertion of muscles distributed in the joints, and the transmission of complex signals across nerves akin but irreducible to electrical networks.

In 1999, while still grappling with McLuhan's theories, I had the opportunity to attend a performance and lecture by the artist Stelarc. On stage, he invited participants to have their bodies taken over by an electrical keyboard that he wired to their muscles. With a simple key press, he passed current through the participant's muscles, allowing him to act as a puppeteer as he raised their arm over their head. They squirmed in discomfort as their body was taken over by the artist's crude machine. When we spoke after the performance, I asked him about the potential of such techniques to restore function to paralyzed limbs. He explained that even his primitive mechanism could effectively solve the control portion of the sensorimotor problem—routing control over the muscles in the paralyzed region of the body to a computer or to a functioning set of muscles was a relatively simple operation. Feeding the complex and variegated data of touch back from the disconnected limbs to the brain, however, proved nearly impossible. The haptic system resisted translation into a machine-legible code.

Our conversation simultaneously reawakened and dashed the hopes of my teenage years, but more significantly, it prompted me to begin kinetically reading about the emerging wave technologies that attempted to transmit tactile sensations over electrical networks. With the impending spike in computer processing power, their engineers promised, these machines would soon work perfectly, allowing a high-fidelity touch to flow effortlessly over the Internet. After several years of following this research, it became clear to me that the technology could not cash the checks written on its behalf. But the narrative being crafted around what I soon came to know as haptic interfaces seemed to echo McLuhan's utopian hopes for touch's elevation by the new electronic media in the 1960s. I soon invested myself in mapping the similarities between the two, what would be the first step in the project that eventually became *Archaeologies of Touch*.

Like superheroes, books have origin stories. The intersection between these two collisions—one fictional and the other, inescapably real—provided the inciting event for my eventual research into the antagonistic relationship between technology and touch. And although the disabling of my sister's body sparked my initial thinking, this is not a book about the social and infrastructural constructions of disability, though it could have been. It is not a chronicling of the efforts made at restoring the connections between the brain and its complex network of nerves broken by collisions, though that, too, provided a tempting pathway for this expedition. Instead, this is a book about touch's impossible complexity: about the dreams of connecting bodies seamlessly through networks, and about the recurrent efforts to unleash a touch transformed by technoscience as a positive, productive, and liberatory force. This Preface, then, serves very much as an acknowledgment that, without vicariously encountering touch's absence through Kristen's condition, this book might never have been written.

Acknowledgments

Archaeologies of Touch took shape initially during my doctoral work at New York University, under the careful and tireless supervision of Arvind Rajagopal, Alexander Galloway, and Allen Feldman. The breadth of knowledge and sharpness of thinking they brought to my work reverberates through each page. My deep appreciation goes to the University of Minnesota Press—especially Doug Armato for the enthusiasm, expertise, and support he brought to the project and Erin Warholm-Wohlenhaus for her guidance throughout the editorial process. The detailed insights provided by Victoria Simon, Jonathan Sterne, and Phillip Thurtle during the time they generously devoted to reviewing *Archaeologies of Touch* proved essential to helping the manuscript realize its full potential. Dawn Martin's skillful work on the index will benefit future researchers interested in this topic.

The College of Charleston provided consistent financial support for this project, including Faculty Research and Development Grants in 2015 and 2016, and annual Dean's Grants from 2010 to 2015 that funded conference travel and research trips. My year as a postdoctoral fellow at the Rutgers University Center for Cultural Analysis afforded an opportunity for dedicated reflection on the project, and I thank Meredith McGill for her support during my time there. Participation in the Texture Matters conference at the University of Vienna shaped much of my argument in Interface 4, and I thank Klemens Gruber,

Jana Herwig, and Antonia Lant for their gracious hosting. I also thank Victoria Fu and Susanne Saether for inviting me to the Touching the Screen conference at the University of Oslo. Larissa Hjorth and Ingrid Richardson's Haptic Play conference at Royal Melbourne Institute of Technology proved immensely valuable as well in prompting me to think more broadly about the ways that touch informs research into practices of play. My sincere appreciation to James Hay and Vincenzo Mele for including me in the Media Archaeology summer school at the University of Pisa; exchanges with the faculty and student participants sharpened my thinking on the relationship between media archaeology and laboratory practice.

As psychologists, haptic interface designers, and academics have lamented for generations, investigating touch can be a lonely and isolating affair, with collaborators few and far between. I was fortunate to exchange ideas with Mark Paterson and Jason Archer during our work on the Haptic Media Studies issue of *New Media & Society*. My appreciation also goes to Susan Lederman and Mandayam Srinivasan, who generously detailed their experiences working on computer haptics during the field's early years.

Since arriving at the College of Charleston, my departmental colleagues and their families have provided consistent intellectual engagement and warm friendship. I thank especially Erin Benson, Mike Lee, Ryan Milner, Sarah Milner, David Moscowitz, Leigh Moscowitz, and Robert Westerfulhaus. A huge and heartfelt thanks to all who offered their ideas and energizing support along the way: everyone from my years in New York, especially Brent Boylen, James Cooney, Jason Schogel, and Katherine Sepp; the last NYU Media Ecology cohort, in particular Sam Howard-Spink, Rob Jones, Devon Powers, and Michael Zimmer; Alex Jager; Rob Shaffer; and the Slacker collective, Stephanie Boluk, Jacob Gaboury, Patrick LeMieux, Laine Nooney, and Carlin Wing.

My deepest gratitude goes to my family for their unwavering encouragement—my mother Anita and my father Peter, my sisters Laura and Kristen, and my niece Ophelia have each inspired me with their continued strength in the face of adversity. Finally, to my loving wife April Bisner, who patiently read every page of the manuscript and talked through new ideas with me at their moments of spontaneous generation: I could not have done this without you.

Introduction

Haptic Interfaces and the Quest to Reinscribe Tactility

> Within the next five years, your mobile device will let you touch what you're shopping for online. It will distinguish fabrics, textures, and weaves so that you can feel a sweater, jacket, or upholstery—right through the screen.
> —Robyn Schwartz, "IBM 5 in 5 2012: Touch"

> How is one to believe that touch cannot be virtualized?
> —Jacques Derrida, *On Touching—Jean-Luc Nancy*

In a 1965 address to the International Federation for Information Processing, computing pioneer engineer Ivan Sutherland detailed his vision for an immersive, computer-controlled "ultimate display" that would present information for "as many senses as possible."[1] Sutherland's talk famously provided the blueprint for what would later be termed "virtual reality," prompting investigations into immersive stereoscopic displays, motion-tracking input systems, along with a host of other human–computer interaction schematics. Assessing the impact of Sutherland's address in a 2009 retrospective, science fiction writer-cum-media historian Bruce Sterling referred to it as "a seed bomb of emergent technologies,"[2] and proceeded to annotate the text of Sutherland's talk with repeated examples of computing devices that had effectively realized Sutherland's immense vision. In the category of "still doesn't exist yet," Sterling placed the most radical component of the imagined display: a complex mechanism to render computational data for tactual senses that Sutherland termed the "kinesthetic display." The kinesthetic display, as he envisioned it, would capture the movements of the human body, and in response, project forces back onto it, effectively simulating the body's physical interactions with matter. Using this "force feedback" system—rudimentary versions of which were already employed in the 1960s to allow for the dexterous remote manipulation of hazardous material—the computer could display complex objects to the user's sense of touch. In its ideal form, the objects presented by the display would have such a high degree of haptic fidelity that "a chair would be good enough to sit in" and "handcuffs . . . would be confining." Even more striking, however, was Sutherland's suggestion that a bullet presented by the display "would be fatal."[3]

Fifty years after Sutherland's address, with his employer poised to release the commercial version of its virtual reality headset, Oculus chief scientist Michael Abrash took the stage at Oculus's second annual developers' conference. He tackled the twofold task of simultaneously laying out the current state of the art in virtual reality research, while also projecting a realistic path forward for the technology's future iterations. In dizzying detail, Abrash explained how virtual reality displays merged technical knowledge with perceptual psychology to effectively and convincingly "drive the senses." Imitating Sutherland's 1965 address, Abrash considered the potential of current-generation displays to create convincing illusions for the different sense modalities. When he came to touch—which computer scientists had taken to calling *haptics* in the late 1980s—Abrash noted its centrality to the virtual reality enterprise, telling his audience that "haptics is at the core of the way we interact with our surroundings, and without it, we'll never be fully embodied in a virtual world." But he then made a curious and startling admission: current-generation virtual reality systems—including the Oculus Rift system that would be released for sale a few short months later—were woefully inadequate to the vital task of synthesizing the felt materiality of virtual worlds. "As important as haptics potentially is for VR," Abrash explained, "it's embryonic right now. There's simply no existing technology or research that has the potential to produce haptic experiences on a par with the real world." Uttered at an event intended to hype virtual reality's purported potential to revolutionize human communication, Abrash's statement seemed an oddly deflating concession. Abrash himself, however, remained upbeat and hopeful, suggesting that the challenge presented by haptics would eventually be solved through "breakthrough research." He promised his audience that "the first VR haptic interface that really works will be world-changing magic on par with the first mouse-based windowing system."[4] Still, the problem seemed insurmountable and enduring.

In the fifty years between Sutherland's 1965 address and Abrash's in 2015, computer scientists, roboticists, engineers, and psychologists set themselves to work on the project of writing touch feedback into computing. Contrary to Abrash's narrative, they have had some significant—and some more modest—successes along the way, formally establishing computer haptics as a new discipline in the 1990s, incorporating vibrating "rumble" motors into more than 500 million videogame controllers distributed around the world, building high-fidelity haptic devices for use in surgical training and remote manipulation, making somewhat effective cybersex machines, developing prosthetic limbs capable of feeding complex tactile sensations back to their wearers, and embedding vibration feedback mechanisms in touchscreen interfaces as a means of approximating the sensations produced by pressing buttons and keys. In spite of the widespread domestication of these haptics applications, the popular and scientific narratives mobilized around haptic interfaces continually portray them as technologies belonging to an imminent but perpetually deferred future, with haptics researchers still questing after an elusive "Holy Grail"[5] of touch interfacing that is only hinted at by the rudimentary forms of force and vibrational feedback present in current-generation technologies.[6] As in Abrash's framing, finding this Holy Grail promises to bring about a drastic upheaval not just in human–computer interaction but also in a whole range of social relationships, consumption habits, labor practices, and aesthetic forms.

Archaeologies of Touch addresses this current technohistorical situation—the consistent efforts made at weaving touch into computing systems, the persistent failures and circumscribed successes of engineers as they attempted to realize this end, and the continued utopian hopes mobilized around touch feedback computing—by describing the gradual emergence, over four centuries, of a formalized technoscientific haptics that provided the groundwork for the twentieth-century project of computer haptics. At the distal pole of this genealogy sits the eighteenth-century cultivation of a practiced epistemology of electric shock, which proved instrumental first to the production of belief in and knowledge about electricity, and shortly thereafter, to the creation of new techniques for studying the functions of the human sense organs. At the proximal pole lies the rapid embedding, beginning in the late twentieth century, of a computational haptics in a range of digital media interfaces, including virtual reality displays, mobile communication devices, videogame controllers, smartphones, scientific visualization machines, wearable computers, and cybersex devices. Between these poles, a new technological *haptic subject* emerged that served to both mark and steer the drastic changes touch underwent as it became increasingly an "object-target"[7] of scientific knowledge, engineering and design practice, bureaucratic management, therapeutic discourses, and commercial investment. This haptic subject embodies the self-conscious efforts scientists, engineers, and marketers made to transform touch, as they sought to give tactility a new utility in a political economy of sensations vital to a society with a growing dependence on the efficient circulation of information through sensing bodies. In 1999, confronting the implications of digital touch technologies that appeared poised to wash over and transform culture, Cathryn Vasseleu grappled with the fundamental question of "under what terms is touch admitted into cybernetic telecommunications networks?"[8] *Archaeologies of Touch,* in positing this haptic subject, takes up Vasseleu's prompt by looking backward to the technohistorical processes and discourses that forged a touch capable of being rendered, if only in selective fragments, through computing machines. In light of this longer history, the recent attempts at engineering a computerized tactility become only one stage in an overarching project pursued by subsequent generations of researchers—the touch admitted into cybernetic telecommunications networks is one that has already been thoroughly reshaped by its repeated interfacing with science and technology.

Five Phases of Interfacing

The story I tell of haptic interfacing's technogenesis is organized around a linear chronology that charts touch's passage through five successive phases of interfacing. Each stage entailed a generative contact between touch and a new set of institutionally and materially grounded discourses. By moving through these phases, I show the technoscientific haptics that underpins and animates the rise of computerized touch to be not unique to the age of computing. Instead it is a product that emerged piecemeal out of a gradually cohering body of scientific and technical research aimed self-consciously at producing an objective mapping of the human haptic system. Locating contemporary haptic interfaces in this overarching tradition calls attention to touch's positioning as both a target and source of scientific knowledge, emphasizing the disciplinary techniques and protocols that allowed it to gain expression as a sense capable of serving a utilitarian function in electronic communication networks. Touch was transformed into haptics first by its deployment as an instrument of scientific investigation, then through its enclosure in the framework of an objective, positivist episteme, and later by its articulation in advertising and marketing discourses. These processes allowed touch—via the new haptic subject—to be "made adequate"[9] to the new demands placed on it in successive epochs.

The first phase, which began with the use of touch to register the charges produced by eighteenth-century electrostatic generators and electrical batteries, involved the cultivation of a practiced tactile sensitivity to electrical shocks. This epistemology of shock underpinned both scientific studies of electrical phenomena and the growing cultural belief in the new and mysterious force. The electrotactile machines of the eighteenth and nineteenth centuries were built with the expectation that their users' bodies would provide not simply a binary registering of electricity as either present (shock) or not present (no shock), but further, that they would acquire a fine sensitivity to variations in the strength and character of the electrical currents that struck them. Users who came into contact with early electrical machines articulated these gradations in experience by developing a detailed semi-standardized language of shock that allowed for the transcription, circulation, and comparison of experimental encounters. By emphasizing tactility's primacy in experiments with machines like the Leyden jar, electrostatic generator, and voltaic pile, I show how the operation of an electrotactile subjectivity fueled both the spreading interest in electrical machines and the knowledge produced as a result of these shocking human–machine contacts. In contrast to the standard account of psychic shock frequently rehearsed in media theory, this model of shock as embodied, tactile, and epistemic—drawn from both scientific and medical discourses around electrical machines—emphasizes the generative function of shock over and against the formulation of shock as a traumatic effect of encounters between bodies and new technologies.

The second phase of interfacing concerns touch's passage into what I term a tactile modernity, as a new set of knowledge-producing apparatuses were set upon the body with the intent of yielding objective scientific knowledge about the operation of the tactual senses. Touch's isolation in the research laboratories of nineteenth-century psychophysicists facilitated the gradual accumulation of data about touch, culminating in the proposal of the term "haptics"—suggested in 1892 by Max Dessoir and defined in 1901 as the "doctrine of touch"[10]—as a response to the accumulating abundance of scientific knowledge that accrued as a product of lab experimentation. Initiated by the "epoch-making"[11] two-point threshold experiments Ernst Heinrich Weber carried out in the 1820s, this new epistemological framework captured touch through a structured adherence to experimental protocols. In the lab, experimenters divided touch into isolatable subcomponents, with pressure, weight, temperature, pain, and movement each sectioned off from one another through the use of increasingly specific laboratory instruments. Where the first interface involved the simple but structured mobilization of touch—a practiced epistemic electrotactility—as a means of producing experimental knowledge about electricity, the second interface turned the gaze of experimentation (with its attendant apparatuses and machines) inward on touch itself. Electricity, which had been an object revealed through tactile investigation, now became a means of investigating touch: applying electrical charges to the skin provided new insights into the mechanisms responsible for tactual perception.

The third interface, dated to the middle decades of the twentieth century, was constituted by a productive contact between touch and technical communication systems, where engineering psychologists designed machines capable of routing data through a touch now reconceived of as a channel for the transmission of information. The overarching aim in this research was to divine what Frank Geldard termed in a 1956 address "the tongue of the skin": a set of machine-rendered vibrotactile or electrotactile signals that the skin would be able to reliably distinguish between and assemble together as the building blocks for a tactile language. Some of the new machines built in the service of this project could pass speech sounds through the fingers, others functioned by using vibrations to project tactual images onto the torso, while yet another category of devices employed Morse code–like schemes for arbitrarily linking letters to vibrotactile signifiers. The psychologists in this third phase of interfacing employed the data and methods inherited from nineteenth-century psychophysical investigations of touch to provide the foundation for their various communication systems, improving on and repurposing many of the apparatuses developed by the prior generation of researchers. Where the search for knowledge about touch as an end in itself defined the second phase of interfacing, this third phase instrumentalized that knowledge, intent on opening touch up to new flows of data. Through the structured solicitation of artificially generated tactual sensations, nineteenth-century touch machines helped quantify and map touch's discriminatory capacities, and by doing so, laid bare its hard-coded ability

to register the differences between unnatural sensations. By assign-
ing values to these stimuli (and teaching their experimental subjects
to concretize associations between machinic tactile signifiers and
linguistic signifieds), touch communication researchers gave the
science of touch a utility it previously lacked. Researchers like
Geldard formulated the new communication systems as necessary
responses to the increasingly taxing burdens that modern media
were placing on the visual and aural channels, with touch providing
an ameliorative means of "transmitting intelligence."[12]

In the fourth phase of interfacing, initiated by the blueprint for
the ultimate display Sutherland laid out in 1965, computer scientists sought to make touch experiences into something that could
be stored, transmitted, and synthesized by computers, building
machines capable of simulating the physical materiality of objects
that existed only in the electronic realm of computer memory.
Appropriating robotic machines used for the remote manipulation
of hazardous nuclear materials, these engineers extended touch
into computer-generated environments. The new interfaces
allowed their users to feel the weight, shape, temperature, and textures of virtual objects. Though they initially proceeded unaware
of the previous century's scientific investigations into touch, the
computer scientists and roboticists working on touch feedback
computing eventually learned of rich research tradition designated
by the term haptics. Owing especially to the efforts of cognitive
psychologists like Susan Lederman and Roberta Klatzky, interface
designers began partnering with hapticians to build more effective
touch feedback interfaces that were informed by the principles of
both electrical engineering and haptic perception. In 1990—nearly
one hundred years after the Berlin psychologist Max Dessoir
had proposed "haptics" as the name for the scientific study of
touch—MIT engineer Mandayam Srinivasan dubbed this emerging field of human–computer interaction "computer haptics,"
effectively fusing the positivist tradition of studying touch and the
practice of building machines that could extend and stimulate the
various submodalities of the human haptic system. Consequently,
the nineteenth-century model of touch—as one broken down
into experimentally isolatable components that could be indepen-

dently stimulated by machines—was inscribed into the design of haptic human–computer interfaces,[13] with separate mechanisms, dedicated algorithms, and coding languages devoted to each of touch's constitutive parts. Although their designers aspired to make touch machines that would create holistic and accurate representations (as Sutherland had called for with the ultimate display), in practice, these machines were only able to act on limited subsets of the haptic system. Any reconstruction of touch, then, entailed strategically selecting which body parts the haptic interface system should interact with, and which of the various haptic submodalities (pressure, vibration, temperature, pain, etc.) it should target for stimulation. Where researchers in the third phase had sought to instrumentalize touch as an information-reception channel, translating audio, visual, and linguistic data for transmission through the skin, efforts in this fourth phase represented a self-conscious departure from the prior tradition—or, as Marvin Minsky explained in his famous 1980 essay "Telepresence," it was time to abandon the goal of transcoding images, sounds, or words for the skin and instead move to the project of "translating feel into *feel*."[14]

The fifth phase of interfacing concerns the efforts made by advertisers in the twenty-first century to produce a demand and desire for touch-based interfaces. In this phase, marketers working for digital technology firms like Nintendo, Apple, Hewlett-Packard, and Immersion Corporation crafted an image of the cultural sensorium in a state of urgent crisis that touch interfaces were uniquely qualified to alleviate. According to the narratives presented in these advertising campaigns—which featured slogans like "Touching is Good," "Touching is Believing," and "Touch the Future"—the sense of touch had been forgotten, left behind, and marginalized by a media interfacing schematic overdependent on audiovisual technologies. In a McLuhanesque maneuver, they claimed that the cultural sensorium could be rebalanced through the active embrace of touch interfacing. Collectively, these ads sought to create a new mode of haptic subjectification that would foster a desire in consumers to reconnect with their lost sense of touch. Touch interfacing becomes instrumentalized simultaneously as a marketing strategy and a means of regaining a lost sense modality. But while these ads portrayed touch as a way to rediscover something ancient, primitive, and pre-rational, they also fetishized touch—in its technologized reincarnation—as a marker of the consumer's passage into a utopic future of fully embodied presence in digital worlds. The attempt to make haptics into a mass-marketed technology involved the ongoing construction and continual reaggregation of a haptic subject—through practices of user-centered design—and the activation of a haptic subjectivity that desired awakening through the adoption of new interface technologies. In the fourth phase, haptics had been confronted primarily as a design problem; in the fifth phase, it became a marketing challenge. At the same time, however, the design challenge persisted, with growing numbers of researchers taking up the task of building more effective haptics applications. By 2010, engineers and developers around the world had established nearly fifty haptics labs spread out across over a dozen countries, with the increasing need to develop new mechanisms for communicating tactile sensations through mobile touchscreens providing fresh infusions of capital into the computer haptics project.[15]

Throughout these five phases of interfacing, I am concerned with mapping a range of interrelated developments around touch's technogenesis: the institutionalized and formalized knowledge-production networks that rose up around touch, the new intellectual and financial resources funneled into the study of touch, the training and regimentation of tactility demanded by the new machines, and the motivations—explicit and implicit—of the various researchers who set themselves to work at the immense challenge of bringing touch under the control of scientific and technical apparatuses. I want to understand what can be thought of, borrowing from Hans-Jörg Rheinberger, as the "experimental situation" and "experimental system" that redefined and reconstituted touch—a "reasoning machinery" consisting of "the dynamic body of knowledge, the network of practices structured by laboratories, instruments, and experimental apparatuses."[16] Through the case studies in each phase, I bring these "materialities of research" and "epistemic practices"[17] into dialogue with the articulated theoretical and practical aims of investigations into tactility. Each trial in the experimental system yielded not only new data about touch but also prompted the refinement of experimental protocols, the formulation of new research questions, and the development of new apparatuses that, taken together, served to specify touch with increasing detail.

While the structure I am imposing on these phases suggests that each constitutes a distinct moment in touch's history, they exist in continuity with one another, linked by direct intellectual, biographical, and institutional connections among the different actors associated with each phase. When the Italian physicist Alessandro Volta first invented his electrical battery in 1799 he immediately suggested that, by applying its currents to the various organs of the human sensorium, the battery could be a tool for the generation of new knowledge about the nature of sensory perception. To illustrate these possibilities, Volta repeatedly applied the battery to his eyes, ears, nose, tongue, and skin, providing detailed accounts of the sensations produced with each contact. Two decades later, Volta's trials proved inspirational for E. H. Weber in his protopsychophysical research on tactile perception, as Weber carried out similar investigations on his own body. By doing so, Weber staged the subsequent development of electrical machines specifically designed for the stimulation of the tactile senses. When Geldard began his search for the tongue of the skin in the 1940s, he had at his disposal a whole range of tools and techniques, acquired during his graduate training in psychophysics at Clark University, where he had studied under several of the figures responsible for experimental psychology's migration from Germany to the United States. Joseph Jastrow and G. Stanley Hall, both of whom possessed deft expertise in the ad hoc design of instruments for investigating tactual perception, were among his mentors. Before Geldard's passing in 1984, he had frequent contacts with Lederman, who by the mid-1980s had begun collaborating with roboticists to help build artificial hands that obeyed more closely the principles of haptic perception established by psychophysics research. Both Lederman and computer science luminary Frederick Brooks—whose GROPE-1 in the late 1960s represented one of the first attempts at realizing Sutherland's vision for a kinesthetic information display—served on the committee for Margaret Minsky[18] in her trailblazing MIT dissertation "Computational Haptics: The *Sandpaper* System for Synthesizing Texture for a Force Feedback Display." MIT's Touch Lab, where the discipline of computer haptics gained its most formal articulation, provided a launching pad for the next generation of haptics researchers during what would later be dubbed the "epoch of haptic interface"—an era marked by increasingly close partnerships among those trained in psychophysical methods, practiced robotics engineers, and computer programmers.[19] Over the centuries, the science of touch gained expression in institutions, too, including the host of psychophysics and experimental psychology labs set up at the close of the nineteenth century, Robert Gault's Vibro-tactile Research Lab (active from 1925–1940; affiliated with Bell Labs), Geldard's Cutaneous Communication Lab (run at Princeton University from 1962 to 2004), Susumo Tachi's lab (established at the University of Tokyo in the 1980s to investigate what Tachi termed tele-existence), Srinivasan's Laboratory for Human-Machine Haptics (began in 1990 at MIT), Hong Tan's Haptic Interface Research Laboratory at Purdue, and Vincent Hayward's Haptics Laboratory at McGill University. All this suggests that touch's technogenesis should not be considered a quasinaturalitic unfolding but instead a project passed down through successive generations of institutionally grounded actors, many of whom were trained in similar sets of established experimental and design techniques (or protocols) for studying, knowing, and synthesizing touch.

The touch that emerges out of these successive phases of interfacing has undergone a radical reformatting and upheaval as a result of its repeated contacts with scientific, technical, and economic practices. These changes are far from finished: the rearticulation of touch through technoscience remains an ongoing project, attracting new intellectual and financial resources, spreading outward through its embedding in new communication infrastructures, and taking on a transmogrified shape in the marketing literature for technologies of digitized touch. Contemporary technical systems, as Bernard Stiegler suggests, are in "an age of perpetual transformations and structural instability."[20] With their market-driven need to rapidly render the present obsolete, this is particularly true for digital media. Regardless of the future forms technologized touch may take, its basic inertia has been firmly established, and its fluctuations will remain bounded by the positivist framework gradually erected around it during these five phases of interfacing.

A Tactile Modernity

The account of touch I provide is organized in its early stages around the idea of a tactile modernity—a way of thinking about touch as an alternative means of registering the impact of rational experimentation and positivist science on the organization of perception. The experimental methods, techniques, instruments, and protocols of what later became known as "scientific psychology" brought with them a new type of machinic tactile experience in the form of artificially induced, precisely targeted stimuli that experimental subjects were asked to carefully attend to and vocalize. In the confines of the lab, tactility became a site for expressing the researcher's fantasy of capturing, controlling, and managing touch. The reorganization of touch through this materially realized fantasy had immediate consequences in pedagogy and medicine (discussed in Interface 2). But more significant for the overarching argument I craft in *Archaeologies of Touch*, the new experimentally derived model of touch laid the groundwork for the eventual engineering and design of haptic human–computer interfaces. By representing touch as something that could be revealed through lab experiment, this model provided an enduring epistemic frame for future investigations into the microprocesses of tactual perception that became increasingly useful to the subsequent generations of researchers who attempted to engineer touch communication systems.

My formulation of a tactile modernity is intended to both complement and expand conventional accounts of technological transformations to the sensorium during the nineteenth century. These narratives depict the senses of seeing and hearing as the primary objects of modern psychosensory science. The quantification and dissection of vision and audition by lab science staged the incorporation of these new models of seeing and hearing in emerging representational technologies. Modern science, modern media, and modern models of perception were each coconspirators in what we might think of as a sensorial modernity characterized by the disruptive application of increasingly structured and formalized laboratory methods to the senses. Through the use of specialized psychophysical apparatuses, the senses were cleaved into discrete, isolatable, and quantifiable objects. In Jonathan Crary's telling of this story, the new image-making technologies developed as a result of vision's subjection to this process worked "to recode the activity of the eye, to regiment it, to heighten its productivity and to prevent its distraction."[21] In conjunction with the veneration of seeing in Enlightenment philosophy and science, this recoded eye aided in vision's ascent to the "master sense" in modernity. A similar set of changes also occurred to hearing as part of what Jonathan Sterne terms the "ensoniment," defined as "a series of conjunctures among ideas, institutions, and practices" that "rendered the world audible in new ways and valorized new constructs of hearing and listening."[22] Touch, however, has generally been treated as external to these developments, framed as a sense left behind by modernity, or reduced to a mere operation of new imaging technologies.[23] In Crary's formulation, for example, the stereoscope indicates the nineteenth century's "remapping and subsumption of the tactile within the optical,"[24] with touching reduced, via an ocular prosthesis, to a function of vision. In arguing for a distinctly tactile modernity, I locate a touch unsubsumed by the optical—irreducible to a mere operation of vision—as the object of the same scientific discourses and practices that were set upon seeing and hearing during the nineteenth century. This alternative account of sensorial modernity provides a parallel formulation of touch as something acted on by and shaped through the new lab science of the nineteenth century. Following medical historian Robert Jütte's observation that "the transformation of the sense of touch in the industrial age is still uncharted territory,"[25] tactile modernity writes touch into a historiography of media that has rendered it absent.

Owing to the prompting of accounts like those provided by Crary and Sterne, the development of contemporary technical media has been linked increasingly back to the psychophysics and experimental psychology programs of the mid-to-late nineteenth century. Primarily, such studies examine the relationship between the quantification of the senses—the expression of the senses through laboratory equipment and experimentation—and their subsequent embedding in media technologies. Friedrich Kittler, writing on media of image and sound reproduction, noted that "media technology must first isolate and incorporate the individual sensory channels" before they can be connected together to form multimedia systems.[26] By "giving a mathematical expression to the data stream of sensual perception,"[27] psychophysics, according to Kittler, yielded a model of the senses as isolated, quantified, and individuated that would prove fundamental to the later development of technical media. Specific knowledge of the threshold between the perceptible and imperceptible—the ability of a sense organ to notice or not notice the difference between a unit of stimulation—allowed for the eventual efficient and rationalized coding of machine-generated sensations by technical media, what Sterne terms a "perceptual technics" used to economize the transmission of sensory data.[28] This tight link between psychophysics and information transmission technology that developed throughout the twentieth century embodies concerted efforts by American experimental psychologists to give their budding science a utilitarian function in industrial society, a multistage project of transforming psychology from a field underpinned by the method of contemplative reflection to one dominated by the mode of laboratory experiment that had proven crucial to modernizing advances in engineering, chemistry, electricity, and physics.

Because of E. H. Weber's influence on his mentee Gustav Theodor Fechner, touch occupies a distinct place—one rarely acknowledged by media scholars, though often rehearsed by psychologists—in this history of psychophysics. Weber's insights, derived from his sustained program of meticulous experiments on the tactual senses, were so influential on Fechner's systematic development of his new science that he suggested, in the opening of *Elements of Psychophysics,* Weber ought to "be called the father of psychophysics."[29] However, outside of touch's primacy in the sequence of developments that resulted in the establishment of psychophysics, the story that emerges around it maps rather neatly onto other historiographies of sensory quantification, such as Alexandra Hui's recent study on the relationship between acoustics, aesthetics, and the emerging psychophysics of hearing,[30] and Jimena Canales's analysis of the tenth of a second as a concept that drove experimental research into the temporal capacities of the human perceptual system.[31] This is not to suggest that touch's history is reducible to these other narratives, but rather to point out that touch did not lie beyond the reach of the scientific programs that media historians have taken to be essential to the development of technical media. Touch, as I show throughout this book, brought its own particularly complex set of research questions, practical challenges, specialized apparatuses, and laboratory protocols, derived in part from the difficulty of confining touch to a singular organ localizable to a specific site on the body. In spite of these peculiarities, touch proved equally capable of being made into an object of structured, positivistic stimulation and observation.

As a space of knowledge production vital to social advancement, the research lab has long played a defining role in accounts of technological, political, and scientific modernity. In Bruno Latour's formulation, the hermetic space of the laboratory established and perpetuated the foundational myth of modernity, providing a site where science could be cleaved off from politics. In the lab, science purportedly became an autonomous form of human activity, free from the contaminating influence of political and religious authority. Building on Stephen Shapin and Simon Schaffer's argument in *Leviathan and the Air-Pump,*[32] Latour claims that the lab's simulated conditions provided a site where nature could be isolated, confined, and witnessed in a space that purported to be value-free through the implementation of experimental protocols. Through the laboratory experiment, nature spoke, and the exper-

imenter merely functioned to record its words. Suspicious of this
conceit, Latour suggests instead that the separation of science from
politics—the basis, he argues, of our modern ontology—never
occurred in the way we imagine it. Science remains shot through
with political considerations and instrumental calculations, as
scientific knowledge remains firmly under the sway of human
interests. Touch's enclosure in the lab, then, entailed a fundamental
shift in its character and status; touch could be observed and laid
bare without the purported contaminations of politics or culture.
During the nineteenth century, touch was made modern through
its enframing within the processes, procedures, and protocols of
laboratory science. This enframing served as a necessary precondition for twentieth-century attempts to incorporate touch into
electronic communication networks: without the establishment of
a modern model of touch, the designers of cutaneous communication systems and haptic human–computer interfaces would not
have had a stable, manageable, and quantified rendering of tactile
processes to embed in their touch machines. In this aspiration to
value neutrality—in the quest to arrive at a purely scientific
account of touch—it emulated features endemic to modern laboratory science more generally.[33]

Finally, a brief note on my choice to identify this as a *tactile*
modernity, rather than a *haptic* one. "Haptics" was adopted to designate the science of touch only in the final decade of the nineteenth
century, only once the new scientific techniques and apparatuses
of experimentation had been set upon touch for several decades.
As such, I use "tactile" to describe this variant of modernity,
because it grounds these practices for knowing the touch senses
firmly in the vocabulary originally used in the nineteenth-century
medical and scientific literature, cementing the position of tactile
modernity alongside the parallel experimental processes for specifying, quantifying, and isolating vision and hearing. Subsequent
formulations of the distinction between haptic and tactile offered
by neuroscientists and physiologists assert that the former involves
an active mode of touching (including the kinesthetic and vestibular senses), while the latter refers to those touch senses housed
in the skin (specifically, pressure, temperature, and vibration). But

while the line between the haptic and tactile senses may appear to be drawn thickly in the pages of psychology textbooks, in practice, even specialists have frequently used the terms interchangeably.[34] As I show by unpacking their relationship in the subsequent chapters, the fuzziness of the two terms often proved generative, with shifts in terminology indexing new connections between previously disparate fields and institutions.

The Haptic Subject

I propose the figure of the *haptic subject*[35] to specify a particular relationship between touch and processes of scientific-technical knowledge production that initially took shape in the nineteenth century. The haptic subject functions as both a driver and an outcome of research on the tactual senses; it is not only a subject who actively touches (consistent with Foucault's questioning, listening, seeing, and observing subjects) but also a subject who was passively touched, poked, prodded, shocked, and caressed by scientific instruments, with the goal of revealing the nature of a touch that transcended the confines and peculiarities of an individual body. In this way, the tactile subject recalls the figure of Crary's nineteenth-century observer: concerning vision, "the idiosyncrasies of the 'normal' eye" became the object of a physiological optics aimed at "determining quantifiable norms and parameters"[36] of human vision. While the methods of investigating touch assumed a similar form, the data given up by the haptic subject were not immediately pressed into service by popular culture machines—there were no zoetropes or stereoscopes for touch during the nineteenth century. Using his aesthesiometric compasses—crude instruments that initiated the emergence of a formalized scientific haptics—Weber aimed at isolating the varying capacities for distinguishing between applied tactual stimuli, probing the subject's capacity for noticing and not noticing the differences between applied stimuli. The perceptual parameters he uncovered gained their most widespread utility in reflexively providing proof of the psychophysical method's concept: the trials, owing to their simplicity, could be replicated with ease by any who wished to learn the rudiments of the new practice. Neophyte psychophysicists undergoing formalized rituals of disciplinary indoctrination were asked to repeat Weber's compass-point experiments, experiencing the various tactile illusions induced by applying the device to different sites on their bodies. The establishment of haptics as a distinct set of techniques for knowing and revealing touch required a distributed network of subjects who would assent to and verify its key suppositions about the quantifiability and isolability of tactile experience.

The haptic subject, then, provides both the epistemological ground for knowing touch and the storehouse of technical knowledge required for touch's artificial stimulation. Its formation remains stable in a set of boundaries, but within those boundaries it constantly shifts, as the haptic subject is reaggregated on a continuing basis, its constitution changing in response to the idiosyncrasies of a given experimental system. Many of its particular features, then, are only ever temporary, as they are reevaluated in the frame of persistently intensifying modes of experimental investigation, and adjusted to meet the shifting demands of information-circulation economies. In short, the haptic subject gives structure and organization to a history of touch. As with Crary's figure of the "observer," the haptic subject exists only at the intersection of institutional, discursive, social, and technical relations.[37] The haptic subject signals the tacit embrace of empirically derived knowledge about touch, while also indicating touch's thorough working over by a set of assumptions about the proper way to generate and organize knowledge of the various tactile processes. It expresses a sequence of wish-images: first, the fantasy that touch can be revealed through empirical investigation, and later, the dream that this same science of touch can allow for tactility's extension into and through digital communication networks.

The production of this haptic subject mobilized a complex network of material processes, training procedures, scientific instruments, and institutionally grounded actors around the goal of revealing, laying bare, and imposing structure on the messy assemblage of human sensations designated by the term "touch." It was only when touch *became* haptics that it could begin to achieve its new, utilitarian function in a society increasingly dependent on the machinic and computational circulation of information. Haptics describes a mode of productively enframing and revealing touch—a way of ordering touch as an exploitable resource in an economy that treated the human sensorium as a calculable network of discrete information-processing channels. Utilizing increasingly specialized and standardized apparatuses, the researchers who executed these lab experiments succeeded in disaggregating and subdividing touch into a set of constitutive submodalities, each with its own unique physiological structures. In the lab, the sense of touch became the *senses* of touch; touch's component parts—heat, cold, pressure, pain, weight, movement, and vibration—became the target of specialized machines, protocols, and experimental programs, all intended to render a detailed and holistic map of touch's ability to notice and not notice the differences between things.[38] Initially, the precision with which the available instruments could stimulate the tactual senses limited the accuracy of this map. But a positivistic, utopian faith in the forward progress of technology—a belief in the power of machines to reveal the secrets of nature—suggested to early experimenters that, for all its fuzzy borders, their extant map of touch possessed an infinite perfectibility. Or, as Frank Geldard suggested in his 1940 treatise on the existence of a distinct vibratory sense, "recent developments in apparatus and method are important since our future facts are a function of them."[39]

Machines, then, were vital to the production of this haptic subject: it was through the various "pieces for haptical work"[40] and the other psychophysical apparatuses that touch began to give up its secrets. The term "apparatus" does productive work in this literature, recurring throughout historical discourse on the psychophysics of touch, to the extent that the production of psychophysical knowledge would not have been possible without the range of explicitly titled apparatuses invented to study whatever aspect of mental life the psychophysicist wished to isolate and quantify. In that context, "apparatus" (appropriated from the German word *Apparat* by experimental psychologists like Titchener, who trained in Germany's newly founded labs) was generally used synonymously with "instrument," as Horst Gundlach explains in his genealogy of the

term "psychological instrument."[41] The intricate late nineteenth-
century pneumatic machine William Krohn and Thaddeus Bolton
dubbed the "apparatus for producing simultaneous touches"
(detailed in Interface 2), for example, allowed the experimenters to
stimulate multiple sites on the subject's skin simultaneously for the
purpose of testing their ability to correctly identify and localize
multiple tactual contacts. In the framework of the apparatus, repeatedly stimulating subjects helped experimenters map the skin's
capacity to perceive touch stimuli both accurately and inaccurately.
Crucially, these psychophysical apparatuses acted on and through
subjects: to generate useful knowledge about human psychosensory processes, they configured and constrained their subjects'
responses to applied sensory stimuli. When experimental subjects
spoke, they spoke through the language of the apparatus. Kittler
described this acquisition of machine languages as fundamental to
the psychophysical project, noting that it was precisely the artifice
of these languages—their differentiation from the test subject's
natural tongue—that made it "possible to isolate the subconscious
mechanisms responsible for the construction of psychophysical
reality from the cultural—that is, language-dependent—functions
responsible for concept formation."[42] These apparatuses, however,
did not assume total control over their subjects; though they
required the subject to assimilate to the apparatus's language, they
did not reduce the subject to a mere function of the apparatus. If
they did, the subject would be superfluous to the experiment.
Instead, these apparatuses aggregated subjectivity, abstracted it,
and made it statistical, disregarding and downplaying the differences among subjects while still depending on those differences
to produce heterogeneous data about the specialized sensory
operations.

   Although the subjects of psychological apparatuses were in theory interchangeable—swappable cogs in a machinery of knowledge production that intentionally imitated the structure of the
nineteenth-century factory—the actual subjects of these experiments (often the experimenters themselves or graduate students
being acclimated to experimental psychology's rigorous methods) had to be trained to carefully attend to the stimuli applied to

their senses.[43] They had to be what Weber called "good observers," capable of maintaining a practiced attention in spite of quite uncomfortable testing conditions (Weber's good observers, for example, had the misfortune of being repeatedly administered enemas of freezing cold water, in an effort to test the temperature sensitivity of the nerve trunks in the bowels). Each apparatus, in other words, depended on the production of a subjectivity—the agreement of the subject to undergo the active disciplining required to fuse with the apparatus in such a way that their articulated experience could yield useful and reliable knowledge about the sense modality or submodality under investigation.

To push on the theoretical implications of this relationship between apparatuses and subjects, I turn to Giorgio Agamben's formulation of the apparatus not as a thing but as a process. Building on and extending Foucault's expansive understanding of apparatuses as mechanisms of control, for Agamben, apparatus—or *dispositif*—designates a set of strategies for controlling and managing the behavior of bodies. Agamben locates apparatuses as definitional components of the "extreme phase of capitalist development in which we live," a phase characterized by the "massive accumulation and proliferation of apparatuses."[44] The lives of individuals seem to be totally subsumed by these formations; with daily existence constantly "modeled, contaminated, or controlled by some apparatus."[45] The power of apparatuses lies in their ability to produce subjectivities, to remake individuals as subjects of disciplinary machines "from tip to toe." But crucially, apparatuses are not imposed from without: at their roots, and thus at the base of all the subjectivities they produce, "lies an all-too-human desire for happiness." Apparatuses clear away the old subject, replacing it with a newly reshaped one, constituted by new desires, which are above all the desire for apparatuses.[46] The double operation of this term—as designating both individual scientific machines and a larger, overarching social machinery that produces, manifests, and manages desires—will lurk in the backdrop of the argument I advance throughout *Archaeologies of Touch*. In early experimental psychology labs, for example, the haptic subject was one who desired a transhistorical account of tactility, one divorced from their own individual idiosyncrasies, which could be revealed through the submission to the lab's sadistic machineries and protocols. The haptic subject conjured by today's designers and marketers is also a desiring one: a subject who understands interfacing with computers through touch as ameliorative and restorative, where a technologized touch can alleviate the suffering inflicted by the subject's continued subjugation to the apparatus of the audiovisual (nonhaptic) interface. Contemporary haptic interfaces promise to *desubjectify*: in claiming to facilitate a mode of interacting with machines that is instinctual rather than unnatural, they tell users that they will be able to communicate with machines in a language of touch that does not have to be learned and submitted to. In other words, these interfaces promise to wipe away the old subject of information machines—the one whose bodily and perceptual habits had to acclimate to the interface's taxing artificiality—and replace it with a new haptic subject, fully embodied through the interface. Haptic interfaces expose what Immersion Corporation frames in its promotional materials as "the exhausted the limits of Information Age technology";[47] in response, these devices offer to revitalize and restore "exhausted consumers" through the layering of touch feedback onto extant computing machines.

Apparatuses, crucially, are enacted through the adherence to *protocols*—through the agreed-on practices of usage that gradually sediment to provide the background consensus informing conventions and ritualized habits of action. Lisa Gitelman identifies a productive kinship between the protocols that structure the habitual use of scientific apparatuses and those that shape the accepted use of media technologies. Both protocols that inform the use of scientific instruments and media protocols require like-minded groups of actors to accept a set of usages associated with a given apparatus.[48] Although communities of actors frequently contest, debate, and refine protocols when they are new, once these protocols have settled—once the new transitions to the old—protocols become "a vast clutter of normative rules and default conditions" gathered around a "technological nucleus."[49] The relationship between protocol, cultural memory, and control is therefore vital: protocols, which Gitelman defines so broadly as to include all supporting structures that underpin the operation of a given medium, exercise their power when they slide into the backdrop of acceptable usage—once they become sedimented in institutions and materially realized in infrastructures. Gitelman's instructive example is the telephone, which is undergirded by habitual uses (such as answering by saying "hello"), a set of business practices (the monthly billing cycle), and physical infrastructures (the wires and cables that connect phones to one another). This process of settling protocols around a medium is frequently messy and protracted, unfolding in a range of spaces and involving conflicts between heterogeneously situated actors vying for power. Where haptic interfacing is concerned, we are presently in a curious moment, as haptics technology is still in the process of being made intelligible, primarily through marketing and popular press discourses about technologized touch (see Interface 5). But part of the argument that I advance throughout *Archaeologies of Touch* concerns touch's prior expression through scientific protocols, which has a long history reaching back into the eighteenth century (Interfaces 1 and 2). As the protocols around haptic interfaces settle—as the meanings and uses of haptics technology are made evident through the discursive construction of the technology—they do so according to a configuration of boundary conditions set forth gradually in the research labs where the haptic subject was first isolated, confined, and subdivided.

Although protocols are structures for enacting control, they also provide the parameters for contesting and interrupting the operation of control. The embedding of protocols in computing machines alters the parameters of protocol formation, control, and contestation. In Alexander Galloway's formulation, protocols—as automated, machinic processes that distribute management and control—configure the exercise of power, while also making possible new forms of resistance and disruption within that configuration. Though protocols are not unique to computing, computing changes the parameters within which protocols are executed and enacted. Accordingly, Galloway draws on Foucault and Deleuze to conceive of bodies as the fundamental objects of protocological control, as they specify the operation of material bodies in particular contexts and spaces.[50] Galloway offers a methodological orientation to protocol that allows it to be analyzed and confronted as a dynamic imbrication of sociopolitical and material processes, rather than a static set of decontextualized and disembodied rules. Protocol enables the "making-statistical" of life forms on a mass scale, functioning "as a management style for distributed masses of autonomous agents."[51] Merging these approaches, protocols can be seen as simultaneously cultural, technical, and institutional, serving as normative mechanisms that regulate, govern, and make productive the behavior of human bodies.

Through this intertwining of subject, apparatus, and protocol, I am suggesting here that the haptic subject be positioned at the intersection of changing techniques of management, control, negotiation, and subject formation. Paraphrasing Krohn and Bolton, the various touch machines described in these pages can be understood collectively as apparatuses for producing haptic subjects—outcomes of individual subjects' tactual experiences in the lab, of material practices enacted by protocological control, and of subjectivities that animate individuals' desire for the desubjectifying power of touch machines.

Recapitulating Touch

Touch's current technohistorical situation makes the intervention *Archaeologies of Touch* provides especially pressing, as questions about tactile relationships with media achieve a new urgency, prompted in part by the touchscreen's homogenization of previously diverse sites of physical interfacing. As we begin to interact with books, banks, music, spreadsheets, films, software, games, stores, and people only through the flat, dedifferentiated surface of the touchscreen, buttons, knobs, keys, shelves, desks, and bodies become nostalgic objects, only appreciated after their erasure. Specific to haptics, vibrational feedback (rendered with increased precision by new algorithms, motors, and actuators) becomes the means of recapturing and recalling the lost materiality of those media and objects subsumed in the touchscreen. The rhetoric mobilized around touchscreens also invests the category of touch itself with a new set of meanings. Media studies, I suggest, finds itself quite unprepared for this situation: while we have excellent, comprehensive genealogies of seeing and hearing that show the senses to be sites where power is expressed and negotiated, to paraphrase Denis Diderot in his call for a tactile language, we have nothing for touch, although this sense has its own distinct history as a technical object. At stake in the present moment is the way that theories and genealogies of media historicize and ideate touch in general and haptics in particular. Considering the long history of technoscientific haptics implicates touch in a broader set of discourses about the relationship between technology, the body, and the senses. It showcases the power of media to order, subdivide, fragment, reconstruct, and reformat the senses, highlighting the new training regimes that the perceptual system underwent to be able to receive, decode, and manipulate machine-generated sensations. *Archaeologies of Touch* aims to get at the microphysical interactions among the material, discursive, and institutional constructions of haptics. It presents a positive empirical response to the present technohistorical situation by actively intervening in the production and reconstitution of what counts as the archive of media history. In this macrohistorical frame, "haptics" becomes as much a political term as it is a physiological and technical one, expressing a biopolitical fantasy where scientific power/knowledge achieves complete dominion over the range of human tactual senses.

Naturalization narratives, like the ones mobilized around haptics, are inherently political: by positioning a given object as an operation of nature rather than of culture, these stories function to obfuscate the operation of power. The repeated claims, taken up in Interface 5, that interfacing through touch represents a more natural and intuitive mode of interacting with information distracts us from all the technical and scientific filters that touch had to pass through before it could enter the computer-mediated sensorium.

It orients us away from the vast institutional, financial, political, and philosophical investments made in this project over its long history. And by doing so, this naturalization narrative forecloses more possibilities than it opens. The strength of the media archaeological approach that *Archaeologies of Touch* takes toward haptic interfaces lies in its establishment of new and productive connections between the past and present, and in its ability to show the scientific and technical imagination mobilized around touch. Ontological arguments about touch—definitive pronouncements about its inherent, unshakable, and enduring qualities—seem not only quite uninteresting when contrasted with touch's rich and textured empirical history, but more problematically, also serve to draw attention away from touch's signature dynamism and flexibility.

Finally, although it is an enterprise that carries the risk of fetishizing what should be its critical objects, there is a demonstrated value to understanding the technical and material features of hegemonic media technologies on their own terms. This approach to media disrupts the power of media technologies by undoing what science and technology studies refer to as the "black boxing" of technology. It allows us to see the processes of negotiation and enunciation that occur at the early stages of a given technology's development. In this way, *Archaeologies of Touch* is not intended as an endpoint but rather as a beginning: a way of orienting media studies to the range of interwoven issues—technical, legal, commercial, historical, aesthetic, epistemic, and political—at stake in the development of haptic interfaces. *Archaeologies of Touch* seeks to make haptic interfaces analogs to visual and aural media not by way of technical analogy, but rather by suggesting that media scholars approach them as analogous critical and historical objects. Understanding the present as an outcome of technogenesis explicitly recognizes the historical contingency of the current situation, and in doing so, opens up new ways of thinking about the future trajectory of media technology. It reveals the ways the body and its senses are specified through the microprocesses of scientific invention, technological deployment, and strategic marketing.

Toward a Haptocentric Media Archaeology

The five phases of interfacing I use to organize the historiographical narrative in *Archaeologies of Touch* provide a means of assembling and shaping a path through the archive of haptic interfacing. Imposing this structure on a diffuse body of materials scattered across four centuries is intended to help contribute to a history of the present. However, this type of linear narrative is anathema to many in the loosely defined field of media archaeology, which is often positioned as staunchly against what Timothy Druckrey calls the "anemic and evolutionary model" that has come "to dominate many studies in so-called media."[52] Media archaeology, as a methodological strategy, is offered as a means of combating teleological accounts of media change, emerging as a response to the deterministic narratives mobilized to explain the evolution of visual media and their affiliated representational techniques. Kittler articulates a similar critique of linearity in his *Optical Media* lectures: "in spite of all beliefs in progress, there is no linear or continuous development in the history of media."[53] The pioneering media archaeologist Erkki Huhtamo also suggests that this approach "runs counter to the customary way of thinking about technoculture in terms of constant progress, proceeding from one technological breakthrough to another, and making earlier machines and applications obsolete along the way."[54] As a theoretical apparatus intended to actively reshape media history, media archaeology positions itself as standing militantly against linear narratives and the ideological association between technical and social progress they imply.[55]

But where touch (and other nonaudiovisual[56]) media are concerned, no linear and teleological narrative exists for media archaeologists to problematize and overturn. Instead, the history of nonvisual modalities and their affiliated media comes into view only by considering their role in fostering vision's seemingly inevitable rise to dominance. The rendering of media-historical time as the history of visual media results in an almost unspoken assumption that these nonvisual modalities have no concrete, independent, or empirically observable historical trajectory. Or if they do, such histories, trajectories do little to complicate the visualist narrative of media history embraced, even if reluctantly, by media archaeologists. Calls to eclipse linear or evolutionary accounts of media history have proceeded against the backdrop of substantial research programs dedicated to constructing an extensive and increasingly sophisticated history of vision and visual media. The linear and evolutionary visualist account of media history therefore serves as the often unacknowledged historical a priori of media archaeology; by attempting to overcome its limits, media archaeologists remain indebted to it, even if only as a point that can be departed from and pushed back against. This is to suggest that, however much linear and evolutionary accounts of (visualist) media ought to be deconstructed and resisted, it is precisely these accounts that provide media archaeology with its cohesive identity. The militant stance media archaeology takes against linear narrative thus originates from a position of disciplinary privilege.

In contrast, those of us interested in the history of touch media do not have the luxury of a linear, dogmatic, hegemonic, progressivist, evolutionary account to push back against.[57] We have no agreed-on canon—the story of touch's imbrication with technical media is one that exists only in fragments, distributed in arcane scientific and technical documents, scattered across centuries and various specialist fields. *Archaeologies of Touch* imposes a necessary ordering on the chaos of this archive, while also recognizing the problems inherent in such impositions. It takes seriously the charge that "media archaeology should be seen as primarily a critique of progress"[58] by portraying haptics not as a value-neutral window to the operations of the human tactual senses but rather as the product of a politically charged positivist research program that repeatedly disavowed and denied its foundational politics. Simply by continuing to historicize and theorize touch only through the visual, media studies has been complicit in circulating this image of touch; to paraphrase Fiona Candlin, the field has continually risked turning touch into a subset and mere operation of vision.[59] It may simply be the case that up until now, we have had no cause to occasion such a systematic engagement with touch's history as an object of technical and scientific research, driving home the point that media change not only remediates old media but also

prompts a revaluation of what counts as media history. Wolfgang
Ernst identified laying bare "the technoepistemological momentum
in culture itself" as one of the animating aims of media archaeology;
the haptocentric media archaeology I present in *Archaeologies of Touch*
exposes not only the technoepistemological momentum embodied
in haptic interfaces, but by doing so, reflexively calls attention to the
epistemological momentum operating in the loosely defined field of
media studies itself.[60]

Perpetually Immanent: The Teleology of Haptic Interface Design

While media studies offers no comprehensive account of touch's
history as a technoscientific object, popular press and technical
chroniclings of haptics have advanced a somewhat cohesive narrative to help familiarize and historicize this exotic new mode of
interacting with computers. The circulation of this narrative is vital
to the propagation of haptic interfacing, illustrating the extent to
which the project is as much social as it is technical and scientific:
the uptake and adoption of haptics technology depends on establishing haptics as a necessary and inevitable response to the limits
of audiovisual media, the outcome of a naturalistic and teleological evolution of the mediated sensorium. The narrative consists of
three strands that work together to present an origin story around
haptic interfaces—a framework in which any new developments can
be slotted without complicating the metanarrative. These narrative
strands are *myths,* in Roland Barthes's sense of the term, that serve to
make sense of an unfamiliar and alien set of technologies—but crucially, as Barthes explains, myth is not a designation that denotes
untruth. Rather, these myths work to conceal operations of power
and ideology by imparting a deterministic and naturalistic inevitability to touch's technogenesis.

   The first narrative strand mobilizes what I term a *logic of analog
medialization*—a discursive framing of changes in the technologized
sensorium that suggests touch can become like the technologically
augmented senses of seeing and hearing through the acquisition of
its own mediatic apparatuses. This narrative framing has its roots

in the beginning of technical media; Thomas Edison, writing in 1888, employed a similar logic when he described the kinetoscope as "intended to do for the eye what the phonograph does for the ear."[61] As if borrowing from Edison, one popular press treatment describes vibration-enabled touchscreens as part of "a new breed of 'haptic' technologies that do for the sense of touch what lifelike colour displays and hi-fi sound do for eyes and ears."[62] As with those prior media technologies, haptic interfaces are situated not only as a way to capture, store, and transmit sensations but also as a way to transform the senses themselves through technical enhancement. As I show in Interface 4, this logic of analog medialization, mobilized by computer scientists in the early years of haptic interface design, was not merely descriptive—it also informed design practice and structured subsequent efforts by marketers to articulate the value of haptics technology.[63] The logic of analog medialization voiced the designer's desire to employ technology as a means of transmogrifying touch: by recuperating a neglected experiential modality, they hoped to regain something lost with the rise of the visualist interfacing paradigm.[64] Designers, in narrating the products of their labor, often foregrounded complex descriptions of electro-mechanical techniques for generating haptic sensations by referencing philosophical arguments about touch's centrality to human experience.[65] The fundamental inadequacies of visual interfacing, according to this argument, could be remedied by adding mechanisms for convincingly synthesizing haptic sensations onto existing media apparatuses. Designers assure an anxious public that the dematerialization accomplished by computational media can be undone by folding haptics technology into the interfacing schematic. A technologized tactility promises to be therapeutic, ameliorating the immense stresses image and sound media have placed on the sensorium. Further, the logic of analog medialization makes a naked and not uncontroversial declaration that touch can be virtualized, countering the common claims that touch, as John Durham Peters puts it, "defies inscription" and "remains stubbornly wed to the proximate."[66]

A second strand locates haptics in a state of perpetual immanence, poised for rapid progress forward and ubiquitous adoption in a wide range of technocultural practices. Since haptic interfaces took center stage in Howard Rheingold's bestselling 1990 book *Virtual Reality,* the technologies have been a source of continued fascination and wonder, simultaneously exoticized by the designation "haptics" and made familiar by the promised intimacy of touch.[67] But as the promised tomorrow continues to never arrive, haptic interfaces are suspended in this state of perpetual immanence, always just on the horizon, always only five short years away, always invested with technoutopian hopes, and always inevitable.[68] Together with the logic of analog medialization, the trope of perpetual immanence works to suggest a natural and deterministic progression to the technologization of the human senses.

Layered on the narratives of analog medialization and perpetual immanence, the third strand foretells the coming of a *master device* that will finally result in the rapid uptake of a standardized, high-fidelity haptic interface. With the arrival of the master device, all preceding efforts will seem like steps toward this single inevitable outcome, rendering them obsolete and irrelevant. Building on the logic of analog medialization, the master device will accomplish for haptics what the adoption of standard recording and playback formats accomplished for image and sound media, giving haptics a unified, stable, and intelligible identity coterminous with the technological extension of touch. The idea of the master device tacitly recognizes what Abrash made explicit in his 2015 address: the challenge posed by haptic interfaces is one of both hardware and software design; the successful proliferation of any eventual master haptic device will require the development and adoption of a shared language for coding computer-generated tactual sensations. And as with image and sound media, the competition to lay claim to a master device and its attendant standards involves a fierce battle between corporate intellectual property holders (described in Interface 5) with each entity vying for control over what Jacques Derrida, in a short passage on haptic interfaces, termed the "algorithms of 'immediate contact.'"[69]

By locating haptic interfaces in the broader political project of productively disciplining the senses, *Archaeologies of Touch* provides an alternative narrative, in which the recent attempts at transforming touch through technology are not unique to the age of computing but instead exist in continuity with the exertions of previous generations of scientists and engineers. This story is not necessarily an unfamiliar one to hapticians—Martin Grunwald's sprawling 2008 edited volume *Human Haptic Perception,* for example, juxtaposes chapters by historians of psychology and medicine detailing touch's scientific and cultural history with essays by researchers actively pursuing the design of new haptic human–computer interfaces.[70] However complex, well-researched, and detailed these histories are, they are situated in a progressivist epistemological framework that treats scientific knowledge as the gradual accumulation of increasingly refined truths (precisely the sort that media archaeology so vehemently calls on its adherents to disassemble). Further, though this origin story may be fairly well circulated among haptics specialists, it is fairly obscure outside of that cohort, remaining nearly unknown to media historians.

As I show throughout *Archaeologies of Touch,* the narrative strands woven around haptic interfaces are not new stories for touch— the teleological suggestion of a finality to research on touch technology, the framing of touch as analog to seeing and hearing, and the notion that one technologized device for tactual communication would emerge victorious over the rest each recur at different points in the history of touch chronicled here. The formulation of haptics as the doctrine of touch, for example, was intended to make the science of touch analogous to the sciences of optics and acoustics. Diderot's suggestion (taken up in Interface 3) that touch could have its own fixed set of signs—a "clear and precise language of touch"[71]—was similarly aimed at making touch like seeing and hearing, and staged Geldard's later attempts from the 1950s on to divine a "tongue of the skin" through the engineering of psychophysically grounded machine languages. In each instance, making touch like seeing and hearing was also understood as a vital, socially transformative project that would bring touch—once it passed through a technological, scientific, or linguistic filter—into a modern ordering of perception that had previously barred its entry. Like the master device, the tongue of the skin was taken to be a singular entity that engineering psychology would eventually reveal. And as with the narrative of perceptual immanence, for several decades, the desired end seemed just on the horizon. Establishing these productive connections between past and present undermines the fetishistic claims of novelty mobilized around haptic interfaces. It resists succumbing to what K. Ludwig Pfeiffer describes as the "vertigo of alleged media revolutions" that media studies and popular culture more generally often relishes, through an exposition of the technoscientific forces that have acted on touch over the past three hundred years.[72] *Archaeologies of Touch,* then, does not attempt to put to lie the narratives of analog medialization, perpetual immanence, and the master device, but seeks instead to connect these stories to longstanding cultural anxieties voiced in debates around the senses and technology.

Genealogies of the Haptic

The term "haptic" has had a curious life in media theory, quite apart from the one it lives in the psychology, engineering, and computer science discourses that I treat throughout *Archaeologies of Touch.* Scholars working with these divergent genealogies have thus far generally been content to allow the tensions between the traditions to persist without resolution. However, given the term's centrality in this book, and the book's positioning at the intersection between media theory and computer science, this brief explanation of how the two genealogies map onto each other helps provide some mutual legibility to the divergent fields. This will not, of course, resolve the tensions between the two genealogies, but it will show them as fundamentally linked at a crucial historical juncture. For media theorists, the term originates in work of Austrian art historian Alois Riegl (1858–1905). Throughout his 1901 *Spätrömische Kunstindustrie* (Late Roman Art Industry), Riegl developed a theory of haptic vision, where the eye, in caressing the visible surface of an artwork, assumed a tactile function—as if it were a fingertip moving across a textured material space.[73] The argument Riegl crafts around this haptic vision is a complex one, invoking a web of relations between the phenomenology of perception, the expression of cultural hierarchies of sensation through works of art, and epochal shifts in reception habits.[74] But crucially, the term "haptic" did not appear anywhere in the book. A year after the publication of Late Roman Art Industry, in a short follow-up essay, Riegl suggested that "haptic" be retroactively substituted for "tactile," not only in that solitary essay but also throughout the whole of the original argument. He justified the strategic repositioning, in the essay's lone footnote, by claiming that the new term "haptic," unlike "tactile," did not situate vision and touch in opposition to one another.[75] It implicated touch in a harmonious rather than antagonistic relationship with the visual; haptic vision, in comparison to tactile vision, indicates a synergistic coupling of the touch and vision, a vision capable of becoming like touch.

Riegl's definition of "haptic" as a nonoppositional mode of touch has been echoed frequently in film and media theory. Gilles Deleuze and Félix Guttari, for example, rehearse Riegl's definition almost verbatim in their claim that "'haptic' is a better word than 'tactile' since it does not establish an opposition between two sense organs but rather invites the assumption that the eye itself may fulfill this nonoptical function."[76] "Haptic" functions consistently as a strategy of sensory dedifferentiation, providing a means of breaking down the barriers between the senses and endeavoring to show how touch can be active as an agent in the process of seeing. As it did in Riegl's formulation, "haptic" serves a strategic and ideological function in these works. Laura Marks makes this explicit in her writings on haptic visuality, defining the haptic as "a feminist visual strategy"[77] that allows her to identify a counter-visual mode of seeing that operates in particular types of cinematic images. In these strands of thinking, the haptic conjures a counter-hegemonic perceptual subjectivity activated through vision.

By his own admission, however, Riegl had not invented the term whole cloth. Rather, according to his footnote, he appropriated it from physiology,[78] where it had been taken up in the 1890s as a way of designating the vast research being carried out on the psychophysiology of tactual perception. In borrowing the term, Riegl had substantially modified its meaning—he steered away from its scientific, doctrinal, and experimental connotations, hinting only vaguely at the new research paradigm it designated. In contrast, in the genealogy rehearsed in psychology, engineering, and computer science, the term's roots are entrenched firmly in the field Riegl poached it from. In this tradition, described above, "haptic" relies on a radically corporeal and embodied touch: it refers to both a body of knowledge about touch and a set of instruments, protocols, and processes used to further specify the components and subcomponents of tactual perception. It is, in short, an epistemic framework for knowing touch. As a consequence of these two traditions, we are left with a bifurcated genealogical account of the term; media theorists,[79] unaware of the scientific tradition associated with the haptic, use it to designate a flexible and not necessarily tactile phenomenology of touching, while haptic interface designers embrace and deepen the haptic's roots in the technicist and materialist practice of experimental psychology. The two traditions imply radically different notions of touch. For media theorists, "haptic" is a model of touch that can operate without touching, where the senses are capable of becoming synesthetically active in one another. For psychologists and engineers, the material act of touching is fundamental to the formation of haptics as an accumulated body of knowledge; they do not seek to differentiate the senses, but instead to radically and intensely differentiate touch itself through the application of experimental techniques and apparatuses.

Through this brief comparison of the two models, I am not seeking to abrogate or resolve their differences. Rather, by pointing to their common origin in the nineteenth-century research lab, I hope to map the opening up of the chasm between them and in doing so, plot possible sites where this chasm might be bridged. The fundamental question here involves touch's capacity to be mobilized—via the haptic—as an agent in a counterhegemonic politics. As I suggest throughout *Archaeologies of Touch,* understanding haptics as an epistemic framework for touch shows it to be the object of a hegemonic—rather than counterhegemonic—instrumental rationality operating in industrial and postindustrial capitalism. At the same time, in the scientific and engineering discourse around touch, the pursuit of technical knowledge about tactility was frequently seen as a means of undoing sedimented, hegemonic hierarchies of perception. Those who took up arms under touch's banner did so with the explicitly stated intent of reversing or ameliorating the harmful effects of a societal overreliance on vision and its affiliated technologies. Although I do not want to suggest to direct a kinship between, for example, Marks's formulation of the haptic as a countervisual strategy and haptic interface designers who posited their practice as a means of reasserting touch's power in a visualist interfacing schematic, there is nevertheless a distinct resistive and disruptive impulse motivating the two mobilizations of the term. By taking some steps toward providing them with a mutual legibility, I hope that *Archaeologies of Touch* will help facilitate future exchanges between the fields.[80]

Groping toward the Future

The promise of haptic interfacing is an enduringly seductive one: to be able to step into a virtual world, or to extend the body via a surrogate into a remote environment, and feel the whole range of wondrous haptic sensations—rain pounding against the skin, wind on the face, the embrace of a distant or lost loved one—that present themselves so effortlessly in everyday life. Achieving this end of a seamless and one-to-one link between the haptic system of the user and their virtual or remote avatar appeared readily at hand in science fiction novels and Hollywood films: tactile variants on cinema (as in Huxley's Feelies and Salvador Dali's *Le cinéma tactile*), the notion of jacking into a cyberspace that provided even more data for touch than it did for vision and audition (a common feature in cyberpunk fiction like William Gibson's *Neuromancer,* Neal Stephenson's *Snow Crash, The Matrix* trilogy, and *eXistenz*), accessing virtual reality through some more messy physical interface (as in *The Lawnmower Man* and Ernst Cline's novel *Ready Player One*), or the establishment of a stable connection between the self and a remote body (Robert Heinlein's *Starship Troopers* novel; the films *Avatar* and *Surrogates*). And although the ease of this feat on page and screen belies its near impossibility in the design lab, the promise of haptic interfacing retains its power. Particularly amid fears over technology's inability to bridge the psychic and emotional

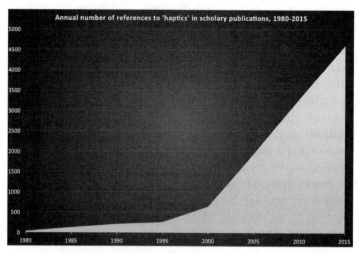

Figure I.1. The number of academic publications that reference haptics by year, from 1980 to 2015. Data based on a Google Scholar search.

gaps between communicative subjects, these touch technologies offer an appealing way of reconnecting through a sense valorized for its ability to provide an authentic and irreducible form of embodied experience.

Over the next decade, the continued proliferation of touch technologies in a range of computing applications will push toward some of these imagined outcomes, buoyed by the recent sharp spike in investments into haptics research. The increasing complexity of the actuators and corresponding control algorithms used to produce vibrational feedback in touchscreens will allow users to feel a greater variety of tactile textures on the screen's flat glass surface. If any of the virtual reality interfaces from Oculus, HTC, or Sony succeed in gaining widespread adoption, the ensuing platform standardization will fuel new investments in haptic interfaces for virtual environments, as developers push toward precisely the type of "breakthrough research" Abrash called for in his 2015 keynote address. Disney's substantial efforts at developing Surround Haptics[81] may begin to bear fruit. Teledildonics devices, no longer tethered to clunky desktop computers by cumbersome wires,[82] are also poised to take some small steps forward after decades of stalled efforts. The dual motor configuration used to produce rumble feedback in videogame console controllers since 1997 may see some minor alterations; its stability across successive generations has allowed for the stabilization of best practices in rumble feedback.[83] The application of haptics in prosthetics research also holds great promise, as neuroscientists team with engineers in an effort to fuse artificial limbs directly to the human nervous system.[84] In his 2015 State of the Union address, U.S. president Barack Obama assured the nation that ongoing efforts by the Defense Advanced Research Projects Agency (DARPA) to create "revolutionary prosthetics" would allow "a veteran who gave his arms for his country [to] play catch with his kids again," with a slide that accompanied his speech showing a robotic hand that could be "moved with thoughts alone, and feel the warmth of touch."[85] Such innovations will continue to be underpinned not only by formalized networks of professionalized actors and ad hoc DIY user communities but also by legal frameworks given shape through ongoing battles over haptics patents.[86] In short, even the most conservative and sober projections of current trends in haptics technology suggest that the next decade will bring the concretization of new standards both in haptics hardware and software—likely not the triumphant rise of any master device but instead incremental continuations of decades-old trends. If haptic interfaces do move toward the widespread uptake of universal standards for hardware, software, and haptic effects design, the present moment will seem in retrospect very much like the early decades of cinema's history, when the new medium gradually cohered around a stable identity with a fixed set of cultural practices, technical standards, and supporting industries. In those formative years, cinema acquired its constitutive protocols through complex processes of negotiation, which quickly became naturalized as definitional elements of the medium.[87] Almost without critical commentary from those in media studies, the past three decades have seen a chaotic swirl of new research on computerized touch, with the number of scholarly publications that reference haptics increasing nearly tenfold, from an average of 443 per year in 1996–2000 to an average of 4,256 in 2011–2015 (see Figure I.1). Regardless of the future direction any individual haptic interface takes, the process of naturalizing and domesticating haptics—and settling the protocols of haptic media usage—promises to continue unabated. In response, *Archaeologies of Touch* calls on media scholars to attend to and engage with its microprocesses, not as an origin of technics but as the intensification of a macrohistorical project explicitly aimed at giving touch a vital new utility in the political economy of communication.

Interface 1

The Electrotactile Machine

When the Phail has been sufficiently electrified . . . the whole Company join Hands; the Operator at one Extremity of the Line grasps the Bottom of the electrified Phial, and the Person at the other Extremity touches the Wire. . . . At that instant, the whole Company receives a Shock.
—Tubervill Needham, "Extract of a Letter from Mr. Tubervill Needham to Martin Folkes, Esq; Pr. R. S. Concerning Some New Electrical Experiments Lately Made at Paris"

Before electricity could be employed as a tool for the storage and transmission of words, images, and sounds, before it could be used, in Siegfried Zielinski's words, to "breathe a new soul into media worlds,"[1] before "electricity itself," as Friedrich Kittler claimed, "put an end to writing's monopoly on the storage of experience,"[2] it was first deployed in the intentional production, storage, and transmission of tactile sensations. The discovery and refinement of a cultivated sensitivity to electricity underpinned the early history of electrical machines, staging the later development of tactile communication systems and haptic human–computer interfaces. As I show throughout this chapter, a practiced tactile epistemology of electricity was crucial not only to the production of scientific facts about electricity's existence and constitutive elements but also figured centrally in the later engineering of machines designed specifically to produce electrotactile effects—what I formulate as *electrotactile media*.

Because media historical research has generally privileged the "archaeology of hearing and seeing by technical means,"[3] it has treated the production of electrocutaneous shock as a sort of accidental or inconsequential effect of electricity, and by doing so, it has also obscured the crucial role tactility played in the design and propagation of early electrical machines. Experiments with the eighteenth-century electrical machines discussed throughout this chapter (such as electrostatic generators, Leyden jars, electric fish, and Voltaic piles) required the cultivation of an epistemically oriented tactility. While electrical discharges often produced optical and acoustic markers (the "sparking" and "cracking" routinely described in accounts of electrical experiments), experimenters devoted the bulk of their attention to describing and comparing electricity's tactile effects. Experimental protocols carefully described the steps to be followed in order to successfully produce shock experience, and then further detailed the tactual effects that should be expected from each experimental configuration. Transcriptions of shock typically attended to its intensity, the areas of the body it spread to, the length of time the charge would remain imprinted on the nerves, and any miscellaneous effects that lingered after the experiment. Electricity, then, could not be separated from its manifestations to the human senses, and of the whole range of senses it acted on, its impact on touch most was the most dramatic. Even with the later development of electrometers capable of measuring the existence and strength of electrical charges, many skilled eighteenth- and nineteenth-century electricians preferred instead to use their own sense of touch and the refined sense of electrocutaneous shock they had cultivated to gauge the qualities of electricity as they manipulated it.[4]

By examining touch's role in scientific, popular, and medical encounters with electricity during the first centuries humans were able to control it,[5] I reframe the electrical machines developed in the eighteenth and nineteenth centuries as a type of tactile media. This remapping is a self-conscious attempt to push the relationship between touch and electricity to the forefront of media history.[6] In these histories, considerations specific to touch are frequently subsumed in the overarching category of the body, and consequently, touch appears to have no agency either as a historical force or as a discursive entity. In contrast, because of their affiliated technical media, the senses of seeing and hearing are constantly attended to in media historical analysis. Even in those media archaeologies that aspire to avoid advancing reductive and progressivist narratives of media change, present-day audiovisual media provide the structuring logic for archival retrieval. The historical reorganization I offer here—designating early electrical machines as tactile media—jeopardizes the typical ordering of the senses provided by conventional media historiographies, supplanting it with a story where touch figured prominently in a cultural sensorium mobilized to apprehend electricity. This move functions as a way to help rethink the conventional timeline associated with the electrification of the senses, while simultaneously grounding twentieth-century attempts to network touch via digital computing in the broader project of eliciting tactile sensations through the structured application of electricity to the skin. In light of the complex relationship between touch and electrical machines, contemporary haptic interfaces should not be understood as signifying a move into an unheralded future. Instead, these machines can be thought of as the continuation of an ongoing cultural training that involved teaching humans to interpret machine-generated tactual sensations—a process of becoming literate in the tactile languages spoken by electrical machines.

Moreover, this haptocentric analysis of electrical machines encourages us to reevaluate critical assumptions about vision's unabated rise to dominance by offering evidence that an electrocutaneous shock figured prominently in the "collective empiricism"[7] of scientific encounters with electricity. Contrary to Foucault's contention that the natural sciences in eighteenth-century Europe involved the increasing exclusion of touch in favor of vision, where "the sense of touch is very narrowly limited to the designation of a few fairly evident distinctions,"[8] the spread of electrical machines depended on and fostered an increasingly refined tactile form of knowledge. Rather than marginalizing and alienating touch, direct and structured appeals to an epistemically oriented touch underpinned Enlightenment scientific encounters with electricity. Though electrical demonstrations frequently produced visible sparks, the eye proved an unreliable and inaccurate register of electricity's movement between bodies,[9] with touch pressed into service to overcome vision's limits as an instrument of scientific investigation.

None of this is to suggest that the body more generally has been excluded from historiographies of electricity: many commentators note its centrality to scientific,[10] cultural,[11] medical treatments,[12] and mediatic[13] treatments of electricity. However, by subsuming considerations of touch in the broader category of the body, these previous approaches obscure the significant role tactility played in giving form to electrical knowledge. As such, they do not attend to the rich tradition of disciplined, practiced touching mobilized by encounters with electrical machines.[14] Through an analysis of the scientific, cultural, and economic protocols necessary first for the production of electrical knowledge and later for the positive deployment of this knowledge via electrotherapy, I demonstrate the importance of a tactile epistemology to the deployment of early electrical machines. Moving chronologically though a series of electrical apparatuses shows how each machine depended on productively incorporating human bodies into ad hoc electrical circuits, with charges passing through organic networks in acts of corporeal spectatorship. These electrical machines were touched in a wide range of contexts by a variety of heterogeneously situated actors. In public demonstrations of electricity, spectators frequently bore witness to electrical facts by becoming part of the experimental apparatus. In the parlors of the eighteenth-century upper class, partygoers were entertained by the curious sensations that penetrated their limbs upon making contact with electrically charged machines. Electricity was applied to bodies not just in lab experiments aimed at yielding scientific knowledge but also in museums, coffee shops, libraries, taverns, and genteel homes, with shocks intended to produce amusement, wonderment, and awe.[15] In each context, the moment of interface was strategically configured and engineered to produce shock effects, with these effects generating the particular scientific and culturally accepted facts of electricity's constitution and behavior. Such electrotactile effects were not incidental or undesirable, but rather, often provided the defining aim of encounters with electrical apparatuses. During the eighteenth century, electricity rapidly became an object of scientific curiosity, a commodity for spectacular consumption, and a medical technology: its successful deployment in each context depended on tactile encounters between subjects and electrical machines that were governed by informal and formal protocols of electric touching. In the chapter's final section, I examine touch's role in organizing electrotherapeutic practice, arguing that an economy of electrocutaneous shock governed and regulated the medical application of electricity. While controversial in the scientific community, medical electricity remained a popular practice; by the second half of the nineteenth century, a host of new machines, techniques, and instruction manuals gave structure to the therapeutic application of electrical charges. The physicians who used electrical machines as curative devices were driven by a techno-utopian hope that administering current to patients would relieve all varieties of patients' maladies.[16] These complex machines also constituted a type of tactile media, providing standardized and carefully calibrated electrotactile experiences while also giving a corporeal grounding to the notion that electricity flowed through bodies as it did through telegraph networks.[17] The currents communicated by electrotherapeutic apparatuses contained veiled messages about electricity's ability to move through nervous networks—messages that touch proved uniquely capable of both receiving and decoding.

My efforts in this chapter also show the emergence and operation of a cultural model of shock that was not metaphorized as psychic shock, but instead designated the phenomenological experience of feeling electrical charges coming into contact with the body. This disentanglement of embodied, electrocutaneous shock from psychic shock is a tricky maneuver, as it involves forgetting or putting aside a good deal about what has been written on the cultural history of shock in the nineteenth and early twentieth centuries.[18] But chronologically, a cultural model of electrically induced shocks that were experienced and interpreted pleasurably, epistemically, spiritually, and therapeutically predated the model of psychic shock commonly discussed in accounts of modernity. As Tim Armstrong suggests, prior to the second half of the nineteenth century, shock described the material impact of physical forces on the body, largely divorced from any connection to mental life.[19] Genealogies of shock frequently emphasize the continuity between these models by linking the notion of physical wounding to an enduring state of mental trauma. Wolfgang Schivelbusch, for example, productively follows shock's transition from physical to psychical, tracing the original meaning of shock to its roots in modernized military conflict. As a term that designated "the clash of two bodies of troops, each of which represented a new unified concentration of energy by means of the consolidation of a number of units into a deindividualized and mechanized unit," shock connoted a move into a mechanized and synchronized model of combat that fused together and combined energies from previously disparate bodies.[20] Citing the damaging impact of these new hybrid energies on bodies, Schivelbusch situates the later adoption of shock by psychoanalysis to designate a type of mental wounding as contiguous with its prior military usage.

However, Schivelbusch largely overlooks the productive intermediary step, where touch came to be commonly used to designate the not *necessarily* traumatic sensation of electricity striking the body. This notion of electric shock as a distinct phenomenological experience makes a brief appearance in Schivelbusch's genealogy, when he quotes a nineteenth-century military surgeon who likened the experience of a firearm injury to an electric shock: "the wounded man does not experience the wounding as such but rather feels the concussion or a shock similar to an electric one."[21] In this account, the experience of electric shock provided the experiential ground for the metaphor of traumatic shock suffered from a battlefield wound. Electricity's impact on the body, then, was not always framed in traumatic terms; frequently throughout electricity's early history, encounters with electric shocks were taken to be simultaneously amusing and enlightening. Further, the case studies in this chapter show that shock was not simply articulated in binary terms—as something either present or absent—but described instead in excessively precise detail. Such rich linguistic transcriptions of encounters with electric shock were taken to be essential components of scientific reportage, fundamental to forward progress in the project of demystifying and controlling electrical forces. Even when they induced pain, these electrocutaneous shocks were understood as *generative,* indicating, for example, the subject's participation in a thoroughly modern culture of scientific experimentation or their exposure to a force believed to have curative effects.[22]

Owing to the curious etymology of the term "communication," the notion that electrical machines communicated with human bodies through touch would not have been remarkable in eighteenth- and nineteenth-century scientific parlance. As John Durham Peters details, earlier uses of the word designated both imparting and transference, often independent of human actors.[23] Consistent with this usage, mid-eighteenth-century accounts of electrical experiments consistently referred to electricity (or electrical *fluid,* as it was initially understood) being communicated from one inanimate body to another. The act of making one's body part of an electrical circuit, a frequent feature in electrical experiments of the day, entailed becoming part of a series of bodies that were thought to communicate electricity sequentially between them. Electrical machines, therefore, facilitated a type of nonlinguistic, tactile communication both between networked humans and the machines they interfaced productively with. It is easy to forget that, until the rise of electric telegraphy in the 1830s, machines capable of artificially generating, storing, and transmitting electricity remained hammers in search of nails—objects of much hope, excitement, curiosity, and wonderment, but outside of their use in medicine, of little practical value. The communicative property of these early machines helped fuel their rapid spread both within and beyond scientific communities.

Tangible Networks

With the invention of the electrostatic generator in 1663, the science of electricity took a significant forward step. The generator allowed electrical charges to be produced by mechanical means, as the machine's spinning wheel produced frictional electricity capable of imparting a charge to all sorts of bodies—human and nonhuman, organic and inorganic. A tradition of carefully attending to the tactile relationship between these generators and their human operators developed from the earliest experiments with the new machines; Francis Hauksbee, in a 1709 trial with a primitive electrostatic generator, noted the peculiar way the machine acted on his body:

> I was surpriz'd with the appearance of a brisk and vigorous Light continued between the point of my Finger and the Glass. It was not only visible on the Finger; but . . . seem'd as it were to strike with some force upon it, being easily felt by a kind of gentle pressure.[24]

Prior to the 1730s, however, most experiments with electricity aimed at demonstrating the attractive or repulsive power of charged objects and bodies rather than making the arcane force directly legible to the human senses. Beginning in 1733, the human body's sensorial responses to electrification assumed a new centrality, as the British natural philosopher Stephen Gray concentrated his experiments not on attracting or repelling bodies but on producing visible, tangible, audible, and olfactory evidence of electricity's behavior.[25] A fascination with the mysterious sensation produced as "electrical fluid" entered or exited the human body structured the design of experiments and the orchestration of experimental subjects' bodies. While this novel experience elevated tactility's importance in scientific data-gathering, it also posed a challenge to researchers, as reporting on experiments required subjects to precisely articulate the sensations felt on contact with electrical charges. Of all the senses impacted by electricity, touch provided the most detail and differentiation—the visible sparks and audible cracks produced by electrical discharges were adequate evidence of electricity's presence, but they provided little insight beyond a mere binary registering. Linguistic transcriptions of tactile encounters with electricity quickly became the most accurate means of experimental reportage: by comparing nuanced accounts of other experimenters' trials to their own sensate experiences touching electrical discharges, these researchers understood themselves to be revealing the essential characteristics of the electrical fluid, with its behavior registered by the shock it produced as it moved between objects and human bodies. While electricity was understood as communicating to all the senses, it spoke most directly to the sense of touch. In the lab, natural philosophers assigned experimental subjects the task of translating the messages the tactile system received from electrical charges into an intelligible language that could be disseminated throughout the nascent scientific network of "electricians."[26]

The body's newfound capacity to act as a storage device for machine-generated electrical charges spurred both scientific and popular curiosity in electrostatic generators. In the late 1730s, Leipzig physicist Georg Matthias Bose invented his own more powerful version of generator, enabling him to carry out a range of amusing demonstrations in courts, salons, and parlors. Bose's most famous trick, the *Venus eletrificata* (the electrified Venus, or the electric kiss), depended on the sensation electricity produced as it passed from its human container into another body.[27] The electrified Venus was simple enough to replicate, provided one followed the precise instructions Bose laid out: a woman stood on a nonconductive pedestal (often made of wax) while an operator charged her body using Bose's electrical machine. An unsuspecting male suitor was invited to try to kiss the wax-perched woman. On attempting to bring his lips into contact with the female lure, the male received a violent shock at his mouth, as the electrical charge passed from the woman's body—using her lips as a conduit—to the body of the man (see Figure 1.1). Under the right conditions, the shock could even traverse the space between male and female without their lips ever meeting. Bose colorfully described his own experience with the kiss in a 1743 poem, noting the brutal materiality of the electrical force:

Figure 1.1. An 1800 reproduction of the electrified Venus. Georg Matthias Bose, Deutsche Museum, Munich, Archive, BN09340.

Once only, what temerity!
I kissed Venus, standing on pitch.
It pained me to the quick. My lips trembled
my mouth quivered, my teeth almost broke.[28]

The male's capacity to endure the painful sensation inflicted by the transferring of electricity between bodies therefore became a marker of his courage: if the man persisted with the kiss in spite of the powerful blow that struck his face, he proved himself successful in the trial.[29] Though the female participant also experienced the shock, hers would have been of a lesser intensity, as charges leaving the body produced less violent sensations than those entering it. Bose prescribed a dramatic cost for defeat, advising any man "scandalized" by the electric kiss to cast himself into the ocean.[30]

Considering the kiss as an electric game—complete with defined participants, formal rule structures, and specified conditions of both victory and defeat—positions it as a quasi-public site of spectacular gendered play with shock at the center of the gamic encounter between the assembled players. In the kiss, the three players each assumed roles determined by their gender, with the male electrician and the female lure pitted against the suitor. The game's parameters were fairly explicit: the male participant had a defined objective (kiss the woman), conditions of success (make contact with the female's lips), and conditions of failure (not making contact with the female's lips, due to the shock of the charge jumping from her mouth to his on his approach). The conditions of victory were reversed for the electrician and his female collaborator; the woman served as a sort of avatar for the electrician, merging with his machine to form an electrified obstacle that the male participant sought to overcome. Particularly with early iterations of the electrostatic generator, producing and containing an electrical charge required a practiced and dexterous skill on the part of the electrician; the kiss provided an opportunity for him to spectacularly display his mastery over this unwieldy force, via the female body-as-battery. For the electrified Venus, her ability to repel the suitor indexed not only her capacity to store what was then known as the "electrical fire" but also served to indicate the painful passion she could unleash (enabled, of course, by the male electrician) upon intrepid suitors.

As a form of popular and spectacular entertainment, the kiss often served as the literal point of first contact between eighteenth-century bodies and artificially generated electricity. Such a positioning emphasizes both the playfulness and sociability of early empirical encounters with electricity, as the successful execution of the electric kiss depended on establishing a hybrid relationship between human participants and the electrostatic machine. The operator, often concealed from both the audience's and the male player's view, provided human power for the generator, while the female participant served simultaneously as a battery for the electrical charge and as a lure for potential male players. Though electrical contacts required the use of a deftly operated machine to charge the female participant, the human–machine relationship remained primarily one between humans and humans; the machine only attained expression by communicating its energy though a human medium.

Bose's curious amusements enjoyed a quick popularity; his 1744 *Tentamina Electrica* included extensive instructions for reproducing the demonstrations, and they rapidly spread throughout Europe, making their way to the British colonies in America by the decade's close.[31] As with Gray's experiments, the electric kiss depended on the tactile system to register electricity's passage out of the body. However, Bose's kiss, by moving the electrical charge between gendered human bodies, altered the signifying function of the electrotactile sensation; shock became both a marker of stored sexual energy and a means of repelling—or at least impeding—the male's amorous advances. Further, shifting the experiential context for the sensation from the lab to the parlor reoriented the subject. Where in Gray's trials shock served as a means of demystifying the strange power of electricity, Bose used it to induce bewilderment, wonder, and a curious sexually charged pleasure. Though both were oriented productively, one aimed at enlightenment and the other at popular amusement. Taken together, these examples illustrate the overarching point that the ability to produce tactile effects drove both the design and use of early electrical machines.

The invention of the Leyden jar in 1745 allowed the electrostatic generator's electrical charges to be stored, enabling electricians to assume increased control over the discharge of electrical fluid while simultaneously providing the means to accumulate progressively stronger charges. Reports of experiments with the jar employed dramatic language to describe the first contact between experimenter and jar—Pieter van Musschenbroek, one of the jar's inventors, reported feeling a shock in the arms, shoulders, and breast, requiring two days to recover from the blow.[32] The sensation disturbed Musschenbroek so deeply that he refused to repeat the experiment. Trumped-up accounts of electric shocks fueled public interest in the sensations produced by interfacing with the Leyden jar; as Joseph Priestley explained in his 1767 summary *The History and Present State of Electricity*: "every body was eager to see, and, notwithstanding the terrible account that was reported of it, to *feel* the experiment."[33] The sense of touch figured prominently in each account of encounters with the jar, as participants described the uncomfortable and disagreeable sensations they experienced in their shared quest to gather new data about electricity. One experiment, which involved simultaneously touching a charged Leyden jar and a gun barrel, resulted in "a great Part of the nervous System [receiving] a Shock so violent, that it would force the strongest Man to quit his Hold, and turn him half-round." A fourteen-year-old boy who tried the experiment described feeling as if "his arms had been broke short off at the elbows, and that he had been cut into two Parts just below the breast."[34] After partaking in the gun barrel experiment himself, Royal Society fellow Tubervill Needham declined to locate the shock as a pain, calling it instead "a mere sudden convulsionary motion, or rather a Shock, which surprises much, and is indeed an uneasy, though not a painful Sensation."[35] While Needham may have differentiated the sensations caused by these experiments from pain, he nevertheless understood them as involving extreme discomfort—subjects were described as "having courage enough to suffer" the various experiments with the charged Leyden jar.[36] Priestley implored his readers to disregard their fears, and rather, to measure "the reality" of the electrical shock they experienced against the descriptions,

exaggerated by feelings of "terror and surprise," that natural philosophers gave on receiving shocks for the first time.[37] Assured by
the aggregated tactile experiences stored within his volume,
members of Priestley's audience could safely engage in autoexperimentation, administering shocks to their own bodies without
being gripped by the same feelings of dread and bodily vulnera-
bility that had seized earlier researchers.

In the years following the jar's discovery, enterprising individuals around Europe began publicly administering its shock for a
small fee. The jar's ability to consistently deliver an electrocutaneous sensation to participants drove its successful commercialization.
Like the electric kiss, the Leyden jar and its affiliated protocols
constituted a medium for transmitting a tactile experience oriented simultaneously toward amusing, educating, and awing audiences. But the jar's appeal also presented a challenge for conventional models of spectatorship: though the public experiment was
a common technique for disseminating knowledge in eighteenth-
century Europe, the Leyden trials hinged on a nonvisual mode of
witnessing. Spectators were accustomed to being tasked, through
viewing these experiments, with multiplying the witnesses to scientific fact. With electricity, however, vision proved an inadequate
register; seeing signifiers of electricity's presence—sparks leaping
between bodies, bodies convulsing as electrical charges moved
between them—only served to fuel the desire for an embodied,
tactile encounter with electricity.

Of these myriad public demonstrations with the Leyden jar,
those carried out by the Abbé Jean Antoine Nollet are perhaps the
most well known to media historians. In a 1745 trial performed for
the delight of King Louis XV, Nollet passed the charge from a jar
through 180 hand-holding French Royal guards, each of whom
leapt into the air with apparent simultaneity as the charge moved
through them. Tom Standage, in *The Victorian Internet,* situates Nollet's later demonstration—which used monks connected by pieces
of iron wire (rather than soldiers connected directly by hand)—as
a sort of prototelegraphy, visible proof of electricity's ability to
move instantly through wired bodies.[38] By contrast, in their work
on electric body manipulation as performance art, Arthur Else-

naar and Remko Scha frame Nollet's demonstration as a "social sculpture," where electrically connected bodies facilitated a previously impossible synchronization of bodily movement.[39] Both interpretations focus on the spectacular dimensions of Nollet's networked bodies, assuming the visible event of the monks jumping when struck by the electrical charge to be the primary aim of the trial. Certainly, given the nobility of the audience Nollet performed the 1745 trial for, it would be a mistake to brush aside the performance's visibility altogether. However, the decision to try to pass the charge through participants was one born out of practicality and necessity—faced with crowds of onlookers who wanted to experience the shock of the Leyden jar, Nollet thought he could make the process of administering sensations more efficient by asking the participants to hold hands (see Figure 1.2).[40]

Figure 1.2. A drawing of Nollet passing the Leyden jar's charge through a circle of hand-holding French guards. From Louis Figuier, *Les grandes inventions anciennes et modernes* (Paris, 1870), 264.

By passing the charge between participants, Nollet effectively succeeded in transmitting tactile sensations through the networked bodies, a type of broadcast tactility that prefigured later attempts by twentieth-century haptic interface designers to transmit standardized, machine-generated tactile sensations through electrical networks. Previously communicated only from machine to human, or between individual humans, shock in Nollet's trials became an experience shared by everyone in the chain.[41] The ocularcentric bias of media historiography conceals crucial aspects of electricity's relation to the body from the view of contemporary analysis—our contact with this archive is structured by a set of ontological assumptions about the constitution of media and the senses they act on. Nollet's ad hoc networks did not merely presage the development of electric telegraphy, as Standage claims. Rather, these experiments with interconnected bodies were a form of electrotactile telegraphy, where the sender could inscribe a message simultaneously onto the nerves of all connected members. Following the protocols of these demonstrations produced a specific set of instrumentally oriented tactile effects that were not accidental by-products of electrically networking bodies. Instead, they were intentional outcomes generated by the adherence to carefully crafted instructions. Shock became a commodity that could be consumed en masse: as experimenters unleashed this new force on them, assembled audiences bore bodily witness to the fundamental characteristics and capacities of electrical flows.

In addition to exhibiting and communicating electricity's embodied effects, this tactile network proved crucial to revealing key aspects of electricity's movement through both human and nonhuman bodies, with felt shocks indicating both the speed at which charges moved and the effects of a charge's passage through different physical bodies. In the absence of suitable instruments, the human tactile system was tasked with registering the behavior of electrical currents. Altering an experimental configuration, while holding the human bodies in the network constant, caused tangible changes to the shock experienced by those bodies. These felt changes were articulated by participants, transcribed as experimental results, and then distributed through printed scientific journals. In 1747, for example, members of the Royal Society attempted to create a circuit composed of several human bodies, two iron rods, a length of wire, a Leyden jar, and the Thames River. Upon closing the circuit, the charge passed through the observers on the side of the river where the electrical machine was stationed, instantly traversed the river, and struck the observers on the opposite shore. Those standing near the machine experienced a strong shock, while those struck after the charge passed through the water experienced a lesser intensity of shock.[42] In the many variants on this experiment, the electrotactile subject remained the one constant—asked to register and report on the charge's strength, the observer's sense of touch figured prominently in the construction of these experimental electrical circuits. Bodies participated in a communicative relationship with the immediate sensation of electricity but also more broadly in a sociotechnical relationship with the other bodies interposed in the circuit. Experimenters organized and oriented the participants' perceptual systems epistemically, seeking to construct a stable body of knowledge about electricity's behavior. Properly configured, the tactile systems in the circuit would allow experimenters to ascertain the speed at which electricity traveled, the materials that best conducted the electrical charge, and the maximum distances over which shocks could be transmitted. In short, these experiments in tactile telegraphy conceptualized the electrical charge as a signal, employing touch as a means of measuring the signal's decay as it passed over various distances and through different substances. The systems of electric telegraphy invented in the nineteenth century can therefore be read as modifications on the electrotactile telegraph. Whereas tactile telegraphs depended on direct contact between machines and human bodies to measure, record, and transcribe the flow of electrical charges, the later devices replaced the human tactile system with machines that made electricity legible to the nontactile senses, allowing electricity, once decoded by a human receiver, to become language.

The fundamental point I want to stress here concerns reliance of forward progress in electrical research on a disciplined, cultivated electrotactile sensibility that would be further refined in successive experiments. Touch's transformation into an instrument for measuring electrical signals and their decay proved crucial to the production of new knowledge about electricity, while also staging the twentieth-century emergence of formalized, machine-rendered tactile languages. Human bodies had to accommodate themselves to the parameters of electrical machines. For both designers of and participants in these experiments, electrical demonstrations depended on a type of craft knowledge only cultivated through the careful repetition of experiments, the sharing of successful techniques, and the building of standardized machines that produced and stored electrical charges. As with other new media, the various electrical machines of the eighteenth century were accompanied by new sets of bodily techniques that circulated through both formal and informal knowledge networks. These machines were, to borrow from Gilles Deleuze, "social before being technical"—the production of facts about electricity depended on contextually situated human bodies to enact, experience, and interpret machinic shocks.[43]

Shock Me Like an Electric Eel

Still gripped by curiosity about the Leyden jar and its variants, European natural philosophers became fascinated with reports from the Dutch colony of Guiana about the strange sensations produced by contact with torpedo fish (more commonly known as the electric eel). The Englishman Edward Bancroft, among the first to note the similarities between the sensations produced by the electric eel and those experienced on contact with the Leyden jar, sparked interest in the eel by publishing accounts of the research he carried out in Guiana.[44] Inspired by Bancroft's report, enterprising electricians on both sides of the Atlantic set out to verify his claim of an affinity between the forces. Those who had contact with both the fish and the Leyden jar noticed a similarity between the sensations they experienced: in spite of emanating from different objects, a phenomenological affinity between the forces that came into contact with the skin seemed to point toward a common origin.[45] Prior to the proliferation of the jar, the torpedo fish had been a species that fascinated cultures for millennia, represented in Egyptian tomb paintings and hieroglyphs and used as a therapeutic in both ancient Greek and Roman medicine.[46] In the seventeenth and eighteenth centuries, the torpedo fish and other similar species captured the attention of natural scientists, who sought to explain its power to deliver painful and numbing shocks on contact.[47] In initial experiments during the late seventeenth century, the fish was thought to produce its effect mechanically by a sudden contraction that so deeply penetrated the tissue of its prey that it impacted the nerves, inducing temporary numbness. This theory, however, could not account for the experiences of those who received the shock at a distance through a mediating agent, such as a metal fishing rod or dense fishnet.

For the electrical researchers fortunate enough to lay hold of one, the fish, like the Leyden jar before it, became a site of generative uncertainty that troubled the assumed division between inanimate and the animate bodies:

> The Eel, touched with an iron rod, held in the hand of a person, whose other hand is joined to another, communicates a violent shock to ten or a dozen persons thus joining hands, in a manner exactly similar to that of an electric machine.[48]

In response to the indeterminate sensations produced by contact with the animal, these researchers developed a new system of experiments intended to expand existing accounts of electricity so that they could accommodate the challenge posed by the fish. The fish, then, can be read as an epistemic thing that had to be managed through the development of experimental techniques and practices. Handling the fish, like handling the jar, required experimenters to assimilate to and hone a set of increasingly specific protocols, as the fish's behavior proved erratic and difficult to manage. These considerations prompt Delbourgo to describe the electric fish as a sort of "organic machine" with its own techniques of proper usage and care.[49] Building on his argument, the fish can be thought of as an epistemic apparatus that provided a means of harnessing animal electricity for the specific aim of generating scientific knowledge through the solicitation of sensory experience.[50] The structured interfacing of human bodies and electric fish produced a shock that could be transcribed, categorized, circulated, and compared by electricians in this nascent knowledge network. Experiments with the fish mimicked experiments with the Leyden jar in their structure and aims—as in Nollet's demonstrations, participants held hands to pass shocks between them, with the fish substituting for the jar as the source of the charge.[51] The experimental protocols required to solicit tactile shocks from the fish thus built on the established practices for working with artificially generated electricity, as the fish was simply swapped into experiments as a replacement for the jar. The arrangement of human bodies and their instrumentalized tactile systems provided the constant in this experiment, with source of shock (either the jar or the fish) serving as the variable. Holding the body constant facilitated the multiplication of scientific witnesses to the fish's powers. Visible evidence of the fish's effects—bodies jerking involuntarily as the force emanating from the fish moved through them—encouraged onlookers to take part in demonstrations, often by holding hands and becoming an active part of the fish–human circuit. Like the jar before it, the fish became a means of transmitting and sharing a particular variety of epistemologically oriented tactile experience.

Though it was the numerous apparent similarities between the spark of the jar and the shock of the fish that prompted European interest in the animal, differences between the two both piqued the interest of researchers and reconfigured the sensory mechanisms needed to investigate electricity. Where bright sparks, sometimes as long as 18 inches,[52] typically accompanied the discharge of electricity from the Leyden jar, "torpedinal electricity" (as Royal Society fellow John Walsh dubbed it) produced no such visible marker.[53] The invisibility of the fish's shocks further shifted the burden of witnessing electricity from vision to touch: in the absence of a spark perceptible to the eye, touch became the only sense capable of directly accessing the existence of animal electricity. Demonstrations with electrostatic generators and Leyden jars were plurisensory spectacles, featuring loud cracks and bright sparks that electricians could intensify by using larger jars or connecting more jars in series.[54] Absent the sparks, and with considerably less dramatic cracks, trials with the fish altered the sense ratios of electric spectacles.

Furthermore, touch facilitated the registering of other crucial differences between the organic fish and its inorganic counterpart. After each discharge, Leyden jars typically needed to be reenergized through the operation of an electrostatic generator; if a second shock could be obtained after the first, it was often of a weak and diminished character, barely perceptible to the experimenter. The fish, however, required no such revitalization: its capacity to repeatedly and constantly issue shocks indicated that it somehow managed to generate its own charge. Consistent with earlier experiments involving electrostatic generators, a refined and practiced tactile sensibility proved essential to the process of gathering the raw data that would inform broader claims about electricity's behavior. But unlike the jar, whose forces needed to be recharged after each trial, the fish's seemingly limitless supply of energy required substantial feats of endurance by those who tested its powers—Walsh, for example, reported receiving "forty or fifty successive shocks from nearly the same part . . . with little, if any diminution in their force."[55] Beyond simply having to suffer these repeated shocks, the experiment required Walsh to maintain sharp enough attention to be able to both count the shocks—to quantify them—and to note any difference in their intensity as they repeatedly struck his body. An instrumentalized tactility again registered the decaying strength of electrical force, but here that energy issued from an animal pushed to its limits by an experimental system intended to map the differences between the realms of the natural and artificial.

The experiments in this system provided a structuring interpretive framework for the fish's shocks, with the fish becoming part of an experimental assemblage that consisted of four components: the experimenter's body (used as a means of summoning and soliciting shock from the fish); the experimental protocols (which served to regulate and configure the performance of experiments); the descriptive language used to articulate the experience of shock yielded by the encounter with the fish; and finally, the fish itself (which had to be kept healthy and alive to produce consistent and sufficient shocks). Each of these interdependent components brought with it a set of challenges related to issues of standardization, networking, and circulation, all revolving around the tactile interface between fish and human. The materiality of human and animal bodies gained expression through the experimental system, as these bodies frequently proved unstable, faltering, and resistant to easy measuration.

While the human body in general may have served as the constant between the earlier Leyden trials and the new investigations into eels, specific human bodies varied wildly in the way that they experienced and registered the shock, making it difficult to divine the objective character of the shock signifier. Further, even the same body did not experience the shock consistently: an individual experimenter's sensitivity to the fish's charge often differed based on the experimenter's own health. During Walsh's trials with the fish, for example, a researcher who had previously possessed a great resistance to the fish's shock became ill. In spite of his condition, he continued to participate in the trials and reported a suddenly increased intensity of sensation upon contact with the fish. The fallibility and inconsistency of a tactile system governed by the idiosyncrasies of the individual body presented an obstacle that these experimenters labored to overcome and factor out. Though the tactile system may have been deployed as an instrument for scientific observation, it frequently proved unreliable, inconsistent, and fallible in that role. And while these subjective variations in reaction to the fish's charge prompted crude speculations into the possible physiological causes of such inconsistencies, the human body was not, as it would later become, the object of knowledge in these electrical experiments.

In spite of these wide subjective differences in experience with shock, electrical researchers still needed bodies to serve as instruments, however imprecise the data they produced. Those European electricians who traveled to foreign lands with hopes of finding healthy fish for their trials often pressed local populaces into service; indigenous, slave, and colonist bodies alike were ordered as raw materials that could be plugged into ad hoc experimental apparatuses. In Guiana, for instance, electricians, colonists, and African slaves all held hands as the fish's shock passed through them, with the increasing number of participants both functioning to multiply witnesses and introduce greater resistance into the electrical circuit. The different ways that bodies experienced shock was explained through the lens of racialized and gendered colonial science, as non-European and nonmale bodies seemed to be responsible for interrupting and decaying the charge that passed between heterogeneous bodies. The same fish that dealt European bodies painful blows appeared to be tolerated by African slaves with little discomfort, prompting investigators to feed forward racialized assumptions about the higher pain thresholds inherently possessed by Africans. Crucially, it was (male) European scientists who transcribed and interpreted the performed experience of African bodies; the Africans were not allowed to self-author, only to function as instruments that were read by the European experimenters. Investigators similarly thought that the gender of participants' bodies could inhibit the transmission of shock: in one trial, an experimenter attributed the interruption to a female participant who had earlier proven capable of handling the fish without receiving its charge, suggesting that the insertion of a single insensible gendered body into the chain would cause a problematic disruption in the circuit.[56] The role of raced and gendered bodies in these trials calls attention to pain's status as simultaneously physiological and political: pain achieves a political function in the governing of both its infliction and its transcription, in the capacity of pain to be registered through language, and managed strategically by the experimenter through the successive refinement of experimental trials.[57]

The development of rigid experimental protocols in some ways helped mitigate these subjective inconsistencies of shock experiences. By carefully circumscribing the conditions under which fish were to be touched, researchers hoped to provide a more consistent, repeatable, and standardized set of electrocutaneous experiences. The networking of observers' bodies aimed to accomplish a similar end: like Nollet's demonstrations connecting soldiers, bodies were often linked to form a circuit that the fish's shock could circulate through, ensuring that the same event could be experienced simultaneously by a network of differently sensing bodies.[58] Just as they had in experiments with the Leyden jar, researchers mobilized and refined protocols of touching to stabilize participants' experiences; these experimental protocols allowed for the control, management, and productive aggregation of encounters with the fish while also anticipating the experiments' replication by future researchers.

To help further overcome and mitigate against the subjectivity of the individual bodies that were called on to serve as experimental witnesses, those who reported on the fish trials gradually adopted standardized language to describe and classify shocks. Much of this process merely involved borrowing the descriptive lexicon from reports of trials with Leyden jars, which provided a readymade vocabulary to index electrocutaneous sensations. In Walsh's experiments, he divided up the sensations according to intensity, using several adjectives to indicate the strength of the experienced shock: "smart," "immediate," "strong," "very strong," "plain," and "weak." He also attempted to quantify intensities "as far as sensation could decide," declaring the shocks the fish gave when touched in the water "not to have near a fourth of the force of those at the surface of the water, nor much more than a fourth of those intirely in air."[59] Quantifying the temporality of shock also provided another distinctive means of measurement, as counting the number of shocks the fish could produce in a given interval allowed experimenters to determine whether rapidly issuing successive shocks depleted the animal's store of electrical charge. For the subjective experiences of shock to be abstracted and segmented as fixed quantities, participants had to carefully hone their attention to electrocutaneous sensations; the generation of a stable body of facts about electricity, then, relied on the experimenter's ability to pass experience through the filter of the shared terminology for describing touch.

The fish also introduced problematic variations into experiments that had to be managed by inventing new standards of handling and care. Fish of different sizes were frequently found to produce different intensities of shock, making the execution of standardized experiments difficult. Though rich descriptions of the physical characteristics an individual fish possessed helped researchers account for and explain variances in experimental findings, producing repeatable results in trials with the electric animals was necessarily more difficult than doing so with the Leyden jar. The intensity of the shock communicated from the fish was also tied closely to its health, meaning that a devotion to maintaining the health of the fish became part of the experimental protocol; any diminution in the fish's state introduced a problematic variance into experimental results. The fish were difficult not just to handle but also to capture and transport—keeping the different species of electrical fish alive and healthy through the arduous journeys from their native lands to the labs of European and American researchers created a challenge both for scientists and for those trying to commercialize contact with the strange animal.[60] In Walsh's first experiments with the fish, conducted on a specimen brought to Paris, he reported receiving only slight, though constant shocks from his contact with the animal. The search for fish with more vigor prompted him to embark on a trip to the French island of Ré, where he had the opportunity to engage with healthy torpedo fish in their native environment. Later researchers, such as Alexander von Humboldt, undertook longer journeys in attempting to bring their bodies into shocking contact with the fish.[61]

Protocols of human bodily comportment, description, and the care and handling of fish were thus attempts to manage and control the variance among shocks received at the moment of experimental contact. Properly configured, the fish–human interface yielded a captivating shock that disrupted nascent theories about the electricity stored in the Leyden jar. These shocks were delivered not just to the individual bodies that participated in the trials; they were also registered by a social body still attempting to come to terms with a newly rediscovered, mysterious force suddenly acting on it. The fish functioned as a type of new electrotactile media, communicating messages—in the form of electric shocks—that disrupted and upset the accepted divisions between the natural and the artificial. The similarity between the shocks produced by the jar and the torpedo fish threw conventional accounts of electricity into disarray while simultaneously solving a mystery that vexed every culture that came into contact with the quasi-mystical fish. Experiments with the artificial electricity stored in the Leyden jar helped natural philosophers decode the mysterious sensation produced by contact with the fish. And the relationship proved reciprocal: "as artificial electricity had thrown light on the natural operation of the Torpedo, [research on the torpedo] might in return, if well considered, throw light on artificial electricity."[62] The debate about electricity, then, centered around the question of how to produce, stabilize, transcribe, and decode shock sensations, with the fish serving as a machine that bodies productively interfaced with by following epistemically oriented experimental protocols.

Volta's New Media Apparatus

In 1800, amid the ongoing debate about the relationship between artificial and organic electricity, the Italian physicist Alessandro Volta (1745–1827) composed a letter to the Royal Society of London in which he detailed his invention of a new apparatus for generating electricity. Volta claimed that this "artificial electrical organ" proved capable of acting "incessantly" on the body through the repeated production of sensible shocks.[63] He framed the purportedly revolutionary apparatus as a merging of the artificial and natural mechanisms of generating electricity, the successful fusion of the Leyden jar and the electrical organ found in the torpedo fish. The letter embodied Volta's attempt to prorogate an apparatus that would serve as a type of new media device, specifically designed with the intent of transmitting a set of machine-generated tactile sensations that would prove disruptive to established theories of electricity. While previous electrical machines tacitly mobilized an epistemology rooted in shocking touch, Volta made his appeal to this tactile epistemology explicit in the letter, bluntly stating "this endless circulation of the electric fluid (this *perpetual motion*) may appear paradoxical and even inexplicable, but it is no less true and real; and you feel it, as I may say, with your hands."[64] As with the experiments on the Leyden jar and electric fish, shock in Volta's trials served a signifying function. However, Volta took aim at providing an interpretive framework for his readers, so that they would come to decode the strange shock signifiers produced by his devices as evidence of his new electrical theory.

The letter provided meticulous instructions for building two different types of electromotive apparatuses, the *couronne de tasses* (chain of cups) and a columnar apparatus (also referred to as a "pile"—a stack of metal plates that alternated between zinc and silver, separated by moistened paper discs; see Volta's illustrations in Figure 1.3). Each apparatus, Volta explained, produced its own distinct effects on the senses. In his discussion of the letter, Joost Mertens argues that the extensive detail Volta gave to the construction of his apparatuses illustrates his intention that they be rebuilt by the letter's audience; as Mertens explains: "the pile must be considered a public demonstration device, part of a strategy to promote general recognition of the fact of 'metallic electricity.' "[65] Mertens attempts to muddy the accepted thesis in histories of science that claims Volta's "epoch-making" letter succeeded in "terminating the debate [about animal and artificial electricity] that had gripped all of learned Europe," while "ushering in a new age of physics."[66] This debate between Volta and the Bologna physicist Luigi Galvani concerned the existence of an "animal electricity" distinct from "artificial electricity." Electricity, according to Galvani's theory, was not confined to shock-generating organisms like the electric fish but was common to all animals. Galvani's theory issued from a series of experiments he conducted, in which a dead frog's legs were made to spasm when they came into contact with metallic arcs. The legs, Galvani claimed, acted not unlike a Leyden jar, storing electricity that was discharged on contact with a suitable conductor. In these well-known and often-imitated trials, animals were again folded into systems of electrical experimentation with the aim of yielding fresh insight on the relationship between organic and inorganic. Although initially swayed by this theory, Volta soon came to believe that Galvani had misidentified the source of the charge: electricity was contained not in the frog's

Figure 1.3. Volta's chain of cups (figure 1 in the drawing) and columnar apparatus (figures 2–4). From Alessandro Volta, "On Electricity Excited by Mere Contact," *Philosophical Transactions of the Royal Society of London* 90 (1800), p. 430.

legs, but in the metals that were touched to them. Volta suspected that the disembodied legs spasmed in reaction not to energy leaving them through the metal, but instead as a result of the charge introduced to them by the metal rods.

The common argument goes that the sparks and shocks issuing from Volta's pile provided him with conclusive proof that metal contained an electrical charge. Mertens, however, suggests that Volta, in earlier lab experiments, had already invalidated Galvani's theory.[67] What he lacked, and what the pile provided, was a means of effectively transmitting and making tangible his proof of concept—what Mertens describes as an "amplifier" for the findings of his earlier experiments. Those proofs that had initially convinced Volta of Galvani's error required meticulous and painstaking laboratory procedures. Witnessing their results also proved difficult: the metallic electricity he found evidence of could only be registered by a custom-made electrometer, locally produced by an expert instrument maker. Volta wanted to design a "more spectacular demonstration." Or, in Volta's own words, the audience for his experimental findings "would like to see sparks."[68] Volta, in constructing his pile, aimed at the production of dramatic effects on the human sense organs, effects that would serve to spectacularly demonstrate electricity's presence in the metals Volta's apparatuses brought the body into contact with. The shocks generated by the pile could serve as powerful signifiers that Volta's new theory of electricity would explain.

Unlike the Leyden jar, which merely stored a charge generated by other means, Volta's artificial electric organ produced its own electricity, capable of continuously issuing repeated shocks. Though the artificial electric organ shared some superficial structural characteristics with the Leyden jar and with the fish's organ, its primary resemblances—those that caused Volta to assert both its kinship to and its departure from those prior epistemic things—could be most productively revealed through tactile inspection, by coming into a shocking contact with his "electromotive apparatus."[69] Volta celebrated his apparatus for its ability to produce unique tactile sensations, akin to the more familiar feelings of touching Leyden jars or electric eels but distinct enough to be considered an amalgamation of the two. Volta could take for granted that his audience would have at their disposal a storehouse of tactile memories gained from repeated experiments with both the fish and the jar. When he introduced the pile to his audience, Volta explicitly asked his readers to compare its tactual effects to those produced both by the Leyden flask and the torpedo fish:

> Is it not capable of giving every moment shocks of greater or less strength, according to circumstances—shocks which are renewed by each new touch, and which, when thus repeated or continued for a certain time, produce the same torpor in the limbs as is occasioned by the torpedo?[70]

Through this linguistic transcription of his own tactile experience, Volta prescribed particular conditions of attentiveness to be maintained when testing the new machine: simultaneously addressed to a social body and the individual bodies that composed it, the audience had to be ready to engage in a specific form of tactual witnessing that would allow them to clearly perceive the distinct character of shocks generated by the apparatus.

But in order to ensure that the instrument would produce in his audience the same effects that he had witnessed in his autoexperiments, the device first had to be built. And, as Mertens argues, the attention Volta devoted to making sure it would be constructed properly suggests that his aims were as much practical as they were theoretical, oriented toward the successful reproduction of a particular flavor of shock. Experimenters attempting to successfully rebuild the apparatus had to follow Volta's meticulous instructions, with proof of a correct build established by comparing their own sensations touching the machine to those Volta had described colorfully in his letter. As such, the specifications for assembling the machine were bound up inextricably with the techniques required for effectively touching it: the experimenter could only learn that the strange apparatus worked by attentively receiving the variety of shocks it communicated.

Once properly assembled, the experimenters could feel shocks of varying sensations by placing their hands at different locations on the apparatus—dipping one hand into a cup of water attached to the pile, with the other hand touched to the top of the column, yielded "a small picking or slight shock." Reconfiguring the positions of the hands resulted in "shocks that affect the whole finger with considerable pain."[71] Increasing the size of the apparatus,

accomplished by adding more discs to the column, brought a corresponding increase in the intensity of the shock, so powerful that
it was capable of extending "to both arms as far as the shoulder."[72]
The disc's placement in the column also altered the strength of the
sensation it produced on contact: the lowest plate in the stack produced a current that almost escaped the toucher's ability to notice
it, with each higher plate in the stack increasing the strength of the
charge perceptibly.

My point here is to suggest that Volta grounded the novelty of
his apparatuses in their capacity to produce distinct and unique
shock effects, legible to his audience as evidence of a new form of
electricity that differed markedly from those yielded by the jar and
the fish. In calling the pile an *electro-motive apparatus,* Volta showed
a keen interest not only in establishing the device's novelty but also
in rooting its newness in its capacity to induce unprecedented
sensations. "We must give new names," Volta declared, "to instruments that are not new only in their form, but in their effects."[73]
The pile functioned not only, as Mertens suggests, as a "demonstration device" for Volta's theory of electricity but also as a means
by which experimenters could produce new forms of knowledge
and pleasure. As Volta made explicit, the shock effects obtained
through the experiments were intended to be both "instructive"
and "amusing" for those willing to receive them.[74] Volta implored
his audience to enter into a relationship with the machines that was
simultaneously playful and epistemic: adding more discs to the pile,
applying charges to different parts of the body, and dipping the
hands into the various cups in the chain each provided a novel,
pleasurable experience of shock that wrote Volta's new theory of
metallic electricity onto the nerves of those who re-created his
experiments. Further, as evidenced by Volta's expectation that
his audience would be able to recognize the novelty of the sensations produced by the pile and the chain of cups, eighteenth-
century electrical investigators proceeded under the assumption
that touch provided the most direct evidence of electricity's existence and constitution. Each set of new electrocutaneous sensations induced by successive generations of electrical machines
served to index forward progress in electrical research. Volta

could only expect his audience to understand the pile as a novel machine because he theorized a link between shock, scientific training, and temporality: the novelty of the apparatus itself was of no consequence unless it strategically interfaced with a body practiced in touching electricity.

In the second half of the letter, Volta's investigations with his new machines took a curious shift inward: where electrical researchers up to that point had generally taken the body to be an (imperfect) instrument for knowing and revealing electricity, they took for granted the sensory mechanisms that facilitated this phenomenological access to the world of currents. In a series of experiments executed by applying the pile to each of the five major sense organs, Volta used the interface between human and electrical machine instead to generate knowledge about the physiological mechanisms responsible for the production of sensory experience. These trials later obtained monumental importance in the history of psychology, staging Johannes Müller's experiments in the 1830s that caused him to posit the doctrine of specific sense energies.[75] Presciently, Volta understood the pile, and the investigations that could be performed using its powers, as having significant ramifications for future anatomists and physiologists interested in the study of the senses:

> The electric fluid, which, when made to flow in a current in a complete circle of connectors, produces in the limbs and parts of the living body effects corresponding to their excitability, which stimulating in particular the organs or nerves of touch, taste, sight, and hearing, excite in them sensations peculiar to each of those senses.[76]

The significance of this move cannot be overstated. Volta effectively transformed electricity from an object made knowable by a stable sensory system into a force capable of disrupting the stability of the sensory system. Through this shift in focus, he instantiated electricity's positive deployment as a means of studying sensory mechanisms. Over the next century, electricity became a vital tool in new experimental studies of the senses, as investigators gradually came to believe that passing electrical current through the body's nerves provided a window into the psychophysiological processes responsible for the formation of sensory impressions. By 1900, batteries that generated electrical current and electrodes specifically designed to allow current to enter through a given sense organ had become common features of the sensory psychologist's toolkit.

The system of experiments Volta carried out on the senses followed the same form as those described earlier in the letter: first, a precise recounting of the experimental protocol; then a pragmatic detailing of the various techniques used to pass current into the body; and finally, a rich transcription of the sensations he experienced in each trial. Again using his own body as an investigative instrument, Volta proceeded to stimulate each organ in turn. Of the five senses, he devoted the most extensive attention and experimentation to touch; taste ranked second, vision third, and hearing fourth. Curiously, it was smell that proved to be the most vexing. When Volta passed current through the interior of the nose, it activated only the sensation of shock rather than any perception of odor, providing an exception to the general theory he advanced based on his trials with the other sense organs.

In following Volta's process—by walking with him as he passed current into his eyes, ears, tongue, nose, and all over his skin—I want to show not only that touch was central in the trials devoted specifically to investigating tactility, but also to demonstrate that touch played a significant role in his experiments with the other four sense organs as well. Volta's results implied that electricity activated in each organ an energy distinct to it (light in the eyes, taste in the tongue, etc.), as well as a distinctly tactile feeling of impact on the nerves, whether current was applied to the ears, nose, eyes, skin, or tongue. These experiments simultaneously *differentiated* the sense organs (by identifying them as possessing distinct energies, and through Volta's decision to inscribe the five-sense model in his experimental process) and *dedifferentiated* them (by identifying feeling as a common sensation that could be activated in each organ).[77]

To test taste, Volta passed electrical currents of different strengths through his tongue, which he accomplished by holding a silver plate between his lips and bringing the tip of his tongue into contact with it. He reported experiencing "very sensible sensations of taste" with a "decidedly acid" character. In contrast, substituting a zinc plate for the silver one produced a "less strong but more disagreeable, acrid, and inclining to alkaline" taste.[78] Positive and negative currents—electricity entering and leaving the body—he deemed to be accompanied by different and distinct tastes. Increasing the charge's strength, however, did little to alter the taste it conjured; instead, it heightened the force with which the charge struck the tongue, resulting in feeling a sharp blow, followed quickly thereafter by "a prickling more or less painful."[79]

Volta stimulated the sense of vision by applying one metallic piece of the apparatus either to his eyeball directly or to his moistened eyelid. To complete the electrical circuit, he pressed another piece of metal to the opposite eye, producing a "weak and transient light." Alternatively, completing the circuit by holding the metal in his mouth caused him to see "a flash much more beautiful" than he had witnessed when the current passed from one eye to the other.[80] In this set of experiments, the effects produced by varying the placement of the inducting electrode aroused Volta's wonder, as he found that moving it about the body while holding the other electrode constantly to his eye caused him to witness a marvelous array of changes in the light's character—via movements of the inducting electrode, Volta choreographed the "transient light" that danced in front of him.[81] This model of vision was, as Müller would later argue, fundamentally subjective, with the stimulus imperceptible to anyone but the experimental subject; although no light was present, Volta nevertheless witnessed it with a convincing viscerality. Divorced from external referents, these sensations seemed to originate in an energy inherent to the nerves themselves rather than communicating impressions from the exterior world. But the stimulus also possessed a distinctly tactile quality; alterations in the location and intensity of the induced current affected "the form as well as the force" of the light.[82] As with his investigations into taste, electrifying the eyes acted on two senses simultaneously. The tactile vision operating in these trials was not merely a metaphor for vision's synesthetic capacity to conjure sensations of touch (as it would be formulated in Alois Riegl's aesthetic theory a hundred years later) but was also a dramatic activation of both senses through the same stimulus.

A similar theme emerged in Volta's limited trials on hearing, where electricity's presence could be felt as well as heard. Of all these potentially injurious experiments forcing electrical current through the sense organs, the tests he carried out on his ears produced the most threatening sensations. On inserting metal probes deep into the canals of the ears, Volta felt a strong shock in the head; after a few moments of silence, he

> began to hear a sound, or rather noise, in the ears, which I cannot well define: it was a kind of cracking with shocks, as if some paste or tenacious matter had been boiling. This noise continued incessantly, and without increasing. . . . The disagreeable sensation, and which I apprehended might be dangerous, of the shock in the brain, prevented me from repeating this experiment.[83]

As with the sensation of light, these sounds were imperceptible to anyone but the experimental subject. However, Volta perceived these sensations induced in the ears in far greater detail than those produced by applying current to the eyes, and he interpreted this intensity as a sign that he might be subjecting his body—a necessary instrument for both present and future trials—to damaging effects.

In sufficient quantities, the electrical current proved capable of activating three sense organs simultaneously. The plurisensory effects of the new machine illustrated electricity's ability to activate new types of phenomenal experience that were divorced from any external referent. A "sufficiently strong" apparatus, connected to a metal plate held between the lips and touched by the tongue, set the sensorium afire, producing "a sensation of light in the eyes, a convulsion in the lips, and even in the tongue, and a painful prick at the tip of it, followed by a sensation of taste."[84] As a device that prefigured the later emergence of media that used electricity to pleasurably deceive the senses, the pile possessed a multimodal capacity that evades even today's most advanced digital media; able to act on vision, touch, and taste through the same point of shocking interface, it suggested new possibilities for both aesthetic and epistemic encounters with electricity.

While Volta attributed significance to his machine's capacity for activating the organs of seeing, hearing, and tasting, he clearly intended the pile to make its most direct, consistent, and productive appeal to the sense of touch. Discussions of the apparatus's effects on sight, hearing, taste, and smell were confined to a short and speculative section at the end of the letter, almost as an afterthought. In contrast, a practiced touch proved indispensable to the successful replication of the trials described in the letter, underscoring the machine's location in a continuum of prior electrotactile media. But unlike the earlier machines, Volta's batteries allowed for a more precise control over the administration and calibration of shocks: increasing the number of discs in the pile, placing the conductors at different points on the skin, and so forth, each provided the experimenter with the ability to customize the machine's tactual effects. Volta hoped that future researchers, in attempting to verify his results, would reconstruct the pile and experience its tactual effects themselves. In these cases, shock evidenced and registered something about the external world of currents rather than revealing the qualities of the mechanisms responsible for perceiving the shocks. Volta's trials probing electricity's effects on the tactile system marked a substantial departure from prior approaches, as electricity went from being the object of tactile knowledge to being an active producer of knowledge about the mechanisms responsible for inducing tactile experience.

Unlike the specialized sense organs confined to a specific bodily locus, the sense of touch was distributed across the entire body. As such, touch had differing qualities that were revealed through the induction of current. The skin acted as a shield that decreased sensitivity to electricity: where it was healthy and undamaged, Volta noted experiencing a "very strong and pricking pain," but passing current into flayed or wounded skin excited an acute pain, so intense that Volta could not endure it for even a few seconds.[85] These trials suggested that touch consisted of varying nerve structures; the skin was not an objective register of current but possessed instead its own peculiar characteristics that allowed it to differently register the same stimulus, depending on its structural integrity, or the configuration of nerves contained in the stimulated site. Carefully applied, electricity had the capacity to expose and specify these characteristics. This finding prompted new investigations into the nature and structure of tactile experience. After Volta, the aim of tactile experiences produced through laboratory encounters with electricity shifted. No longer oriented outward at revealing the character of the electrical fluid encountered through touch, the focus instead turned inward, directed at uncovering and revealing the fundamental characteristics of the tactile system that electricity acted on. Electrical machines, as epistemic things, served a double role: initially objects of an instrumentalized tactility, they gave new knowledge about the interlinking of animal and artificial electricity that proved essential to collapsing the border between nature and machine. Following Volta, their later deployment as instruments for investigating the production of tactile sensations allowed touch's constitutive and variegated elements to be charted and plotted at a progressively increasing level of specificity. Future machines capable of productively deceiving touch—or communicating through the newly quantified tactile channels—built on the principle of electrotactile interfacing established by eighteenth-century experimentation.

The Art of "Electricization"

While public displays of electricity had captured the wonder of anxious audiences, those early spectacles also begged the question of its utility; by demonstrating human control over electricity, electricians tacitly suggested that their new machines could someday be put to a positive practical use.[86] Audiences often phrased the utilitarian question as a belittling taunt to exhibitors—and defiant natural philosophers refused to engage it, opting instead to insult anyone who suggested that they instrumentalize their grandiose research. Benjamin Franklin, for example, famously countered these challenges by asking "What is the use of a baby?"[87] Halle doctor and professor Joseph Gottlieb Krüger was among the first to address this problem, noting in 1743 that, as no application had been found for electricity in theology or jurisprudence, "where else can the use be than in medicine?"[88] Krüger urged his students to experiment with electricity as a therapeutic, and in response, they began applying charges to bodies in hopes of treating all manners of maladies. News of electrical cures quickly spread through European medical journals, encouraging enthusiastic and skeptical physicians alike to try their hands at replicating the successes others reported.[89] From the earliest days of their practice, electrotherapists were constantly hounded by claims of quackery (charges sometimes leveled by competing physicians trumpeting the superiority of their own methods), but medical electricity attracted growing numbers of devotees, eventually culminating in a "Golden Age of Electrotherapy" generally dated from 1880 to 1920.[90] Electrotherapy's subsequent decline can be attributed to a host of cultural changes that rapidly positioned electrical cures as obsolete and resulted in the quick discarding of the many therapeutic devices that had accumulated during the practice's extended run of popularity.[91]

Moving quickly through electrotherapy's long history, I highlight its dependence on a technics of both touching and being touched by electricity: knobs, buttons, dials, brushes, rollers, and electrodes used in the application of medical electricity configured the tactile experiences of patients and physicians alike. They also prefigured the later development of computing technologies capable of dynamically rendering and synthesizing tactile sensations. Considerations of touch and of electrotactile sensibility were primary concerns in the design of therapeutic apparatuses, in proscriptions for applying current, and in the perceived efficacy of electrical treatments. Here I suggest that electrotherapeutic practice entailed specifying, constructing, and mapping the body as an electrosensitive organ in need of both management and cultivation. Such mappings were not necessarily confined to the skin; electricity's capacity to penetrate and move through the body's interior space, evidenced by the sensations felt as electricity passed through it, yielded detailed depictions of the body as a conduit of and register for electricity. Further, these electrotactile sensations helped establish analogies between organic and technological communication systems, as practitioners likened the feeling of electricity moving between electrodes pressed against the skin to the transmission of messages along telegraph lines.[92] Therapeutic electrical touch not only evidenced the general claim that "electricity is life" but also served as a means of embodying specific lessons about the force's capacity as a communicative agent.[93] While the model of the body as a communication system had achieved some prominence by the middle of the nineteenth century, the feeling of currents moving between electrodes positioned at different points on the skin provided a bodily grounding for that metaphor, propagated through the increasing standardization of instruments used in therapeutic practice.

For more than a century and a half, the immense cultural investment in the truth of electrotherapy's efficacy trumped its physiological reality. Electrotherapy therefore presents a curious case study in the relationship between science, medicine, and consumer culture: as de la Peña argues, the desire affixed to electrical treatments served as a powerful placebo, inspiring patients to believe in its curative efficacy. Electrocutaneous sensations provided tangible evidence, or material signification, of electricity's utopian power over the human body—the shocks and gentle currents that accompanied electric cures, framed by celebratory rhetoric hailing electricity's success at relieving the body's infinite ailments, indicated the passage of both the patient's body and social bodies into a new era of harmony between humans and technologies. Electric belts, worn to charge the body, "generated a light buzzing, or tingling, sensation under the skin where it came in contact with the current-conducting wires."[94] Specific to my reading of electrotherapy machines as instantiations of tactile media, electrotherapeutic discourse can be understood as a formalized field of scientific utterances that treated the tactile system as the primary witness for and register of electricity's curative powers.

Throughout the era of electrotherapy, attitudes toward the tactile sensations brought about by treatment sessions shifted substantially. Early instructions for administering charges often understood pain as a necessary and fundamental part of the therapeutic process, while later texts, advocating instead a process of negotiation between physician, patient, and electrical apparatus, urged physicians to use discretion in conjuring painful sensations. The changing sensations of electrical treatments gradually came to signify forward progress in medical research. Before detailing numerous methods for applying electrical charges as a medicinal in his 1784 *Essay on Electricity,* George Adams declared electricity's success "in relieving the sufferings of mankind." Recent advances in both techniques and apparatuses made its efficacy more "sensible."[95] Patients therefore served to register electrophysicians' newfound abilities to better modulate intensities of charge and to localize the application of current at the tactile level, as the sensations that accompanied therapy sessions began to take on an increased diversity. Adams differentiated his own methods by the sensate experience they induced in the patient; one technique worked "by a sensation between a shock and the spark, which does not communicate that disagreeable feeling attending the common shock," while another induced a "mild and pleasing" sensation "resembling the soft breezes of a gentle wind; generating a genial warmth."[96] Each distinct sensation could be understood as a "species of shock" with a specific set of corresponding effects. Some species had the advantage of not "inducing that pungent sensation," communicating instead a "quick vibratory sensation"; others brought with them intense feelings of pain.[97] Later formulations employed similar tactile taxonomies. For example, Celia Haynes named the different methods of skin faradization for the sensations that accompanied them: "the electric nail . . . produces a sensation like a red-hot nail pressing into the flesh"; "electric cauterization," where an electrified wire brush that moves across the skin, "produces a sharp, burning sensation, that has been compared to that caused by a cautery."[98]

Though disagreeable sensations were not to be avoided altogether, attempts to minimize patient discomfort illustrate touch's centrality to electrotherapeutic practice. In 1787, Francis Lowndes advocated a radical split from prior practices, claiming that "the modern practice of Electricity rejects those violent modes, such as strong shocks, etc., which accompanied its first introduction into medicine." Electrotherapy's passage into a new era, for Lowndes, was marked by the substitution of painful sensations with "more gentle applications."[99] Similar claims echoed throughout electrotherapy's long history. In 1872, the French physician Ernst Onimus referred to a break from the "barbarism" associated with previous practices of electricization, which involved torturously "holding a copper cylinder in either hand and receiving tremendous shocks though both arms."[100] So-called modern electricization consisted instead of the "rational application of electricity," obeying one crucial dictate: "avoid all unnecessary pain."[101] These attitudes toward electrotherapeutic practice specified electrocutaneous sensitivity at a material level: like Adams and Lowndes, Onimus refined apparatuses and their affiliated techniques of use with the intent of making the application of current as agreeable to the patient as possible without compromising the treatment's purported efficacy.

At the same time, physicians did not want to altogether eliminate the sensations caused by electrical current, precisely because the feeling of electricity entering or exiting the body signified electricity's therapeutic benefits and effects. The success of electrotherapy depended on keeping the patient's felt experience of shock within certain parameters; an unnoticeable charge would fail to indicate the treatment's efficacious nourishment, while an exceedingly intense shock signified a barbarism or lack of skill on the part of the physician. In marketing and commercializing electrotherapy, physicians made a direct appeal to the sense of touch, seeking to determine a sort of golden mean between electrical sensation's absence and its overwhelming abundance. However, such a mean remained elusive, varying based on the sensibility of the individual patient, being resisted by the unreliability of the therapeutic apparatus, and evading the grasp of the untrained physician. The many handbooks devoted to electrotherapeutics evidence this struggle, as their authors intensified and rigidified proscriptions for administering therapy. Physicians (or "operators," as they were often called) were urged to augment or diminish the "force of the shock" based on "the strength and sensibility of the patient."[102] They regulated the dose of electricity "like that of a drug" according to the "age, race and habits of the patient under treatment."[103] Electrotherapy assembled patient, physician, and electrical machine in a circuit of mutual legibility: the patient felt and articulated the shock sensation; the physician (whose own body often served as a conduit through which electricity passed from electrical machine to patient) regulated the passage of shock into the patient's body by dexterously manipulating the unwieldy electrical mechanism; and the machine encountered what was termed the "resistance of the body" as different patients and their different body parts provided variations in conductivity, slowing and impeding current as it flowed from the machine through the physician–patient amalgam and back.[104] Skilled therapists were expected to be masters of inducing electrocutaneous sensations and were advised to become intimately familiar with electricity's sensible effects through repeated autoexperimentation. A physician with "no personal experience of the sensations produced by the currents" was ill prepared to administer them to others. Through this self-education, the physician could become acclimated to "the different sensibilities of different parts of the body, the motor points of muscles, the mode of applying the rheophores to the skin, [and] of graduating current."[105] By applying currents to their own bodies, electrotherapists tracked the felt paths electricity traveled as it moved through body; they noted the resistances it encountered, the obstacles it faced, and the tactual effects its passage generated. Proficiency in what was dubbed "the art of electrisation" could be obtained only by cultivating a refined sensitivity to subtle variations in administered electrical currents.[106]

To help manage the sensations caused by treatments, the incongruity between the bodies of the physician and the patient had to be accounted for in exhaustive detail. This typology of electrocutaneous sensibility grouped bodies by their resistive capacities, allowing the physician to anticipate the type of apparatus to be employed as well as the strength of the charge needed for therapy to be effective. As Haynes explained:

> The skin of the aged is firm, frequently dry, and is less easily penetrated; therefore it will require not only larger electrodes, but that they are placed as near as possible to the organ or muscle through which the current is sent. The skin of dark

races does not permit a current to pass as readily as lighter one; therefore they need to have the electrodes large and near together. Those accustomed to laborious pursuits, involving severe muscular exercise, offer great resistance to the passage of current, on account of the firmness of muscular tissue, and will require the same arrangement of electrodes as the preceding.[107]

As this passage illustrates, patients' tactile experience in treatment sessions—the strength of the current that delivered shocks to their skin—was configured according to the codified, purportedly scientific anthropological narratives circulated in medical texts. Subjects whose skin possessed "good" conductivity were young, light-skinned knowledge-workers, while any aged, dark-skinned, or laboring patients were made to endure strong shocks communicated to their bodies, with physicians turning up the dials on the electrotherapeutic apparatus or increasing the size of the electrodes used to induce current. Vague generalizations about differing sensitivities were underpinned by sets of quantitative data about the amount of current specific subjects could withstand, the size of electrodes to be applied, and precise distance applied electrodes ought to be placed from one another.[108]

The flow of electrical current from battery to patient was mediated simultaneously by human and nonhuman actors, with each bringing its distinct structures to bear on the body of the therapeutic subject. The therapist's newly electrified hand became an essential component of the electrotherapeutic apparatus, used to channel electricity into and through the body. When administering current to various regions of the body, physicians often had to manually probe deep into patients' fleshy crevasses; while designing electrodes tailored to individual body parts aided in this labor (see Figure 1.4), it was frequently the physician's naked hand that served as the most easily controllable electrode. By slightly increasing or decreasing the pressure placed on the patient, or by adding or subtracting fingers to the point of contact, physicians could regulate the strength of the charge applied to the patient. In these electrical massages, "the operator's own body . . . became the means to gauge the necessary strength of electrical stimulation."[109] This

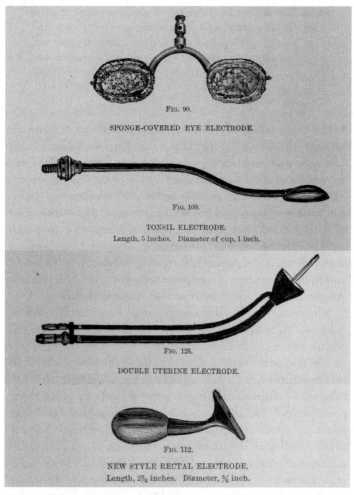

FIG. 90.

SPONGE-COVERED EYE ELECTRODE.

FIG. 109.

TONSIL ELECTRODE.
Length, 5 inches.   Diameter of cup, 1 inch.

FIG. 126.

DOUBLE UTERINE ELECTRODE.

FIG. 112.

NEW STYLE RECTAL ELECTRODE.
Length, 2¾ inches.   Diameter, ¾ inch.

Figure 1.4. Eye electrode, tonsil electrode, double uterine electrode, and rectal electrode. From Celia Haynes, *Elementary Principles of Electro-Therapeutics* (Chicago, Ill.: McIntosh Galvanic & Faradic Battery Company, 1884), 286, 314, 373, 341.

cultivated electrodexterity required the therapist to carefully attend to the precise sensations of current passing from their hand to the specific site on the patient's body in need of treatment. As with spectacular demonstrations of electricity, a well-disciplined human body, equipped with a practiced electrocutaneous sensitivity, frequently proved a more effective instrument than the machines that would later displace it. Physicians' trained bodies regulated the electrotherapeutic economy of shock that underpinned therapeutic practice.

Throughout the nineteenth century, as the range of conditions that electricity purportedly relieved or cured expanded, electrodes grew in sophistication and specialization. One compendium depicted more than one hundred electrodes, each configured to apply current to a different body part. The eyes, throat, ears, urethra, feet, uterus, tongue, teeth, and rectum each had their own dedicated electrodes, with corresponding techniques for inducing current and descriptions of the acceptable range of accompanying electrotactile sensations. The electrodes designed for insertion into the uterus could be used to terminate fetuses; those placed on the eyes were thought to relieve wearied vision and restore a sharpness of sight to the retina. Applying electricity to the tonsils, according to some reports, provided an effective treatment for tonsillitis, while rectal electrodes could temporarily alleviate paralysis of the sphincter. These electrodes were designed to be interchangeably attached to standardized connectors on electrotherapy kits, indicating the harmonious relationship that developed progressively in the latter half of the nineteenth century between itinerant electrotherapists and the medical instrument industry. In contrast, using "electric baths" allowed currents to be diffused over the patient's whole body, spreading their curative effects across the skin's vast expanse. Submerging the patient in electrically charged water, with the strength of its currents regulated by a "beautiful and elaborate apparatus" (see Figures 1.5 and 1.6), was thought to treat a whole range of generalized afflictions while subjecting the patient to only a "very slight sensation" of electrification.[110]

From a medical perspective, electrotherapy involved forging a technics of electrical touching and organizing an economy for

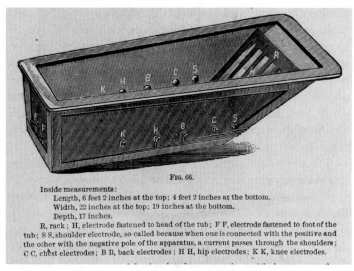

Fig. 66.

Inside measurements:
    Length, 6 feet 2 inches at the top; 4 feet 2 inches at the bottom.
    Width, 22 inches at the top; 19 inches at the bottom.
    Depth, 17 inches.
    R, rack ; H, electrode fastened to head of the tub; F F, electrode fastened to foot of the
tub; S S, shoulder electrode, so called because when one is connected with the positive and
the other with the negative pole of the apparatus, a current passes through the shoulders;
C C, chest electrodes; B B, back electrodes ; H H, hip electrodes; K K, knee electrodes.

Figure 1.5. Electric bath, with electrodes for targeting specific body parts. From Celia Haynes, *Elementary Principles of Electro-Therapeutics* (Chicago, Ill.: McIntosh Galvanic & Faradic Battery Company, 1884), 227.

regulating the dispensing of shock. But crucially, the experience of electrotherapy also gave embodied form to new ideas about the interchangeability of organic and technical communication systems.[111] In Laura Otis's reading, such ideas were visualized both in graphic depictions of telegraph lines and in medical illustrations of human nerve systems,[112] while de la Peña considers the body as a phenomenal register of electricity's movement, as electrodes penetrated its orifices and currents surged into its depths. Taking both arguments in conjunction, I suggest that, in the second half of the nineteenth century, the tactile experience of electrotherapy helped illustrate the principles of telegraphic communication through felt sensations: as current left a positive electrode, passed jarringly into the body, and then exited out a negative electrode, therapists and patients alike bore witness to the concept that information moved along wires strung between telegraph poles. Alfred Garratt's formulation of bodies as communicative networks in need of repair

Fɪɢ. 68

McINTOSH COMBINED OFFICE AND BATH APPARATUS.
Size of horizontal base, 18 × 20 in. Size of upright base, 18 × 20 in.

Figure 1.6. McIntosh Combined Office and Bath Apparatus. Source: Celia Haynes, *Elementary Principles of Electro-Therapeutics* (Chicago, Ill.: McIntosh Galvanic & Faradic Battery Company, 1884), 228.

through the therapeutic application of electricity suggested an affinity between the labor of telegraphers and therapists that could be witnessed by both physical and social bodies:

> . . . hence we say that the conductors, and whatever body is included between the poles of those conductors of a battery, are *traversed* by a current. . . . The works that can be accomplished between these two tips of wires, called poles, already telegraphing nations with a harmony of intelligence, besides

performing every day a thousand other useful works, must
also be brought to the aid of our enfeebled bodies, or
deranged nervo-telegraph systems, to vivify, to restore communication, to clear them of their debris, and enable them
again to regulate their more normal and constantly-renewing
supplies.[113]

Drawings of electricity's path through the body (see Onimus's diagram in Figure 1.7 and Erb's illustration in Figures 1.8 and 1.9)
visualized this narrative portrayal of electricity's passage through the
body, with both depictions reinforced by the tingling electrotherapeutic treatments induced. This sensational analogy rendered
the body as a less efficient transmitter than the telegraph line,
with signals diffusing through it in an arc rather than moving on a
direct path between nodal points. Nevertheless, by constantly repositioning the electrodes (or telegraph poles, as Garratt described
them), the therapist painted a powerful message onto the patient's
nerves. Combined with narrative framings of this experience that
described it as analogous to telegraphic communication, such
embodied experiences helped tangibly render electric telegraphy's
fundamental principles. The perceived capacity of electrotherapy to clear nerves of their debris—and to recharge and replen-
ish the nervous system's decaying energies—transposed the utopian
imagination around telegraphy onto the patient's body. Where
Nollet's tangible networks had demonstrated that electricity
could instantly traverse space exterior to the body, the ad hoc
networks electrotherapists distributed across the skin inscribed
the lesson that the interior and exterior worlds were mutually
informatic: each depended on electrical technology to facilitate
the easy flow of messages across distances, and each proved
capable of being rehabilitated through the rational application
of electricity.

The electrotherapeutic economy hinged on the capacity to commercialize, ration, and regulate the dispensing of shock. At the
height of electrotherapy's golden age, the traveling physician, able to
treat maladies both physical and psychical with their complement of
batteries, variegated electrodes, and electric belts (see Figure 1.10),
was the central component in the operation of this economy, as

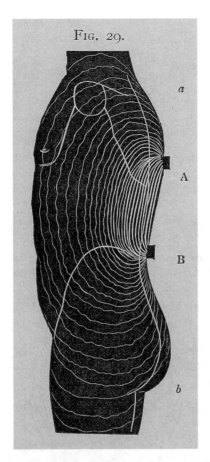

FIG. 29.

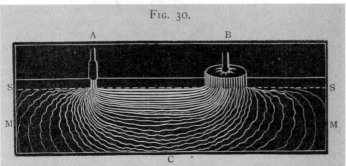

FIG. 30.

Figure 1.7. Two illustrations of the path electricity travels through the body. From Ernst Onimus, *A Practical Introduction to Medical Electricity*, trans. Armand de Watteville (London: H. K. Lewis, 1878), 32.

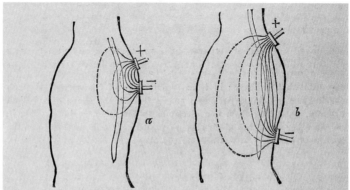

FIG. 14.—Schematic representation of the distribution and density of the threads of current with regard to their entrance deeply into the tissues (in this instance, into the spinal cord). *a*, when the electrodes are in close proximity, *b*, when far removed from one another.

Figure 1.8. Density of current as it traverses the body. From William Erb, *Handbook of Electro-Therapeutics* (New York: Wood, 1883), 29.

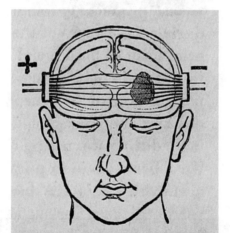

FIG. 15.—Schematic representation of the best method of application of the electrodes in order to bring a lesion, situated deeply in the left cerebral hemisphere, into the field of the most dense and effective threads of current.

Figure 1.9. The induction of electrical current through the skull. From William Erb, *Handbook of Electro-Therapeutics* (New York: Wood, 1883), 31.

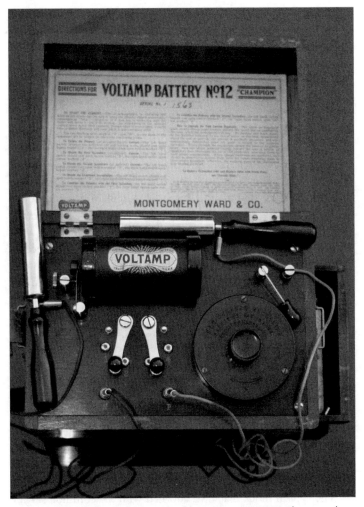

Figure 1.10. Voltamp no. 12 medical battery, circa 1907. Photograph by the author, 2016.

patients grew accustomed to receiving the vivifying effects of electrical treatments, indicated by the shock the therapist applied. Beginning in the 1880s, the rise of arcade electricity began to fracture the stability of this arrangement. The therapist's role in

administering restorative currents started to decline, as they were
gradually displaced by interactive, coin-operated machines that
dispensed customizable amounts of current for a small price.
Machines like the *Volta Electric Automaton* (circa 1920) and the
Electric Sailor (circa 1910) promised operators the ability to regulate
the dispensed currents "at will,"[114] and by doing so, suggested that
the human physician had become superfluous to the treatment process. Coin-operated electrotherapy collapsed the role of operator
onto the patient by simply automating the application of medical
electricity. In addition to being "good for the nervous system" more
generally, the *Silent Physician* (Mills Novelty Company, circa 1904)
claimed to be suited to treat "all forms of muscular ills" without the
intervention of a human electrotherapist. Predictably, physicians
resisted automatic machines and do-it-yourself electricization kits.
Cautionary tales advised patients against using electrotherapeutic
instruments without the aid of a trained operator, warning potential users that they would fall prey to the temptation to overadmin-
ister current. "The popular idea seems to be," as Haynes observed,
"that 'if a little electricity is good, more must be better.'"[115] Shock
served to indicate the consumption of current in normal and acceptable doses while also marking excessive, pathological consumption
habits. But the inviting promises of arcade machines proved so
alluring that they countered the advice of physicians and even
goaded patients into partaking greater quantities of current. For
example, the Peterson Medical Electro Battery (circa 1904) claimed
to "strengthen your nerve by a mild invigorating electrical treatment" while asking "how much you can stand," effectively daring
the operator/patient to deliver the maximum intensity of shock in
the self-treatment session.

The typical shock machine, adorned with the common slogan
"electricity is life," featured metal handles that could be grasped
and turned to adjust the strength of the current that passed into
the operator. Others added numeric dial displays to visibly indicate and quantify the intensity of the shock experience. Machines
frequently signaled the player's feat of enduring the maximal
currents by loudly ringing a bell. An 1886 machine went so far as
to dispense a card that indicated the voltage that the patient had

subjected him- or herself to, effectively providing a transcription of the shock experience.[116] Although they frequently acknowledge the medical lineage of these machines, arcade historians have tended to lump shockers together under the general category of strength and endurance testers, where shock machines were games that urged players to show off their capacity to withstand the still-novel force of electricity.[117] In this interpretation, under-pinned by assumptions about the gendering of technological mas-tery, these devices acquire a decidedly masculine connotation, as enduring the discomforts that shock machines subjected the body to spectacularly signified, in the public space of the arcade, an agonistic triumph over the electrical machine and the forces it generated.

While such readings are certainly justified, they distract from the sincere hopes metropolitan residents placed in medical elec-tricity: the modern force provided both a means of alleviating the urban body's sufferings and technique for restoring the vital energies drained by city life. The popular *Imperial Shocker* (Figure 1.11), inlaid with a photograph of a smiling young woman framed on each side by a pair of ornate cast iron female figures, implored operators to "take a shock and look pleasant." The *Imperial Shocker* linked cabinet electricity to regimes of beautification and a more general therapeutic ethos that fueled the emerging consumer cul-ture, offering "electric treatment great for one-night jags" to any-one who could slide a coin into the machine's slot. No longer confined to the physician's office, these machines opened up the distinctly modern benefits of electrotherapy to new populations by simply lowering the price of admission, while simultaneously undercutting the expertise of trained therapists. The specialized technics of electrotherapy required for the operation of earlier machines, which had included the ability to apply specialized elec-trodes to all spaces of the body, was reduced to the simple motions of inserting a coin, grasping handles, and turning knobs.

Figure 1.11. The *Imperial Shocker* (Mills Novelty Company, circa 1902). Other iterations of the Shocker featured the popular slogan "electricity is life." Photograph by Chad Boekelheide; courtesy of Chad Boekelheide, www.chadscoinop.com.

The Twilight of Electrotactile Machines

While this move to conceptualize early electrical machines as tactile media may appear to be just semantic jockeying, I insist that it constitutes an important initial step in actively crafting a new genealogy of tactile media, helping destabilize narratives that locate seeing and hearing as the primary sense modalities impacted by the emergence of electrical machines. Further, it troubles the iron-clad Enlightenment linking of vision to knowledge by showing how the progress of Enlightenment electrical science depended on a practiced touch, structured by protocols, and employed as a means of gathering data about a force inadequately registered by the privileged sense of vision.

Moreover, contemporary touch-based interfaces are frequently considered to be the first attempts at incorporating tactility into electronic communication systems. However, the evidence offered in this chapter suggests that they are rather the latest in a series of instrumentally oriented contacts between electrical apparatuses and the human tactile system. The development of a cultivated sensitivity to machine-generated tactile sensations, therefore, is not unique to the present moment. Instead of marking the passage into an unheralded future, contemporary haptic interfaces simultaneously recall both touch's primacy in knowing electricity and the process by which electricity helped make touch calculable. By upending the conventional set of associations between electricity and the senses, this approach opens up new possibilities for thinking about the relationship among sensation, technology, and mediatic modernity, extending backward by more than a hundred years what Carolyn Marvin referred to as "the starting point for the social history of Anglo-American electric media."[118]

Moreover, by situating touch as the original mode of communicating with electrical machines, this chapter helps reverse—or at least, call into question—axiomatic claims about touch's fundamental incompatibility with electronic communication networks. As Tom Standage argues, by connecting bodies and circulating electricity through them, Nollet offered proof of concept for the electric telegraph nearly a hundred years before its commercialization.[119] An ontological understanding of electric telegraphy and the senses it acted upon has obscured from view what we can understand, to appropriate a phrase from Mark Hansen, as the electric telegraph's "primordial tactility."[120] It was touch that initially allowed bodies to become part of electrical networks; the tactile sensation of shock as it passed instantly between wired bodies signified membership and participation in a new type of networked sensational community. In light of this history of hard-wiring together of tactile systems, the "haptocentric intuitionalism" that Derrida claimed leads scholars to believe that touch resists virtualization is exposed as an imposition of ideology on media history.[121]

By locating early electrical machines next to contemporary haptic human–computer interfaces, I do not mean to suggest an unbroken continuity between the two developments—I do not want to imply that modernity involves electricity progressively taking over and enveloping the sense of touch. Instead, even in spite of twentieth-century attempts at developing media that specifically act on touch, tactile encounters with electricity have declined substantially after reaching a high point during the so-called golden age of electrotherapy. With the discrediting of electrotherapy in the early twentieth century, electrotactile contacts dwindled substantially. Since the ebbing of this tide, once-common tactile encounters with electricity have been confined to a small handful of fairly extraordinary circumstances—behavioral research (Stanley Milgram's controversial electric shock machine),[122] extreme psychiatric treatments (electroconvulsive treatment), torture (as in the iconic image of the Abu Ghraib prisoner connected to electrical wires),[123] state-sanctioned killing (the electric chair), law enforcement (the use of tasers by police as a purportedly more humane means of managing disorderly bodies), medical treatments (defibrillators, both implanted and external),[124] transcutaneous electrical nerve stimulation (TENS, which has recently enjoyed a resurgence in popularity and public visibility), and other miscellaneous or accidental encounters (testing a battery on the tongue, static discharge, etc.). Recreational or performative applications of electricity directly to the body are limited to still fewer situations, such as sadomasochistic, sexual electricity play (the Violet Wand and its many variants), shock games (*Lightning Reaction* and its knockoffs), and performance art (Eddo Stern's "Torture Tekken";[125] Stelarc's networked body[126]). Recent attempts to incorporate electrotactile sensations in information display[127] suggest this trend may be reversing itself, but such technologies have yet to widely proliferate.

Most contemporary tactile media are discontiguous, both formally and technically, from the electrical machines described in this chapter. As I detail in Interfaces 3 and 4, contemporary tactile media typically employ different mechanisms for dynamically acting on the body, using electricity to activate motors that spin in controlled frequencies to generate vibrations apprehended by the tactile system. This move to *electromechanical* tactile media entails a curious reversal. Where initially, tactile media depended on the translation of mechanical energy into electricity via the electrostatic generator and then subsequently of electricity into electrotactile shock, today's tactile media frequently depend on the reverse mechanism—transforming electrical energy into motion—to create and render tactual effects. This would eventually be the operative principle in Geldard's Vibratese system of tactile communication, and continues to be expressed by vibration feedback mechanisms for touchscreens, videogame controllers, and electric massagers.[128] The chronology essentially follows this sequence: from mechanoelectric (the electrostatic generator and the Leyden jar) to organoelectric (the torpedo fish) to chemicoelectric (Volta's pile), and then finally to electromechanical (the electric motor and its aforementioned range of deployments in various tactile media systems throughout the twentieth century). Each change brought with it a shift in the experiential quality of administered tactile sensations, along with a new range of possibilities for modulating the configuration of applied sensations. The development of computers in the second half of the twentieth century, for example, allowed these sensations to be more precisely controlled, administered, and targeted, and as such, brought with it a new set of communicative possibilities and potentials. The category of tactile media, then, can be understood not as a stable and constant formation, but rather as a shifting assemblage composed of technical elements, embodied sensations, and cultural practices. This formulation provides a means of tracking persistent changes in the way technology has been used to generate dynamic sensations for touch.

Kittler's claim that media are fundamentally about "the deception of sensory organs"[129] directs us to consider the development of a media technics as dependent on mobilizing or producing specific knowledge about the capacity of the sense organs to be deceived. The design of eighteenth-century electrical machines, however, did not aim to deceive touch. Rather, it was touch's reliability—its stability as an epistemic modality—that allowed it to serve such a vital role in registering electricity. These electrical machines hailed a unitary and undifferentiated touch. In touching electricity, natural philosophers used their tactile systems to uncover new knowledge about electricity, to delve into a wondrous world that touch both provided access to and promised to demystify. But they did so without questioning or reflecting on the perceptual mechanism that facilitated these encounters. Gradually, some shifted their attention away from electricity's fundamental characteristics and started haphazardly applying current to the different skin regions. These sorts of increasingly structured experiments were aimed at inquiring into the character of the sense organs themselves, and they eventually contributed to Johannes Müller's doctrine of specific nerve energies, which severed the link between sense experience and its external referent.[130] During the nineteenth century, electrical machines were modified and repurposed with the explicit aim of studying tactile processes. These apparatuses, and their corresponding experiments, shattered touch into fragments, effectively deterritorializing what had been a roughly mapped dimension of human experience. The further application of electricity to the skin facilitated the reconstruction of these shards back into a holistic unity, but its composition remained fundamentally altered by the disruptive contact between electricity and skin. This reterritorialization provided a specific, technical body of knowledge about touch necessary for the twentieth-century development first of tactile sensory substitution systems, then of electromechanical tactile displays, and eventually of haptic human–computer interfaces that I describe in the final three chapters of the book. The electrotactile sensations generated by contacts between machine and skin underwent a key transition: in eighteenth-century physics and popular amusement, they were employed to generate knowledge about electricity itself, while in nineteenth-century psychophysics and physiology, they were used to produce knowledge about tactility and its multitude of components. In Interface 2, I map this transition in thinking about touch as it shifted from being an epistemic agent to being the object of a self-conscious scientific episteme. Once its capacities were quantified and specified—once touch's processes themselves became the object of experiments and instruments—tactile media obtained new abilities to both deceive and educate the newly variegated organs of tactile perception.

Interface 2

The Haptic

First of all we used the instrument called *Stangenzirkel* in German [beam compasses]. This consists of a long metal beam in the shape of a right-angled prism, with two points fastened at right angles (to the beam). Because these points can be moved towards and away from each other in a straight line, the instrument seemed suitable for touching the skin with the two points a certain distance apart. To prevent the points from puncturing and hurting the skin, we fixed a cork stopper on each point. Using these stoppers, we touched various parts of the skin of the forehead, face, neck, back of the neck, chest, stomach, back, hip, upper arm forearm and hand, and finally the thigh, skin and feet. We compared the judgment of the distance between stoppers with the true distance, the judgments being made while looking away.

—Ernst Heinrich Weber, *De Subtilitate Tactus*

In the late 1820s, Ernst Heinrich Weber (1795–1878), a young anatomist working out of the University of Leipzig, set out to quantify the relationship between applied tactual stimuli and his subjects' mental experience of it. A practiced empiricist, Weber, frequently assisted by his brother Eduard, attempted to fix his experimental subjects in a controlled space, adjusting for any environmental factors that would contaminate the consistency of his results. With

meticulous attention to detail, likely attributable to the influence
of childhood friend and acoustics pioneer Ernst Chaladni (1756–
1827), he instructed his subjects to remain still while he applied
the two points of a beam compass to various parts of their bodies.
Each time he pressed the points against their skin, he asked his
blindfolded subjects to report the number of contacts they felt.
Intent on determining the smallest distance apart the blunted compass points could be placed while still being perceived as distinct
from each other, Weber gradually moved the points farther apart
(see the compass in Figure 2.1). Situated too close to one another,
the subject would experience the points as a single stimulus rather
than two separate stimuli. This minimum distance at which the
points could be perceived as distinct Weber designated as the two-
point threshold (or two-point *limen*)—the justice-noticeable difference between the two stimuli. Weber's investigations were based
on what he understood as touch's fundamental ability to *misperceive,*
to yield inaccurate data about the stimuli acting on it. The two-
point threshold experiments produced a map of the skin as a surface

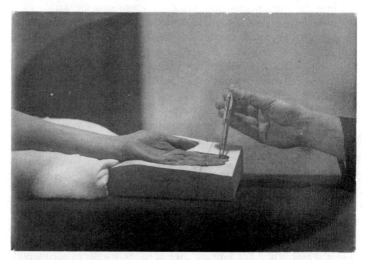

Figure 2.1. An aesthesiometric compass being applied to the finger of
a test subject. Photograph provided with the friendly support of the
Adolf-Wuerth-Center for the History of Psychology, Germany.

with varied acuity for discriminating between stimuli; at the tip of the tongue, subjects could perceive the compass points as distinct when they were separated by only 1.12 millimeters, whereas on the chest, the two compass points were perceived as one until a gap of 45 millimeters stretched between them (his findings are illustrated in Figure 2.2).

Using this simple experiment, Weber measured the skin's heterogeneous capacity for localization, crafting a method for quantifying both noticed and unnoticed stimuli. In soliciting his subjects' experiences with tactile stimuli, Weber thought himself to be simply revealing crucial structures of the tactile system that had remained hidden from the view of less adventurous anatomists.

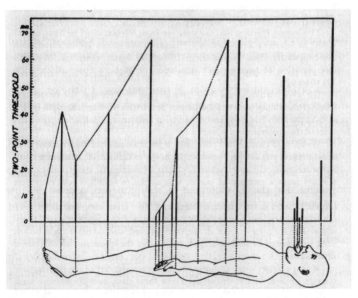

Figure 2.2. Data from Weber's research on the two-point threshold mapped onto an image of the body. Threshold distances are listed in millimeters; Weber's were originally in Paris lines (subdivisions of the Paris inch). Based on Weber's data in C. S. Sherrington, "Cutaneous Sensations," in *Text-Book of Physiology*, E. A. Schafer (London: Pentland, 1900), 936. Theodor Ruch, "Somatic Sensation," in *Neurophysiology* (Philadelphia: W. B. Saunders, 1965), 316.

But Weber's investigations into these just-noticed differences had implications that exceeded mere anatomical description: rather than simply getting at the materiality of the nerves responsible for touch sensations, he was helping establish a methodological framework for two related projects—psychophysics and the science of touch—that would be taken up by subsequent researchers during and beyond the nineteenth century.

Given its most systematic articulation by Weber's one-time student Gustav Theodor Fechner in his sprawling 1860 manifesto *Elemente der Psychophysik* (Elements of Psychophysics), psychophysics involved quantifying the specific relationship between the physical and the psychical worlds. Fechner suggested that Weber, in his experiments on touch, had inadvertently uncovered an experimental technique for determining the generalizable laws that govern the relationship between stimulus intensity and the subjective mental experience the stimulus induced. But where Weber had focused mainly on touch, Fechner broadened the scope of his investigations to accommodate other sense modalities, asserting that each was underpinned by a similar unifying law.[1] By the end of the nineteenth century, the psychophysical method—a means of fixing, measuring, dividing, and quantifying psychosensory processes—became the dominant paradigm in psychology, displacing philosophical psychology as the primary method for studying mental processes.[2] In the wake of Kittler's genealogically grounded argument that the development of technical media depended on forward progress in the measurement of sensation, the history of psychophysics has assumed an increased centrality to contemporary media theory. Kittler suggested that, by "giving a mathematical expression to the data stream of sensual perception," psychophysics provided a positive body of knowledge about the specific thresholds of the individuated senses essential for the subsequent emergence of media that could target and exploit these thresholds.[3] For Kittler, the medialization of the senses occurred not in the invention of technical media like cinema and phonography, but rather in the laboratories of experimental psychologists, where the psychophysical interface—using a range of apparatuses and corresponding techniques—treated the senses "as if they were technological media."[4] The "experimental quantification" of the senses rendered mental experience as the convergence of psychophysiologically distinct mechanisms for registering changes in both the external and internal realms. This emphasis on psychophysics as a science of sensation shows the model of the senses embodied in media technologies to be historically constructed, the outcome of turning scientific investigation inward on mechanistically conceived sense organs. However, media theorists (including Kittler) have predominantly taken image- and sound-reproduction media as their starting point in looking backward to examine the experimental programs that had to be executed for technical media to speak to the senses. Consequently, those working in this tradition have downplayed or ignored touch's initial centrality in structuring investigations on the measurement of sensation.[5] The vast and programmatic research carried out on touch comes into view only obliquely, and it is generally taken to be of little consequence for the historical trajectory of media systems thought of as predominantly involving a technics of seeing and hearing.

The second project to emerge from Weber's investigations on tactility—the development of a formally organized, systematically articulated science of touch—involved touch's gradual partitioning into smaller, experimentally isolatable subcomponents. Although touch's story here echoes in some ways the more familiar narratives crafted to historically contextualize the psychophysical investigations of seeing and hearing, of all the senses isolated, broken down, and reconstructed by psychophysical methods, touch underwent perhaps the most radical transformation. Where the processes of seeing and hearing had long been the subject of extensive scientific study, when Weber began poking his subjects with blunted compass points, comparatively little attention had been devoted to empirical examinations of touch.[6] Weber's experiments are identified, in retrospect, as the start of an unbroken tradition of research on touch, to the extent that those working on the design of touch-based human–computer interfaces continually position him at the paternal root of their family tree, as both the "father" of the science of touch and as the "godfather" of Fechner's later psychophysics. For example, Martin Grunwald and Matthias John, describing the German pioneers of haptics in the 2008 anthology *Human Haptic Perception: Basics and Applications,* bluntly summarize Weber's position in the field's history: "the scientific and systematic examination of the human sense of touch began with . . . Ernst Heinrich Weber."[7] Akin to the "perceptual technics" Jonathan Sterne identifies as fundamental to the economization of signal transmission in sound-reproduction technologies, Weber's methods for quantifying touch achieve a new utility in the practice of haptic interface design, with the just-noticed difference providing a means of determining which computer-generated tactile sensations users will perceive, and which ones will exist below the thresholds of tactual perception.[8]

By the end of the century, as a result of continued experimental research, tactile experience came to be seen as so variegated it could no longer be contained under the unitary designation of touch. In a sprawling 1892 essay that outlined and furthered the vast experiments conducted on touch during the nineteenth century, the Berlin psychologist Max Dessoir (1867–1947) suggested that the term *haptisch* (haptics) be taken up as a new banner under which to unite the varieties of tactile experience.[9] Dessoir's neologism formally expanded the category of touch to include sensations generated by nerves in the muscles and joints (such as movement and weight) that did not originate solely in the skin.[10] As Grunwald and John suggest, Dessoir's nomination of the term originated in a dissatisfaction with the "classic terminology for the sense of touch"; "haptic," by contrast, provided "an encompassing generic term which included the different aspects of the sense of touch" and facilitated "the scientific teaching" of touch.[11] Researchers in experimental psychology quickly adopted Dessoir's preferred nomenclature; by the close of the century, they had begun to use haptics synonymously with touch, effectively subsuming the latter category in the former.

Although this move from tactile to haptic did not displace "tactility" with "haptics" entirely, in the lexicon of physiological psychology, "tactile" was to be reserved for those sensations that originated solely in the skin, while "haptic" was proposed as a more general designation, encompassing sensations that arose from both the skin and bodily movement (sometimes reduced to a division between tactility as passive touch and haptics as active touch). In practice, however, even the most precise psychologists frequently slipped into the habit of using the two words interchangeably (a trend that continues today, both in the popular and specialist discourse on haptics). In spite of these repeated and sometimes indiscriminate oscillations between the two terms, the stamp that the lab pressed onto touch remains the enduring legacy of haptics. From the nineteenth century on, psychophysical investigations of touch continued to generate specialized knowledge about the subdivided senses of touch. Though the divisions between the various modalities that comprise the touch senses shifted, they did so according to the epistemic parameters of the lab experiment. As with sensory psychology more generally, progress in haptics research became affixed to and dependent on refining the instruments and methods employed in experimentation. In the mid-twentieth century, for example, developing machines that could more precisely target controlled bursts of electricity to the skin yielded positive, quantifiable new knowledge about the skin's capacity to perceive differences between stimuli (taken up in greater detail in Interface 3). Haptics, then, constituted a declaration of power over tactility; rapidly embedded in the design of research labs and psychophysical instruments, it signaled touch's enclosure within a new epistemological framework. As a field of knowledge, haptics depended on the active solicitation of test subjects' experiences in the human-built laboratory environment. A new type of perceiving subject—a haptic subject—touched and was touched within the carefully configured parameters of lab space.

Accordingly, this chapter examines the scientific processes that facilitated touch's folding into the epistemic framework of haptics. Following Dessoir's suggestion of the term, Edward Bradford Titchener, one of a handful of experimental psychologists who brought the German techniques for measuring sensation to the United States in the late 1800s, defined haptics as "the doctrine of touch with concomitant sensations and perceptions—as optics is the doctrine of sight, and acoustics that of hearing."[12] From a genealogical standpoint, the emergence of haptics should be understood as a response to the interface between the laboratory experiment and touch. Haptics—as the doctrine of touch—was simultaneously the product of new laboratory techniques for managing tactile experience and a normative model that implied a forward path for future studies on touch. From Weber's comparatively crude experiments with beam compasses in the 1820s onward, touch became the object of an increasingly complex set of apparatuses and operations intended to accumulate knowledge about the specific parameters of tactile perception, as that critical border between the known and the unknown—between the governable and the ungovernable—expanded outward through touch's enframing within the apparatus of haptics.

To illustrate this process of touch becoming haptic, I open by examining several preconditions that laid the foundation for Weber's initial investigations. From there, I move to a detailed treatment of Weber's experimental system, showing how it allowed him to cleave touch into a variety of component sensations (including localization, temperature, and pressure). In addition to this fragmentation of touch, Weber's trials also succeeded in disentangling touch from a host of sensations that it had traditionally been associated with. Most notable among these was pain, which Weber understood to be separate from touch, originating in a physiologically distinct, nontactile network of nerves. In the confines of his lab, tactility became purified, freed from the sensations that formerly clouded its ability to accurately distinguish among stimuli, and was rendered as a set of intertwined systems whose primary function was to discriminate accurately among things. The experimental interface provided a means of partitioning and segmenting sensory qualities off from one another, while also serving as a mechanism for creatively reassembling those qualities under carefully controlled conditions. One experiment could isolate the pressure sense from the temperature sense by normalizing the thermic qualities of the pressure-stimulating apparatus; another could vary the temperature of the apparatus as a means of testing the interaction between the (instrumentally segregated) senses of pressure and temperature. Once his methods for quantifying the tactile system had become accepted and standardized practice, the instruments Weber used in the lab began to spread,[13] as the science of touch obtained a positive use value in both clinical and pedagogical contexts. I close the chapter by considering the ramifications that folding touch into haptics had for the overarching process of remaking touch as a communicative sense, situating the haptic subject as the necessary ground for future attempts at coding, storing, and transmitting the data of touch.

A Tactile Modernity

Before touch became haptic in 1892, it first had to be assimilated by regimes of experimentation. This assimilation process constitutes a distinct historical formation—a tactile modernity—that signifies and indexes touch's response to the demands made by a new science of the senses. In the eighteenth century, touch served a crucial epistemic function in electrical science, used to provide access to a dimension of electricity hidden from the other senses. But the experimenter's touch in the eighteenth century remained a stable formation: as a means of probing, revealing, and registering electricity, its constitutive processes were taken to be biologically given and unitary. In the nineteenth century, however, touch itself became an object of rational experimentation. Like electricity in the century before, it was made to answer to the demands of particular methods and their accompanying experimental protocols. By folding touch into the framework of the rational experiment, it became fully implicated in a modernization project, well in advance of late-twentieth-century attempts to incorporate touch into computational media. Both the rational experiment and the lab where these experiments were executed play central roles in accounts of modernity. In Max Weber's formulation, rational experimentation participated in the modern "disenchantment of the world" by providing "a means of reliably controlling experience" necessary for the operation of empirical science.[14] Bruno Latour situates laboratory space at the center of a fictionalized split between the modern and premodern; knowledge generated in the confines of the lab purports to be free from the operations of political power. In the lab, nature can be made to give up its secrets through structured, methodical, and repeatable displays. This modernization project entailed efforts by Enlightenment scientists to develop an objectively scientific understanding of the natural world according to its supposed internal logic. During the seventeenth and eighteenth centuries, nature was treated as the object of a new epistemology intended to "demystify and desacrilize"[15] knowledge, with experimenters calling on a particular brand of empirical science—forged in the artifice of the laboratory—to displace received, traditional truths. The human body and its senses occupied a liminal space in this field: by refining its mechanisms for synthesizing perception, empiricism rendered the need for human perceivers superfluous, while providing unprecedented methods for quantifying and arresting a perceptual system that had previously served as the epistemic ground for truth claims.

Outside the intertwined fields of psychology and haptic interface design, however, awareness of touch's reworking by nineteenth-century empiricists remains scant. As a consequence of this sustained inattention, the ideas embodied in the scientific doctrine of touch exercise a hegemonic power over contemporary understandings of touch, embedding in their theories a set of commonsense truths about the purportedly value-neutral physiological functioning and composition of the tactile system. Those who frame touch as the ground for an antimodern or anti-ocular politics often unconsciously base their conception of touch on a psychobiological and psychophysical model of scientific tactility that is the product of the quintessentially modern experimental scientist's lab. Rather than challenging the political anthropology of touch that developed in the lab space, failing to acknowledge its power serves to reinscribe it. To counter narratives that periodize the senses by ascribing to vision the role of the "master sense" in modernity, I position touch as the object of a scientific method that, by the century's end, enabled tactility to be portrayed as rational, predictable, and manageable. From Weber's initial experiments on touch in the late 1820s until the century's close, physiologists, anatomists, and psychologists pioneering a coherently articulated science of touch laid the ground for its successive inclusion in a variety of bureaucratic apparatuses. As part of an overarching upheaval in psychology, the doctrine designated by haptics was pragmatically and positively deployed—used to quantify and monitor workplace fatigue, to ensure the proper cultivation of the senses in pedagogical programs, and to measure the pain thresholds of criminals and other deviants in phrenological practice. Taken together, these applications signal the sedimentation of the methods and apparatuses that constituted haptics across a range of social spaces, as touch became swept up in the "transformation of the senses by industrialization and technology"[16] associated with modernity.

While scholars in sensory studies have engaged in valuable work on touch's historical, philosophical, aesthetic, and political dimensions, my analysis of touch's shifting constitution will remain focused on the scientific and technical discourses that enveloped touch during the nineteenth century.[17] These modes of knowledge production took on increased importance first with the rise of industrial capitalism and later with the emergence of an economy grounded in the circulation of sensory data through electrical networks, as they generated normative models of the body and its senses that influenced the constitution of objects far beyond the lab's walls. As Foucault argued in *Discipline and Punish,* changing modes of political and economic power in Europe during the eighteenth and nineteenth centuries required new techniques for managing subjectivity. What he refers to as "the quantitative assessment of sensorial responses" pursued by the nascent scientific psychology accorded tactile experiences the possibility of being specified, numerically rendered, and normalized.[18]

In Crary's thesis on subjective vision, the shifting circulation of visual signs in the nineteenth century rendered the embodied tactility of the eighteenth century obsolete. Tactility, Crary claims, was only "made adequate" to the demands of this new perceptual environment through new visual prosthetics, with the stereoscope being the most notable among them. The stereoscope provided vision with an "immediate, apparent tangibility" that severed the reciprocal assistance between sight and touch.[19] What had been a relationship characterized by the mutual dependence of the two sense modalities became one defined by vision's dominion over

touch, as the stereoscope indicated the "remapping and subsumption of the tactile within the optical."[20] However, in presenting a
strictly optical account, not just of modernity but also of tactility,
Crary casts aside the possibility that touch had both a function
and a history independent from its role in apprehending images.

Approaching this reorganization of the perceptual field from
the standpoint of tactility rather than vision shows touch's modernization to be dependent on its coming under the dominion of
evidentiary techniques similar to those that had been set upon
vision. Tactile modernity, then, designates a new set of disciplinary
structures and discursive formations that facilitated touch's articulation as a distinct psychophysical system. These discursive formations and disciplinary structures were accompanied by a corresponding array of subjective tactile experiences in the confines of
the experimental psychologist's lab, where individuals assimilated
new regimes of sensation, judgment, and discrimination. In short,
the doctrine of touch demanded the forging of a redefined relationship between sensory experience and its linguistic transcription. As
Kittler argued, psychophysical methods necessitated the application
of artificially generated stimuli to humans treated as "guinea pigs"
in the psychophysical laboratory. In response to the new sense
experiences imposed by psychophysical apparatuses, subjects had to
acquire a responsive language divorced from their natural tongue.
Absent this artificial new responsive language, Kittler claimed, "it
would not have been possible to isolate the subconscious mechanisms responsible for the construction of psychophysical reality from
the cultural—that is, language-dependent—functions responsible
for concept formation."[21] Articulated through the language of
Weber's two-point threshold experiments, subjective tactile experiences could be aggregated as a statically modeled, normally functioning tactile system that transcended both individual experimental
subjects and their cultural contexts.

Due to its historically unique position, including its ambigu-
ous relationship to sexuality, pain, emotion, and other amorphously
mental and physical states, fitting touch within the framework of
modern experimental science required techniques and instruments
that conformed to its physiological materiality. The mode of

measure became inextricably linked to the measured object. Guided by a radical devotion to empiricism, nineteenth-century physiologists and experimental psychologists engaged in a materialist reconstruction of touch from the ground up, discarding a priori knowledge about touch in favor of that which could be revealed through the structured application of mechanical and electrical stimuli to the body. Contemporary haptic interface design, informed by the same devotion to empiricism, depends on the strategic recall of the haptic subject that emerged from the nineteenth-century research labs. These experimental apparatuses served to order touch, in Heideggerian terms, as a "standing-reserve" that would later be "challenged-forth" in the service of the new economic need to circulate information efficiently and effectively through subjects.[22]

Like the electrical experiments described in Interface 1, the validity of Weber's quantified account of tactile acuity depended on its dispersal through a network of scientists willing to repeat his tests. Crucially, when Weber summarized and circulated his findings through the publication of two significant works on touch—*De Subtilitate Tactus* (On the Sensitivity of the Touch Sense) in 1834 and *Tastsinn und Gemeingefühl* (The Sense of Touch and Common Sensibility) in 1846—he detailed the methods by which his results were obtained, so that future researchers could attempt to replicate his results.[23] Weber's trials filled a vacuum of knowledge concerning touch: in the absence of any hegemonic, experimentally informed account of tactility, his methods were greeted by the exposed and anxious bodies of anatomists and physiologists, all eager to have compass points pressed repeatedly against their bodies. By the century's close, any textbook published on the newly emergent experimental psychology not only detailed Weber's "epoch-making" two-point threshold experiments[24] but also provided instructions on how to replicate them, often followed by an account of the author's own experiences when doing so. Though researchers later in the century disputed and honed Weber's results, they did so on the methodological terrain he established, indicating the extent to which the lab experiment progressively territorialized tactility during the nineteenth century.

Toward a Psychophysiology of Perception

Prior to Weber's experiments, the scientific study of touch had unfolded in a piecemeal fashion, absent any coherent, systematic, and unifying account. Whereas a good deal of empirical study at the time had been devoted to understanding the "higher-order" senses of vision and hearing, before Weber's interventions, the study of the "lower senses" (including touch) had been "one of the most neglected areas of physiology."[25] Christian doctrines limited the transmission of knowledge about touch, causing Renaissance and Enlightenment studies of touch to proceed with scant awareness of older investigations and theories.[26] While touch assumed a new importance among what Jessica Riskin terms "sentimental empiricists" of the Enlightenment, many of their assumptions about the tactile processes remained grounded in speculation rather than systemic examinations of the mechanisms that produced tactile experience. In contrast, new anatomical studies attempted to provide physiological explanations for phenomenal tactile experience, accounting for differing tactile experiences and acuity through a mixture of visual examinations of the skin, extrapolations from clinical cases of tactile disorders, and rational inference grounded in phenomenological experience.

From the mid-1700s to the end of the nineteenth century, the understanding of the human nervous and perceptual system underwent a drastic shift, attributable in part to the chaos that resulted from discoveries about the relationship between electricity and the body. Having applied to the body the same methods applied to the study of nature, physiologists and philosophers turned their attention to the processes by which the senses transmitted sensations to the mind. In Germany, the interpenetration of physiology, physics, and philosophy prompted the development of new ideas about the relationship between nature and human biology. As an attempt to resist the more mechanistic model of Newtonianism, German nature philosophy positioned the human as a coequal part of nature, explaining biological functions in relation to natural—rather than mechanical—ones.[27] When Weber began investigating touch in the early 1800s, he positioned his efforts against the vitalist belief that humans were animated by irreducible and fundamental natural forces. Up to that point, Enlightenment philosophers had extensively debated touch's status in the overall configuration of the human, entering into broader conversations about the relationship between and primacy of perceptual modes. For example, Johann Gottfried Herder located touch at the center of his epistemology of the external world. Friedrich Schiller defined touch only as a force that was acted on, distinguished from the active, distal senses of seeing and hearing that allow the subject to engage in self-construction.[28] Both Weber's methods and his conclusions represented a significant departure from these treatments: rather than engaging in philosophical reflections on the role of touch in the formation of the self as a political and moral subject, he opted instead to isolate touch as a physiological process, differentiated based on its anatomical structures as they revealed themselves in the simulated conditions of the lab. His decision to locate touch in anatomy can therefore be understood as a reaction to, and an attempt to move away from, the speculative method of nature philosophy.[29] The relationship between the interior and exterior could be uncovered through dedicated attention to nerve structures and their limits. By quantifying what the nerves both captured and failed to capture, by charting differences between noticed and not noticed stimuli, and by identifying sensations both present and absent, Weber built a new theory of perception from the ground of the lab experiment up.

Of these myriad changes that swept over the study of the nervous system in the nineteenth century, two important developments served as preconditions for Weber's science of touch. The first concerned the division of nerves into sensory and motor components, and the second involved developing the doctrine of specific nerve energies. Scottish physiologist Charles Bell's (1774–1842) finding that distinct nerves were responsible for transmitting information to and from the extremities cleaved functions that had previously been considered linked. Bell published an account of his new theories in an 1811 pamphlet titled *Idea of a New Anatomy of the Brain,*[30] where he described, in chillingly detached detail, a series of rabbit vivisection experiments. In the trials, he alternatively stimulated the anterior and posterior portions of the animals' spinal cords. The results of these gruesome experiments, in which the rabbits' reactions to Bell's probings were viewed as a model for human sensorimotor function, suggested that the nerve structures responsible for motion were distinct from those responsible for sensation. In 1822, Francois Magendie built on these findings with his own experiments, severing the posterior nerve roots in the spinal cords of several puppies. After stitching the wounds up, he noted that the animals had no sensation, but were able to move without difficulty. He repeated the experiment, this time severing the anterior roots. The animals' limbs were "immobile and flaccid" but "retained an unequivocal sensibility."[31] Together, these two works established the Bell-Magendie law of the spinal nerve roots, rendering a model of the nervous system where the structures necessary for action were distinct from those necessary for sensation. Magendie, "in pinching, plucking, or pricking" the posterior (sensory) nerves at their roots, noted that "the animal manifests pain."[32] This pain "is nothing compared to the intensity that develops if one touches, even lightly, the spinal cord at the origins of these roots." With the delight of a puppeteer, he described contractions produced in the animal when "pinching" and "jabbing" at the anterior (motor) nerve root, noting that the animal did not display any apparent sensitivity when the muscle convulsed.[33] The role of touch in this process should not be downplayed: the vivisecting scientist took control over the motor and sensory functions of the subject through tactile manipulation, either directly or mediated by the scalpel (here functioning as a puppet's string). The motor and sensory functions of the nervous system were reduced to basic input and output systems, each transmitting information through a dedicated set of wires revealed through lab experiment.[34] By the time of Magendie's trials, electricity's animating effects on muscular tissue had been extensively demonstrated through lab experiments and public performances, with the new science of electrification showing that the autonomy of the human body could be co-opted and compromised through the application of electrical current. Electricity could both control the bodies of the living and—as Giovanni Aldini demonstrated in his experiments on human and animal corpses—animate the bodies of the dead.[35] Magendie attempted to further validate Aldini's findings using this technique. By applying galvanic current to the severed nerves, he induced strong muscular contractions in his subjects. The living animals' touch, when fused with electrical current, became a mechanism for producing positive knowledge about the segmentation of sensation and motion in the human nervous system.

The second precondition for the science of touch involved rendering the senses as carriers of discrete energies, a move generally credited to Johannes Peter Müller's doctrine of specific nerve energies (first offered in 1826 and published in his 1838 work *Handbuch der Physiologie des Menchen*).[36] Bell, however, had posited a similar theory in his 1811 pamphlet, where he situated his findings as a counter to "the prevailing doctrine of the anatomical schools" that took the whole brain to be a "common sensorium." The sense organs, according to the doctrine Bell attempted to displace, were only differentiated by their degree of sensibility: the eye was essentially only a highly sensitive form of touch: "the nerve of the eye . . . differs from the nerves of touch only in the degree of its sensibility."[37] Sense information was taken to be of a common type that was differentiated only upon entering the brain. Bell and Müller disputed this position, claiming that each sense organ was equipped to process only a single type of sensation. As the theory of differentiation by heightened sensation suggested, a needle piercing the eye produced light rather than an extreme and excruciating pain. A blow to the head, Bell noted, produced distinct sensations in each of the affected organs, with these sensations arising from the irritation of the sense organs themselves rather than from any external stimulus they were presented with. As a result of the vibrations from being struck in the head, "the ears ring, and the eye flashes light, while there is neither light nor sound present."[38] For Müller, the external world did not produce sensations but rather excited specific energies contained in the nerves themselves. Applying the theory to touch, he observed:

> The sensations of the nerves of touch (or common sensibil-
> ity) are those of cold and heat, pain and pleasure, and innu-
> merable modifications of these, which are neither painful nor
> pleasurable. . . . All these sensations are constantly being pro-
> duced by internal causes in all parts of our body endowed
> with sensitive nerves; they may also be excited by causes act-
> ing from without, but external agencies are not capable of
> adding any new element to their nature. *The sensations of the
> nerves of touch are therefore states or qualities proper to themselves,*
> and merely rendered manifest by exciting causes either exter-
> nal or internal.[39]

In this formulation, Müller treated sensation as something that could be *solicited* by external stimulus, but resided ultimately in the specific nerve stimulated. The sense organs did not serve to medi-ate between the external and internal world, but instead conveyed to the sensorium only the character of their own state, which itself was only incidental to the stimulus. Sensation therefore possessed a self-referential quality, providing knowledge "not of external bodies but of the nerves of sense themselves."[40]

Müller and Bell's experimentally derived theories prompted an upheaval in the way physiologists treated the senses. Moreover, in conjunction with Weber's nascent work, their collective efforts indicated the coalescing of a new study of perception, where the varying capacities of the individual sense modalities became the primary objects of scientific investigation. Once the atomist doc-trine of specific nerve energies became accepted dogma, the senses were isolated from one another with a validity accorded by natural science and then subjected to new modes of instrumental interro-gation. Experimentally isolated, the individuated sensory pro-cesses could be quantified, abstracted, and represented using novel graphical methods. As part of the drive to master the unknown—to shed light on the dark corners of human experience—physiological psychologists developed a comprehensive range of techniques for measuring sensory processes. The premise underlying Müller's claim called into question the senses' ability to provide reliable information about the external, physical world. An inherent insta-bility constituted subjective perception; as Müller argued: "the

immediate objects of the perception of our senses are merely particular states induced in the nerves."[41] Scientific observation and
experimentation rendered the sensory system—the same system
that had previously been deployed as an instrument for revealing
the external world—an untrustworthy mediating apparatus. Whereas
earlier movements augmented the senses by suturing new instruments onto them, nineteenth-century experimental psychologists
used instruments instead to reveal the specific capacities and limits of
the individuated perceptual organs themselves. The electrical apparatuses discussed in Interface 1 illustrate this transition: after being used
in the eighteenth century to investigate the physical laws that govern
electricity's behavior, in the nineteenth century, following Volta's
suggestion, anatomists turned these instruments inward on the senses,
measuring and mapping the ability of instruments to deceive the
senses of their experimental subjects.

On its surface, this disruptive paradigm shift seemed to imply
a negative and nihilistic attitude toward the senses. However, the
response to this newborn epistemological skepticism entailed what
Crary describes as "a positive reorganization of perception and its
objects,"[42] where the senses themselves—rather than the objective
world they provided access to—came to be the focus of intense
scientific scrutiny. Müller's theory of nerve energies (described by
several commentators as a theory of "specific 'sense' energies"
rather than of nerve energies[43]) would eventually be furthered by
his student Hermann von Helmholtz to account for the depletion
of these nerve energies in his law of conservation. Though some
have claimed that Müller and Helmholtz, in focusing their experiments primarily on vision and hearing, failed to attend to touch,[44]
Müller's theoretical speculations about the sense energies were
informed by a latent theory of tactility, although it never became a
priority in his experimental agenda. The eighteenth-century
model of vision depended on being embedded in a tactile body,
what Crary describes as an "anti-optical notion of sight" rendered
as analogous to touch.[45] The nineteenth-century model, informed
by Müller's doctrine, grounded visual experience in an eye that
possessed its own unique energies, activated by external stimuli but
governed ultimately by its psychophysiological characteristics. In

divorcing the subject's visual sensations from the external world, Müller did away with the fixed, objective experience of light represented by the camera obscura model, decoupling the internal world of the perceiver from any stable link to external referents. Although radically corporealized through its new location in the individual and specialized nerves, optical experience in the nineteenth-century model became detached from the body's other sensory registers. The doctrine of nerve energies decontextualized visual experience; it no longer needed to refer to an external world of objects, nor did it need to be linked to other sensory experiences. For Crary, this formulation of subjective visual experience mirrored the development of nineteenth-century image-making technologies. The eye's ability to be deceived and to generate its own experience (powers not unique to it, according to Müller's doctrine) became the subject of an investigation that yielded a body of quantified knowledge, which was immediately mobilized to produce a range of devices for inducing optical illusions. Techniques for mediating vision, then, inducing an evolving set of theories to account for the new visual experiences these techniques produced.

However, this detachment of visual and tactile sensations also proved generative for a touch that was no longer defined in relation to vision, as touch quickly became the object of dedicated empirical analysis.[46] The physiological separation of the senses accomplished by Müller's doctrine (and foreshadowed by Volta's electrical investigations) appears in retrospect a necessary precondition for the isolation of the sense organs as discrete objects of experimental study. In nineteenth-century physiology, touch—like seeing and hearing—became radically implicated in the trajectory of scientific modernity: demystified, quantified, and *mediated* by new empirical techniques for measuring perceptual processes. By subjecting touch to analogous experimental processes, instruments of measure, and investigative techniques, these psychologists understood themselves to be actively making touch like the senses of seeing and hearing.

Cartographic Tactility: The Two-Point Threshold Experiments

In the scientific disassembly and reconstruction of touch, no system of experiments was more significant than Weber's two-point threshold trials. The particular positioning of his experimental subjects allowed Weber to generate purportedly objective knowledge based on their articulated perceptual experience. Seated at the inception of what is referred to as "scientific psychology,"[47] Weber's numerous and meticulous experiments, not just on touch but on the whole range of sensory processes, helped establish a methodological ground for future research on the psychophysiology of perception, and staged the execution of a macrohistorical project dedicated to the empirical study of the tactile senses.

The protocols for the threshold discrimination experiments expressed Weber's desire to produce constant, reproducible, and predictable results that would serve as a way to reveal the dermal nerve structures responsible for the subjective experience of space. As these nerve structures were inaccessible to the microscopes of his era, the two-point threshold experiments can be understood as a response to the limits of visual inspection and instrumentation. The search for objective and stable knowledge about touch in this context involved controlling, fixing, and disciplining the subject's tactile experience while simultaneously isolating the specific subcomponents of the tactile system that Weber sought to examine. The threshold that Weber asked his subjects to articulate referred to a sort of magic minimum: the moment at which the subject was able to perceive the two compass points as distinct from each other. Initially spaced closely together, Weber moved the compass points incrementally apart until the subject could discriminate between them.

The imperative to produce stable and reliable knowledge about the structures of the tactual system had ramifications for the specific configuration of the experiment, as Weber sought to standardize both the experimental subjects[48] and the stimuli that were applied to them through the use of experimental protocols. Subjects were to be rendered docile during the trials, as Weber insisted that they sit as still as possible and in a fixed position while the two compass points were pressed against them.[49] This protocol allowed subjects to carefully attend to the points of contact between the apparatus and their skin, free of any tactual noise that might be produced by bodily movements. Other variables also threatened to taint the consistency of experimental results and therefore had to be eliminated. Concern over what he understood as fatigue in his subjects' senses prompted Weber to suggest that the experiments be carried out "at a time when our attention is not distracted by too many other sensations, i.e. particularly, in the evening or at night."[50] Constant exertion, he believed, left the senses bereft of energy, resulting in the subject's inability to discriminate between stimuli that they would have perceived as distinct under optimal conditions; standardizing the time during which he conducted the experiments helped control this variable.

Concerning the compasses (see Figures 2.3 and 2.4), Weber made similar efforts at homogenization. In laying out the proto-

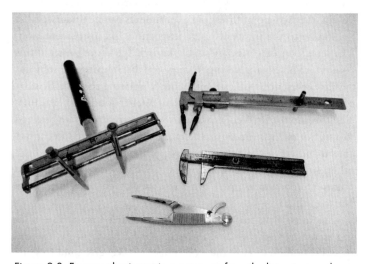

Figure 2.3. Four aesthesiometric compasses from the late nineteenth century, used in two-point threshold tests. Photograph provided with the friendly support of the Adolf-Wuerth-Center for the History of Psychology, Germany.

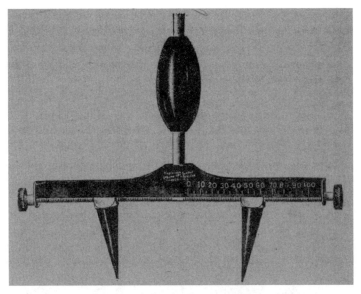

Figure 2.4. Jastrow's Improved Esthesiometer (1910), a device developed to refine Weber's two-point threshold discrimination experiments. From Guy Montrose Whipple, *Manual of Mental and Physical Tests* (New York: Arno Press, 1910), 246.

cols for these experiments, he stressed that the objects applied to touch-organs were to have a constancy and uniformity; they "should be identical in material and shape . . . have the same temperature in the various experiments . . . and [be] placed on the touch-organ at the same time with the same force, but in such a way that the organ is not at all damaged and affected by pain, which would certainly impair the sense of touch."[51] His attempt to standardize the character of the applied stimuli mirrored his efforts at standardizing the subjective conditions of his participants. Normalizing the sensations generated by the compass's points as they were pressed into the skin allowed Weber to isolate the feeling of pressure from the other tactual sensations. The blunted compass points ensured that pain, excluded in Weber's model from the tactile system, would not cloud the subject's tactual faculty. Homogenizing both the temperature of the points and the pressure with which they were applied served a similar function. Disentangling the tactile senses required adherence to these rigid protocols; any errors in the execution of the experiment, including the failure to properly care for instruments, risked contaminating the sense of pressure with sensations of temperature and pain.

Weber's attempts to reduce variables were more than just sound experimental design; considered in the more expansive project of modernizing the sensorium, isolating, confining, and quantifying the skin's tactile acuity can be understood as part of the process of making the senses adequate to new techniques of measurement. Weber's cartography of the skin, produced through these meticulous experimental protocols, emerged as part of a drive to produce a body of objective knowledge about aspects of perception previously thought to be subjective. Guided by Weber's instructions, devotees and challengers alike could repeat his experiments, articulating a model of touch coequal to the protocols used to measure, fix, and subdivide it.

A chart Weber included in his text broke down his results by skin region. For each site on the skin, he listed the distance at which the compass points were applied, the clarity of the distinction between the two points (the categories were "clear," "not very clear," "yes," "no," "poor," and "obscurely"), and the degree to which the subject could tell whether the compass points were oriented horizontally or vertically (using the same categories). Weber carried out the experiment on 190 different skin regions from the top of the head to the bottom of the feet, and from the tips of the fingers to the ends of the toes. In plotting the skin by this method, Weber provided a coordinate system for the skin—a map of its regions according to their capacity to accurately assess the distance between the compass points. The experiment rendered space as an objective quantity correctly judged by the compass (what Weber termed the "true distance") and necessarily judged incorrectly by the experimental subject. The skin became a matrix of discriminatory and resolving capacities, with each region possessing quantifiably differentiated abilities to notice and not notice the differences between things.

His findings in these experiments would lead him eventually to posit the existence of "sensory circles," where a nerve sat at the center of each circle, and the perception of stimuli weakened the farther the stimuli moved from the circle's middle. Touching, Weber claimed, involved gradually gaining a "dim awareness of the number and position of our sensory circles" through "long usage and repeated motor activity of our limbs."[52] This theory emphasized the subject's active role in constructing felt experience through the use of a *Muskelsinn* (muscle sense). Based on this tacit, unconscious awareness, Weber claimed that the tactual perception of an object's size increased relative to the density of sensory circles in the dermal area used to perceive it:

> The more sensory circles lying between the two points of a compass touching us, the further apart the points seem to be and vice versa. If the sensory circles are small and numerous, as on the finger-tips or the palm of the hand . . . then the compass-points touching these areas seem to be further apart than if they touched part of the dorsal skin, in whose surface only a few elementary fibres are found to terminate.[53]

According to these results, the perception of distances through touch was relative to the nerve structures distributed in different densities throughout the skin. This claim provided the kernel of what would later come to be known as Weber's illusion, which holds that the distance between two fixed compass points appears to grow as the compass is moved from a region of the skin with a lower density of nerve receptors toward a skin region with more tightly clustered nerves. Weber's illusion, then, highlighted and specified touch's capacity to be deceived—its inability to accurately register the space of the external world. Similarly, another set of experiments prompted Weber to assert that the perception of shape varies based on the density of nerve fibers in the skin region where we encounter a given shape.[54] Each region of the skin provided a relativistic dermal image unique to its structure.

The nerves that perceived pressure were not solely responsible for distorting the skin's perception of space. Through yet another battery of experiments, Weber found that stimulus temperature—which

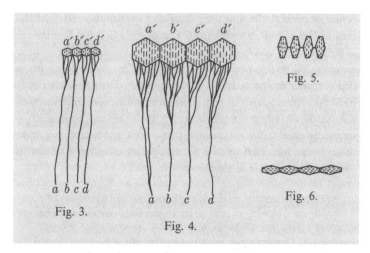

Figure 2.5. Weber's diagram of sensory circles on the skin. Smaller circles are found on the areas with greater discriminatory capacity, such as the fingertips. Previous thinking about touch attributed sensitivity to skin thickness. In the diagram, a', b', c', and d' are the sensory circles; a, b, c, and d are the nerve roots. The images labeled Figures 5 and 6 (on the right) represent the different shapes and orientations that Weber posited as existing in the sensory circles. From Ernst Weber, "Ernst Heinrich Weber on Sensory Circles and Cutaneous Space Perception," in *Source Book in the History of Psychology*, ed. Richard Herrstein and Edwin Boring (Cambridge, Mass.: Harvard University Press, 1965), 147.

he had initially worked to normalize in his early experiments with the compass—also impacted the subject's perception of space. Warming the stimulus caused it to feel lighter on the skin, whereas a cooler stimulus felt heavier and could therefore be more accurately localized. Weber's structured experimental system, which materially expressed what he viewed as the crucial divisions among the various touch senses, allowed him to first disentangle the perception of space from the perception of temperature, and then to reintroduce the variable of temperature under carefully controlled conditions. It was not just that he succeeded in disaggregating the tactile senses, but that he took the further step of productively and generatively *reaggregating* them, through the experimental reconfiguration, to accurately plot their various interdependences.

Against the backdrop of a scientific culture that increasingly valorized the production of stable sensate experience as the grounds for knowledge, quantifying and mapping the tactile system's capacity for the differentiated perception of the same object embodied an effort to normalize and manage what had been a confounding and unreliable mode of perception. Weber, and those who later adopted his methods, sought to accomplish this task using the same modes of experimental confinement that, over the preceding two centuries, had yielded increasingly useful representations of the natural world. Tactile experiences and their mental correlates became part of a mechanistic and quantifiable universe. Whereas in the Enlightenment depiction of the relationship between the sensing subject and the external world, "the senses deceive, and reason corrects these errors,"[55] in the new Weberian paradigm, the rational experiment's interface with the senses modified reason's corrective faculty. Reason did not fix or eliminate the errors of the senses but instead provided the means to construct a normalized psychoanatomical model of the tactile system, in which its various capacities to provide deceptive and illusory sensations were accurately mapped.[56] This rendering of the tactile system gained nuance and refinement through continued innovations in both experimental method and instrumentation (see Pillsbury's map of errors in localization in Figure 2.6, for example), articulating a model where touch operated as a data reception system. Though technical media capable of productively exploiting this map did not arrive until more than a century after Weber crafted it, touch would not have been able to achieve its later integration into computing systems without first being rendered as a web of distributed data-reception nodes. Before data could be rendered for the flesh, flesh first had to be rendered as data.

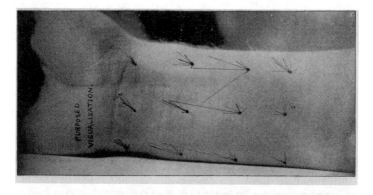

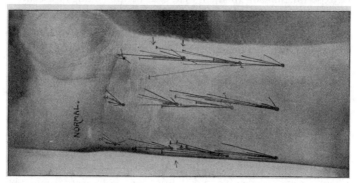

Figure 2.6. Pillsbury's technique for testing the relationship between visual and tactual acuity. From W. B. Pillsbury, "Some Questions of the Cutaneous Sensibility," *American Journal of Psychology 7* (1895): 55–56.

Purifying Touch: Pain and the "Common Sensibility"

While threshold perceptions and discriminatory capacities proved stable under the artificial conditions of Weber's lab, their stasis remained haunted by the constant threat of pain. In the lab, Weber took careful steps to avoid inducing painful sensations, guided by the assumption that pain interfered with touch's capacity to make accurate judgments about the external world. He had gradually barred pain from his trials through the rigorous refinement of experimental techniques, cordoning off the nerve mechanisms responsible for tactile perception (location, pressure, and temperature, in Weber's model) from those that induced sensations of pain. In spite of these efforts, the close link between pain and touch remained a constant source of imprecision throughout his trials. To more thoroughly confine the tactual senses, he had to isolate this deficiency completely. The psychophysiologically distinct category of the "common sensibility" provided him with the means to cleave off a bundle of bodily sensations that had traditionally been grouped together with touch. Moreover, the common sensibility served a strategic function in the overarching project of rendering touch rational, knowable, and ultimately suitable to be used as a reliable information-processing channel. The sensations linked under the designation of touch were a jigsaw puzzle that Weber recut, and then, under the naturalistic guise of rational experimentation, assembled anew. Unlike conflicts over touch in the eighteenth and nineteenth centuries, touch's relationship to vision was at most a tertiary concern for him; instead his understanding of touch radically realigned a group of sensations prior thinkers had been content to axiomatically lump together.

Though Weber had first formulated the common sensibility, or coenaesthesis, as distinct from touch in *De Tactu* (1834), it assumed a more prominent role in the later and more speculative *Der Tastsinn* (1846), where he defined the common sensibility concisely as "the ability to perceive our own sensory states."[57] This understanding of the common sensibility did not deviate substantially from other formulations circulating at the time; Christian Friedrich Hübner offered a similar theory in his 1794 doctoral dissertation *Coenesthesis,* written at the University of Halle.[58] However, when read in relation to Weber's theory of tactility, the common sensibility achieved a functional utility for Weber it lacked in other contemporaneous models. As with his other theories, Weber's splitting off of touch from the common sensibility was informed by a rigorous program of experiments. In those trials, he sought to distinguish between those sensations that were oriented inward toward the perception of the body's own state and those directed outward toward the state of the external environment. Pain, laughter, and nausea were each understood to be part of a nervous network designated by the common sensibility. Prior thinkers assumed that the common sensibility was derived from input from all the senses (smell, for example, was thought to contribute to nausea, while touch contributed to pain), but Weber modified this theory by claiming that humans were given access to the common sensibility through a distinct set of inwardly oriented nerves. And though this assertion would be disproven by subsequent studies, embracing it had positive implications for Weber's theory of touch.[59]

Articulating the split between the common sensibility and the external senses, he offered the following speculation:

> Our sense-organs are directed outwardly not inwardly, in order that the mind may receive impressions from the external world: it would become very confused if internal processes were persistently demanding its attention. One intestinal canal touches and rubs against another, lungs rub against the skin of the pleura covering the chest cavity, muscles press and rub against each other: but we have no sensations of these.[60]

To test this hypothesis, Weber carried out a battery of bizarre and inventive experiments intended to probe the body's interior spaces. If his subjects could accurately perceive the quality of stimuli applied to the inner regions of the body, their experience implied the existence of nerve structures capable of generating these perceptible sensations. In one particularly uncomfortable experiment, he administered enemas filled with cold water to himself and two other "good observers."[61] In the first trial, which employed a 14-ounce enema of water cooled to 18.7° C, Weber noted that the water excited "a strong feeling of cold in the anus upon its introduction and later expulsion."[62] Once it entered his bowels, however, he felt a "very faint, almost unnoticeable sensation of cold, which gradually seemed to spread toward the middle of the abdomen."[63] Unsure of what this unstable sensation implied, Weber repeated the trial, holding the volume of water injected inside him constant while lowering its temperature to 7.5° C. The colder water, he speculated, would sharpen any sensations induced by trial. The second trial, however, yielded similar results. With greater focus, he was able to localize the source of the cooling sensation not to the interior of the body but to its exterior. By placing a thermometer against the skin of his distended abdomen, he found that the water had succeeded in cooling his skin by roughly 1 degree. In other words, the instrument confirmed Weber's intuition: the sensation he felt originated not in the nerve fibers that travel throughout the body's interior, but rather in the specialized nerves distributed throughout the touch-organ. The lessons Weber drew from these tests, however, are of less interest here than the methods by which they were obtained. Weber derived the segregation of the tactile senses from those senses that comprised the common sensibility—his breaking off of interiority from exteriority, and of touch from pain—from dozens of similar experiments, as he attempted to build a science of neuroanatomy up from the ground his trials provided.[64] Because of its physiological differentiation, the common sensibility could be split off of touch, with its affiliated, inwardly directed, irrational component—feelings—banished to another realm of sensory experience.

With pain, Weber's concern focused on its poor functioning as an information-gathering system. Whenever a stimulus activated the pain sense, his experiments showed that subjects were consistently unable to make accurate judgments about its qualities. As his experiments in *De Tactu* had shown, the constitutive elements of the tactile system (including the senses of location, temperature, and pressure) functioned impressively as discriminating mechanisms, frequently able to make judgments almost as fine as the instruments used to stimulate them. Pain, Weber lamented, stripped touch of these powers:

> We can even grade general sensations, e.g. pains; but how underdeveloped we are at this, compared with the numerous degrees of temperature or levels of pressure which we can observe and to some extent measure with the sense of touch. If a warm body excites no pain, we can clearly distinguish . . . a difference of only 0.3 or 0.2 degrees C; but if the temperature of the warm body does excite pain, we can no longer achieve such a fine discrimination. We cannot even make coarse discriminations.[65]

The physical realm, in Weber's protocybernetic formulation, consisted of data that could be accurately measured by a range of instruments: the thermometer for heat; weights for pressure; and, in the two-point threshold trials, the compass for space. As it hindered touch's discriminatory faculty, pain had no place in this instrumentalized conception of tactility. The problem pain posed was emblematic of Weber's concerns about touch's capacity as an information receiver: as a type of noise, pain interfered with touch's ability to reliably take in information from the external world. His experimental method embodied a philosophy of sensation that required the senses to enter into a mimetic relationship with the instruments used to assess their capacities. As no instrument yet existed to quantify pain, it was banished from the newly reconstituted realm of tactile experience; it was only brought in as a way to measure the moment at which it would overwhelm the touch senses.

As with many of his other hypotheses, Weber derived his account of pain from a comparison of others' findings with the data yielded by his own experiments. In the experiments he designed to investigate pain, Weber's own body frequently provided the testing site. Attempting to observe the difference between touch and pain, he repeatedly hammered his fingers at different points, noting a brief gap between when he felt the sensation of touch and when he felt pain.[66] In his efforts to separate the temperature sense from pain, his trials were more systematic. One set of experiments involved dipping different parts of his body into scalding hot water, noting the time that it took for the pain to become so intense that he had to withdraw the body part from the water (unsurprisingly, the joints on Weber's fingers were more tolerant of water heated to 62° C than his tongue; where he left his finger in the water for six seconds, he withdrew his tongue after only two seconds). In another battery of tests, he pressed a freezing cold iron key against different points of his head and recorded its effects. At some points (the ear lobe, for example), the key induced only the sensation of cold, without an accompanying sensation of pain. At others (such as the eyelid), the key quickly produced a painful sensation akin to burning. Further, he experimented with electricity's ability to induce pain, but found that, although the sensations produced by electrical current were "unpleasant," they stopped short of causing pain.[67] The drive to specify with increasing precision the relationship between exterior and interior undergirded each new set of experiments, as Weber doggedly probed the borders that separated sense modalities.

Based on the immense and variegated data he gathered through this system of experiments, Weber concluded that fluctuations in sensations of pain were not as great as the differences humans perceive in other forms of sense input: "so much is certain, that qualitative differences between pains and other feelings of general sensation are much less numerous than feelings from the special senses."[68] As Fechner would later formalize in his positing of Weber's law, Weber had found that the perception of differences among stimuli—for touch and for the senses of seeing and hearing—involved a proportion of the original just-noticed difference. The perceptual modalities that constituted the common sensibility were exceptions to this generalizable finding: the structures that make up the common sensibility were thought to be inaccurate and easily deceived. Pain proved particularly distracting, precisely because its activation called attention from the exterior to the interior world, and by so doing, short-circuited the capacity for rational, detached, and autonomous perception:

> One effect of the intensity of many feelings provided by common sensibility is that the mind is prevented from calmly contemplating them in the manner necessary for the sensation to be referable to objects. Instead, the attention of the mind is driven by the pains to its own state of suffering, and its own body. The effect of this is that the sensations excite not so much the cognitive faculties as the faculties of desire, so that we are driven to avoid pain by instinctive or intentional movements.[69]

The effect of pain was distraction: by refocusing the mind on the body's own state, pain caused the subject to avoid rather than interrogate the offending stimuli. Conjuring the pre-rational and instinctive, pain interfered with the accuracy of the sensory system, activating an animalistic, unrefined, and nonhuman subjectivity. The capacity for rational and reasoned behavior, under the command of the so-called cognitive faculties, defined the human in the scientific culture Weber inhabited. Enlightenment anthropology demarcated humans by their capacity to reason; as long as touch and pain were collapsed onto one another, the tactile would remain linked to a nonhuman and evolutionarily primitive mode of subjectivity.[70] In aesthetics, touch had been confined, along with smell and taste, to the lower order of the senses, due in part to its close association with the instrumental function of pain: Arthur Schopenhauer, a contemporary of Weber's, articulated precisely this concern in *The World as Will and Representation*.[71] Weber's attempt to disentangle touch from pain provided a technique for sanitizing touch, a means of freeing tactility from the confusing and irrational screams of the body and its sense organs in pain.[72] If pain were part of the tactile system, then it would signal a deficiency in touch's ability to function informatically in its capacity to reliably mediate the external world. The separation of these systems can be understood as emblematic of the attempt to transform touch into an efficient epistemological instrument, one tasked with accurately and predictably measuring the external world.

Although in the lab, instruments artificially freed touch from the risk of unintentional pain, outside the lab's protective walls, touch increasingly became subjected to the new vulnerabilities that accompanied industrial capitalism. Work in textile mills, railroads, and mines, for example, exposed the body to the harsh, injurious conditions of mechanized labor, with machines constantly threatening to sever limbs and digits.[73] Urbanization, too, pressed bodies closely together, shifting the scene of touch from the calm serenity of the countryside to the chaotic and frequently squalid conditions of the city.[74] Tactile modernity, from this perspective, designates a strategy of protection—a method of shielding touch from pain by rendering the latter as an experimentally and experientially distinct category of sensations. The lab allowed for the production of a tactility innocent of any painful experience, severing touch from centuries of accumulated associations and memories. This complex new experimental apparatus regulated the mirror desires of seeking pleasure[75] and avoiding pain—desires that had been previously managed through various forms of religious disciplining and corporeal training. In these models, pain often served as a strategic resource that could be mobilized in the service of religious and political programs (the practice of redemptive flagellation in the Catholic tradition; and juridical torture in the late Middle Ages).[76] Such calculated invocations of pain did not cease, of course, with the rise of experimental psychology, but the mechanistic detachment of pain from touch allowed pain to be banished to a realm where it could not contaminate the purity of true tactile experience.[77]

Diagnostic and Pedagogical Aesthesiometry

The new science of touch, like the "new psychology" more generally, affixed itself to utopian hopes about the capacity to educate and improve the sensory capacities of individuals. As a positive and pragmatic formulation, it employed rigorous methods and increasingly sophisticated instruments to precisely specify the relationship between the physical and the psychical worlds.[78] The apparatuses used to fragment touch (grouped together in the category "haptical" in Baldwin's *Dictionary of Philosophy and Psychology*) included test weights for quantifying the pressure sense, thermaesthesiometers for mapping spots of hot and cold sensitivity, algometers for measuring pain thresholds, and rotation tables for inducing and recording the sensation of movement (see Figure 2.7 for a partial list and Figures 2.8 and 2.9 for representative instruments).[79] The degree of differentiation among these instruments indicates the substantial commercial investment made in the project of further isolating the psychophysiological components of bodily experience. The market for what Titchener et al. referred to as "pieces for haptical work"[80] expanded as psychophysiologists intensified their push to break touch into increasingly finer units.[81] Attending to such dynamics counters hegemonic assertions about touch's historical naïveté, demonstrating that, like the other modes of perception, it too was reconstituted in response to the emerging biopolitical imperative to monitor, regulate, and quantify bodily processes. In the wake of Weber's work, the sense of touch was dissected; once cut into manageable pieces, each unit could be explored and experimented on independently as a discrete field of enquiry. For

C. HAPTICS AND ORGANIC SENSATION.

I. Haptical, etc., Sensations: Intensity.

1.	Pressure balance. Scripture. Willyoung.		$8.00
2.	Minimal weights. Scripture. Willyoung.		$3.00
3.	9 weights for method of right and wrong cases. Jastrow. Garden City Model Works.		$8.50
4.	16 weights. Scripture. Willyoung.		$4.00
5.	30 Weights. Galton. Cambridge Instr. Co.		£5.0.0
6.	2 glass funnels, with weights of shot.		
7.	6 wooden cylinders for loading with shot.	each	$ .25
8.	Wooden egg for loading with shot.		$1.00
9.	Set of 100 cartridge weights. Sanford. Made in Ithaca.		$2.50
10.	Set of 120 envelope weights. Sanford. Made in Ithaca.		$1.00
11.	Pressure balance. Von Frey. Zimmermann.	Mk.	40.00
12.	Algesimeter. Cattell. Brown.		$15.00

II. Haptical, etc., Sensations: Quality.

13.	4 pressure pencils. Scripture. Willyoung.	each	$1.35
14.	8 pressure pencils. Made in Ithaca.	each	$ .05
15.	Apparatus for exploring cutaneous surface. Washburn. Krille.	Mk.	50.00
16.	2 atomizers for inducing anæsthesia.	each	$1.00
17.	Menthol pencil.		$ .10
18.	Improved kinesimeter, with attachments and armrest. Hall. Yale Lab.		$110.00
19.	4 temperature tubes. Scripture. Willyoung.	each	$2.00
20.	12 temperature cylinders. Goldscheider. Made in Ithaca.		$3.00
21.	1-gal. copper vessel, fitted with two Roux regulators and 3 Friedburg burners.	ca.	$15.00
22.	2 thermometers, graduated in degrees,—24 to+200° C. Eimer & Amend.	each	$1.75
23.	Thermometer, graduated in degrees,—25 to+250° C. Eimer & Amend.		$2.00
24.	Thermometer, graduated in tenths of degrees,—7 to+100° C. Eimer & Amend.		$4.00
25.	Thermometer, graduated in degrees,—10 to+250° C. Eimer & Amend.		$1.75
26.	Upright physiological inductorium. Du Bois-Reymond. Petzold.	Mk.	120.00
27.	1 bipolar electrode; 1 unipolar electrode; 1 plate electrode. Chloride of Silver Dry-plate Battery Co.		$5.00

III. Haptical, etc., Perception.

28.	Apparatus for perception of movement by the elbow. Sanford. Willyoung.		$8.00
29.	Set of blocks and points for filled and open space. Titchener. Krille.	Mk.	10.00
30.	23 rubber strips for estimation of extent by the skin. Titchener. Made in Ithaca.		$ .50
31.	Interrupted-extent apparatus. Titchener. Willyoung.		$20.00
32.	Set of glass and rubber forms for determination of cutaneous form-limina. Major. Eimer & Amend.		$5.00
33.	12 surfaces for cutaneous impression. Made in Ithaca.		$1.00
34.	Set of charcoal points, with sharpener, for localization experiments.		$1.00
35.	4 rods, 5 handles, 2 cups, etc., for study of eccentric projection.		$5.00
36.	Stationary apparatus for study of eccentric projection. Made in Ithaca.		$5.00
37.	Rectilineal arm-movement apparatus. Münsterberg. Elbs.		$45.00
38.	Combined tilt-board and rotation-table. Titchener. Willyoung.		$50.00
39.	Set of 7 needle-æsthesiometers. Washburn.		
40.	4 simple æsthesiometers. Scripture. Willyoung.	each	$2.00

Figure 2.7. Partial list of haptical instruments used at the Cornell University Psychological Laboratory, including instruments for the dedicated investigation of pain, movement, temperature, pressure, and weight. From Edward Bradford Titchener, *The Psychological Laboratory of Cornell University* (Worcester, Mass.: Oliver B. Wood, 1900), 8–9.

the science of perception, touch ceased to be one solitary category of experience and was recognized instead as a group of senses. Before Weber, "the belief that the skin might house a variety of sense modalities had won only a handful of converts."[82] In the post-Weber studies on the subject, the "unitary experience of touching" was fragmented into differentiated considerations of how we perceived "roughness, warmth, cold, pressure, size, location and weight."[83] Some argued for the existence of specific nerve structures responsible for distinguishing between warm and cold; others grouped the perception of hot and cold together as a single temperature sense. They debated the existence of a unique sense for detecting vibration, differentiated between the tactile capacities of hairless and hairy skin regions, and split touch into active and passive modalities.[84] Some included pain in the tactile senses, whereas others followed Weber in locating it outside tactility's bounds.[85] By the early part of the twentieth century, the categories of proprioception and kinesthesis had become discrete areas of psychophysiological research, as psychologists, guided by empirical study in the framework of the lab, redrew the borders of mental experience.[86]

It is not my intent here to catalog the extensive research psychologists carried out on the tactile senses. Rather, I want to focus on two characteristics linking these quests for practical knowledge about tactility that together provided the basic consensus behind the many divergent claims psychologists, anatomists, and physiologists made about touch. First, the participants in this debate agreed on the validity of rational experiment as the method for knowing touch. The myriad controversies and disagreements centered on the constitution of the specific experiments and their efficacy in isolating the fragment of touch they sought to measure. Second, and related, these researchers understood touch as a purely psychophysiological operation, equating their object with their epistemic frame. As with contemporary media, the separation of the senses became embodied and embedded in a set of external apparatuses. Psychologists divided textbooks into sections on audition, vision, and haptics, with specific tests applied to each individual mode of perception. The organization of instrument catalogs also

Figure 2.8. G. Stanley Hall and H. H. Donaldson's kinesimeter (1893). Pictured in the photograph are R. C. Hollenbaugh (left) and Fletcher B. Dressler (right). Courtesy of the Clark University Archives.

Figure 2.9. Ensemble of dermal and kinesthetic apparatuses (1893). Courtesy of the Clark University Archives.

reflected this new recutting of the senses, with apparatuses grouped together according to the modality or submodality the device was designed to measure. Further, the structure of lab space inscribed these divisions among the senses; in Edward Bradford Titchener's map of his psychology laboratory at Cornell, he devoted separate rooms to optics, acoustics, and haptics, with taste and smell being grouped together in their own space (see Figure 2.10).[87] Psychology, then, divided the senses not only in abstract formulations, but also in the physical spaces of labs, and through separate institutionally grounded research programs devoted to each differentiated modality.

Over the last decades of the nineteenth century and into the twentieth, the instruments and tests developed in the lab took on uses outside its walls.[88] The improvements psychologists and anatomists made to Weber's crude beam-trammels allowed them not only to measure spatial limen with more precision, but also to evaluate subjects' capacities for the perception of shapes, to test reaction times, and to quantify pressure sensitivity. But the positive accumulation of knowledge in psychology also implied a normative function, consistent with the application of human sciences Foucault described in *The Order of Things*. "The human sciences," for Foucault, "laid down an essential division within their own

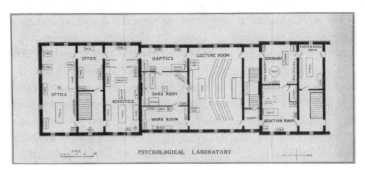

Figure 2.10. The Psychological Lab at Cornell. Note the placement of separate rooms for optics, acoustics, and haptics. From Edward Bradford Titchener, *The Psychological Laboratory of Cornell University* (Worcester, Mass.: Oliver B. Wood, 1900), n.p.

field: they always extended between a positive pole and a negative pole; they always designated an alterity."[89] With the science of touch, diagnostic aesthesiometry—the comparison of an individual's tactile acuity to an experimentally derived statistical mean—allowed the identification of a new negative pole: an abnormally functioning touch that served to indicate a host of psychosensory pathologies. Diagnostic aesthesiometry effectively reversed the function of Weber's compass; in originally seeking to uncover the structure of cutaneous nerves, he encountered fatigue as an obstacle to be factored out and controlled for in his trials. If the same subject perceived the same stimulus applied to the same part of the body differently, according to the time of day at which the experiment was conducted, the experiment was functionally useless, as sensory fatigue distorted the image of the nerve structures Weber hoped to reveal. But clinicians at the end of the nineteenth century employed this method to an alternative end, measuring patients' tactile acuity against the map Weber's research had provided. Failure to perform responses to the two-point threshold experiments became a sign of nervous exhaustion, an indicator that vital nerve energies were depleted, and the patient was in need of rehabilitation. This "blunting of sensitivity" signified a diseased or disordered neurological system, attributable to a range of conditions (including, in one account, neurasthenia, hysteria, insanity, and idiocy).[90]

Neurasthenics were thought to suffer from depleted nerve energies, as the repeated perceptual exertions required to exist in hyperstimulating urban environments exhausted the storehouse of reserve sensory energy needed to accurately discriminate between things.[91] Measuring the two-point threshold became a diagnostic technique: uninterested in determining the absolute structures of the nerves, clinicians aimed rather at exploring precisely that notion of sensory fatigue that Weber had been so careful to factor out. Just-noticed differences became just-*not*-noticed differences; the sensory world produced by the exhausted nervous system was characterized by uniformity, particularly in the case of the two-point threshold. To the skin of the fatigued subject, what had been two became one. An increasing distance between the two points on the aesthesiometric compass indicated sensory fatigue. The two-point threshold grew, the circumference of sensory circles expanded, and the subject's overall tactile resolution shrank as a consequence of such exhaustion. Initially, these spatial limen tests were the only measures of decaying tactile sensibility, but quickly, the subdivided senses of touch, and the new apparatuses used to apprehend them, provided further means of registering nerve maladies. Electrical machines, too, were repurposed for use in this diagnostic apparatus: a patient's sensitivity to electrical currents ("faradization") administered to different body parts became an indicator of nervous fatigue, with clinicians measuring the patient's responses to quantified units of applied electrical shocks.[92] The newfound psychoanatomical distinction between the structures responsible for sensations of pain, touch, and temperature allowed fluctuations in their respective sensitivities to be tracked independently of one another. Each mode, now isolated, acquired its own trajectory of statistical normalization and "mathesis";[93] the subjects touched in earlier experimental contexts gave up knowledge about themselves that could be used to construct previously unknown parameters of sensory deficiency and deviance. Aesthesiometry achieved a juridical function that depended on the continued aggregation of haptic experience through the use of diagnostic instruments.

The condition of exhaustion itself was closely tied to changing labor practices and the increased need to monitor workers in new spaces of work; it not only indicated the loss of physical energy (and thus, as Anson Rabinach explains, the human motor's decreased capacity to supply energy to machines), but more crucially, also functioned to render the worker as a sensing and discriminating machine. Although historiographies of neurasthenia generally ignore tactility's role in informing notions of sensory fatigue, the continued use of aesthesiometers into the twentieth century demonstrates that tactile discrimination was of paramount concern in studying the effects of modern life on the individual's sensory apparatus. The science of touch gained positive deployment in response to fears that existence in modern society entailed an assault on the senses inevitably accompanied by a corresponding depletion of

nerve energies. The fixed map of tactual acuity that haptics provided defined the parameters of both normally and abnormally
sensing subjects.

In spite of this proliferation of increasingly complex apparatuses
and techniques for measuring touch, Weber's original method
retained its validity. The German psychologist Hermann Griesbach
made extensive use of aesthesiometers to test the tactile acuity of
schoolchildren, with practitioners around the world quickly adopt-
ing his methods.[94] By the early 1900s, compasses similar to those
Weber had used in his trials were adopted by psychologists in Germany, Italy, Japan, Belgium, France, Bulgaria, the United States,
and Canada. Echoing Weber's two-point threshold tests from nearly
a century before, a 1910 guide instructed educators to carry out the
following test using a compass-style aesthesiometer:

> Instruct all S's [students] in the same words substantially as
> follows: "I'm going to touch your arm with points something
> like pencil points. They won't hurt you at all. You are to give
> careful attention to what you feel and tell me if you think
> I'm touching you with one point or with two points. You
> will have to watch very carefully. If you feel only one point
> say one if you feel two say two."[95]

The point I want to press here concerns the constancy of the experimental protocols used to isolate tactile experience. As in Weber's
trials, subjects' eyes were to be shielded from the stimulated skin,
out of fear that visual experience would contaminate the results of
the test. The instructions insisted that the subject, despite being
blindfolded during the trial, "watch" carefully, suggesting a distinctly tactile form of attentiveness that compared the embodied
visual act of fixing one's gaze to the mental process of attending to
pressures placed on the skin. To avoid activating ticklish sensations
(which Weber had dumped into the nebulous category of the common sensibility), the protocols also cautioned the administrator
against placing the instrument on hairy skin. This confinement and
isolation of a nominally pure and innocent tactility allowed touch
to acquire a new use-value in monitoring and surveillance. Severed from its traditional associations with emotion, desire, sexual-

ity, pleasure, and pain, this ascetic touch was tasked only with the labor of registering the differences between things, with discriminating accurately and attentively between stimuli as they were applied to the skin.

Adherence to the protocols of the two-point threshold trials had to be learned and mastered, both by experimenters and their subjects. Assimilating these protocols conveyed a normative message about touch's proper conditions of use: beyond simply monitoring and gathering data about an individual's sensate experience, aesthesiometric tests signified the new conception of the tactile sense as a necessarily imperfect information-processing mechanism. Subjects could never perceive the accuracy of the compass points with exact precision; the batteries of tests that subjects underwent conveyed a fundamental lesson about the superiority of instrumental judgments, as they were designed to be failed, but failed within certain parameters. By showing the tactile senses to be prone to gross distortions, test subjects absorbed crucial and practical messages about "the limitations and deceptions of their own senses";[96] their repeated failures were, therefore, simultaneously generative and instructive.

However, these haptical apparatuses also took on a pair of significant positive functions. First, their repeated use by experimental psychologists demonstrated, concretized, and legitimated the new psychology to students studying at the recently founded labs. They provided material expression for the animating idea that intensities of sensation, and mental life more generally, could be rendered numerically. Second, the measurements made by these instruments quantified the gains that could be achieved by repeatedly and carefully practicing touching, lending a verifiable credence to late-nineteenth-century pedagogical programs that emphasized sensory cultivation. The notion that touch could be educated was not a particularly new idea, but haptical apparatuses provided both standardized mechanisms for training touch and the means to quantify any gains achieved through this dedicated drilling.

Touch's centrality in the work of Italian physician Maria Montessori (1870–1952) illustrates this point. Montessori located the instruments of experimental psychology at the core of her educational program. In her pedagogical treatise *Il Metodo della Pedagogia Scientifica,* she stressed the changes that followed from the findings of those in the Leipzig school: "physiological or experimental psychology which, from Weber and Fechner to Wundt, has become organized into a new science, seems destined to furnish to the new pedagogy that fundamental preparation which the old-time metaphysical psychology furnished to philosophical pedagogy."[97] Where the "esthesiometer carries within itself the possibility of *measuring,*" Montessori's tactile media "often do not permit a measure, but are adapted to cause the child to *exercise* the senses."[98] Montessori described a plan of tactile education that made use of touchable "didactic material" to hone the discriminatory capacities of schoolchildren. As in experimental psychology textbooks, the protocols for creating and using didactic materials were meticulously specified: children were conditioned in the proper care of the hands and nails, in preparation "for a life in which man exercises and uses the tactile sense through the medium of these fingertips."[99] She hoped that, through these repeated rituals of care, students would come to understand their fingertips as vital instruments for knowing and encountering the world. Separate materials and exercises aimed at fostering the tactile, thermic, baric,[100] and muscular senses, as the division of touch into component parts continued to gain material expression in the design of instruments and apparatuses. With proper training, accomplished by blindfolding her students during their encounters with the tactile materials, Montessori hoped she could offload some of the labor of perception from the eyes onto the fingers, effectively imparting the ability of "seeing without eyes."[101] Following in this tradition, the Italian futurist F. T. Marinetti advocated the use of "tactile tables" that consisted of different categories of palpable materials not only to train the fingertips, but to usher in a new set of "tactile values" capable of disrupting and undoing what he understood as the arbitrary distinction between the five senses.[102]

In recalling these various deployments of haptics outside the lab's walls, my goal has been to show how various actors pressed the new science of touch into service both to assist in the construction of the normal and the pathological and to help cultivate a refined sense of tactile discrimination. Psychophysicists did not simply discover the parameters of haptic perception, only to have their findings adopted uncritically by clinicians and educators. Rather, the techniques for isolating and quantifying touch were folded into larger social projects that reflected new instrumentalizations of the senses. Diagnostic aesthesiometry embodied a growing concern with the subject's ability to maintain healthy levels of nerve energies in the face of modernity's constant assaults on the senses. The mobilization of haptics in educational programs became the groundwork for the subject's perceptual development throughout life, training that would prove essential for their insertion into the emerging industrial workplace. Both were part of what experimental psychologists trumpeted as a general homology between the new psychology and the demands of a rapidly industrializing economy, where the standardization of machines and bodies transformed labor, medicine, education, and mental life.[103]

Haptics: Enframing the Tactile

In the waning years of the nineteenth century, Dessoir proposed the term *haptic* to designate this new range of experimentally generated facts about touch.[104] A result of his frustration with the limits of available terminology for touch, the term "haptics" allowed experimenters to mark the transition from an older way of thinking about touch to a newer, more modern one made possible by folding touch into the lab. Haptics declared the experimenter's power to capture, order, map, and represent the variety of sensations formerly grouped messily together under the heading of "touch." Titchener's definition of haptics, offered in Baldwin's *Dictionary of Philosophy and Psychology* shortly after Dessoir proposed it, as "the doctrine of touch with concomitant sensations and perceptions—as optics is the doctrine of sight, and acoustics that of hearing"[105] rendered it as an analog of optics and acoustics, suggesting that the scientific study of touch would imbue the sense modality with a legitimacy it had previously lacked. Mainly owing to Titchener's efforts, those in the emerging field of experimental psychology quickly adopted the neologism, and by the close of the century, it became fashionable to use the terms "touch," "tactile," and "haptics" synonymously, effectively subsuming touch in its doctrinal variant. Haptics implied not just a new way of conceptualizing touch as a scientific object, but a new mode of *doing touch* through instrumentally aided experimentation.

William Krohn's Apparatus for Simultaneous Touches (Figures 2.11 and 2.12) embodied the mode of confinement employed to study touch in the nineteenth century. Krohn's technique deviated from Weber's, employing pneumatically activated corks capable of pressing against up to ten skin regions at the same time.[106] By opening or closing stop-cocks on the air box, the operator could control the number of corks that would press against the skin. Using the bellows, the operator briefly activated the corks and then quickly asked the subject to indicate where they felt contacts. Consistent with other experiments on the senses, the apparatus provided an

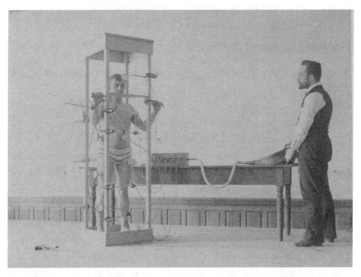

Figure 2.11. Thaddeus Bolton and William O. Krohn with the Apparatus for Simultaneous Touches. As Krohn (right) pumped air through the pneumatic tubes, the device activated corks positioned across the skin. The subject (Bolton) was asked to identify points on his skin where the apparatus stimulated him. Photograph courtesy of the Clark University Archives.

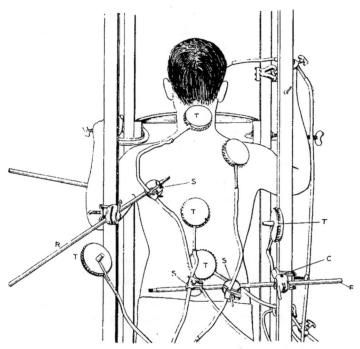

Figure 2.12. Apparatus for Simultaneous Touches, rear view. During Krohn's turn in the frame, he experienced 512 touches of the stop-corks on his joints and correctly localized 78 percent of them. The other two subjects performed in a similar range. From William Krohn, "An Experimental Study of Simultaneous Stimulations of the Sense of Touch," *Journal of Nervous and Mental Diseases* 18, no. 3 (1893): 169.

objective reality against which subjective perception was measured. Haptic experience became located in a frame within a frame: Krohn's device served a function similar to Boyle's vacuum chamber in the seventeenth century, perpetuating the illusion that the experimenter had succeeded in confining and isolating his object beyond the reach of any political considerations. Both Krohn and his subject, however, were encompassed in a broader structuring apparatus that promised to allow unrestricted and unprecedented access to the mechanisms from which mental life emerged. Experimental scientists could only uncover the body's natural tactile

processes by embracing the new psychology's modern methods.
Owing to advances in both techniques and instruments, the new
psychology allowed sensory experience to be "studied with as
much specialization of both field and method as modern astron-
omy."[107] The incorporation of increasingly mechanized and standardized devices for applying stimuli to test subjects progressively
factored out the possibility for human error. Like the Krohn's
Apparatus, the refined aesthesiometer Jastrow crafted in the 1880s
ensured that stimulus points would "touch at once and with equal
intensity,"[108] while also allowing the experimenter to easily vary
the character of the applied stimuli (see Figure 2.13; in addition to
pressure, the device also allowed for thermic and electrical stimulation). As with today's new media, a celebratory and fetishizing
rhetoric announced the field's arrival.

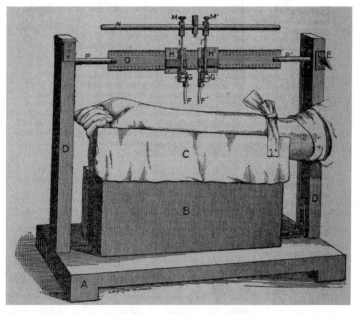

Figure 2.13. An aesthesiometer developed by Jastrow to mechanize the
simultaneous application of stimulus points, reducing the possibility of
human error. From O. T. Mason, "Notes," *American Journal of
Psychology* 1, no. 3 (1888): 552.

This move to enclose touch within haptics marked an important genealogical shift, as twentieth-century conversations around touch began to refer to haptics without marking its point of origin in a specific scientific metaphysics. Whether unconsciously or intentionally, psychologists gradually erased the distinction between the two categories, allowing even those familiar with such theories of tactility to claim that touch is persistently shrouded by a veil that obscured it from scientific and technical gazes.[109] This body of ongoing research can be seen as constitutive of a new, historically distinct type of perceiving subject—a haptic subject that registers touch's territorialization by a haptics defined as the doctrine of touch. This haptic subject emerged as the result of a strategic interface between touch and the apparatuses of experimentation. As such, its existence is contingent on the productive aggregation and quantification of tactile experience according to a particular set of structuring scientific protocols and machines. The haptic subject encompasses a matrix of sensory operations, atomistically isolated in the lab and then assembled into a new set of interlinkages that specify the mutual dependency of its various components—a network of systems distributed throughout the body's depth and across its surfaces. Although the haptic subject exists in a necessary state of flux, the borders established by lab protocols constrain its oscillations.

By suggesting that the twentieth century saw the progressive territorialization of touch by haptics, I am not suggesting a retreat from this model or that we attempt to recuperate something like premodern touch. Rather, I want to correct a misguided set of assumptions informing historiographies of mediated perception that portray touch as scientifically naïve, grounded in a transhistorical body, and thus beyond the reach of scientific and mediatic modernities. This representation of an innocent touch colors our imagination of touch's historicity and consequently steers our treatment of the contemporary relationship between touch, media technology, and the political economy of the senses. When confronting the present-day diffusion of haptic technologies, media historians often rely either on a set of common-sense, intuitionist assumptions about touch or refer back to phenomenological theories of touch that built on (or in some cases, explicitly challenged) the psycho-physiological model established in the nineteenth century. As the doctrine of touch, haptics then becomes reified not just in the practice of interface design but also in media theory, where the scientific formulation of touch appears stripped of its human origins. As a product of nature rather than of culture, it loses both its scientificity and its historicity. The existence of discrete nerve structures distributed throughout the joints that allow the perception of the body in space, the configuration of mechanoreceptors in the skin that facilitate judgments about the pressure and texture of external objects, and the isolation of nociceptors that cause sensations of pain[110]—each of these developments helped form the historical a priori for our contemporary understanding of touch, and each existed as unknown continents before the psychophysical paradigm Weber initiated in the early 1800s.[111]

Contemporary machines designed to synthesize touch inscribe this mapping of touch-as-haptics in their material configuration, employing separate mechanisms to act on the different components of the haptic system. In doing so, they prepare touch to take on a new role in the economy of information circulation. As I show in Interface 4, haptic interface design, dubbed in the 1990s the "new discipline" of "Computer Haptics,"[112] participates in a process of recalling and repurposing the nineteenth-century model of touch while simultaneously erasing it—the tactile system it reconstructs purports to be a natural object rather than a technical one. Reconnecting with the lost and alienated sense of touch can only be accomplished through the embrace of new perceptual prostheses united under the technofetishistic banner of haptic interfaces. While recent attempts to synthesize touch in computer-simulated environments have fueled financial and intellectual investments in fixing a statistically normalized mass haptic subject, this subject's origins are grounded firmly in the nineteenth-century project of quantifying human psychosensory processes. Treating such developments as external to the lineage of contemporary media inscribes a technologically determined audiovisualist bias into media historiographies.

While the nineteenth-century science of touch embodied an effort to isolate the sense of touch through experiment and apparatus, it is crucial to recognize that haptics also entailed the imposition of visual and mathematical abstractions on tactile experience. The various mappings of touch produced by haptics research rendered touch visible according to new graphic logics. Diagrams of tactile acuity, maps of hot and cold spots on the skin, drawings of the body that plotted differences in electrocutaneous sensitivity[113]—each depiction of tactile processes represents an attempt to bring the experimenter closer to touch through a process of making it visible. Such maps recall Sander Gilman's suggestion that "the iconography of touch . . . seems to be alienated from its sensory organ, the skin."[114] A tactile modernity therefore does not serve to counter ocularcentric and auralcentric accounts of modernity. Instead it supplants and challenges several key assumptions informing these narratives by describing the unique path that touch followed as experimental psychologists attempted to construct an account of touch independent from its former associations with optics. The instruments used to facilitate this isolation helped touch become adequate to the new epistemological ordering and pragmatic deployments of the senses: tactual experience could be divorced—however artificially—from any irrational and instinctive urges through careful adherence to lab protocols. The experimental method provided a technique for interfacing with touch that facilitated the strategic exclusion of its noninformatic functions. By submitting to the normative dictates of laboratory science's protocols, this new haptic subject facilitated the intensification of knowledge about touch's constitutive functions. As a result, its capacities could be further specified and its abilities calculated with ever-greater precision. The haptic subject provided researchers with a foundation on which to layer new experimental techniques. Armed with this increasingly sophisticated model of a subdivided and quantified tactual system, engineering psychologists in the twentieth century began to imagine touch as an informatic network capable of interacting with information transmission machines. These machines could only be grafted onto a matrix of tactile systems that had been rendered

capable of noticing and not noticing differences; the labor of scientists in the nineteenth century that had transformed the tactile
into the haptic served as a necessary precondition for the progressive mediatization of touch throughout the twentieth century.
The tactile was remapped and subsumed, not within the optical
but within a new frame of disciplinary methods, doctrinal dictates,
and instrumental measurement techniques designated by the modern category *haptics*.

Interface 3

The Tongue of the Skin

Our eyes and our ears are assaulted so continuously, such
frequent and insistent demands are placed on them, that the
visual and auditory channels are seriously overburdened at
times. Such oversaturation leads quite naturally to the
question of whether it is only vision and hearing that can
serve in communication.
      —Frank A. Geldard, "Adventures in Tactile Literacy"

Early in the 1920s, against the backdrop of the prior century's
investigations into touch's psychophysical operations, Robert Gault
began experimenting on the possibilities for utilizing a technologically enhanced touch to route speech through the skin. Intended
as a means of helping those with severe hearing impairments understand spoken language, Gault's research proceeded from what he
understood as a demonstrated kinship between the ear and the surface of the skin; "the human ear," as Gault noted, "has evolved
from the organs of touch and vibration."[1] Pressing this structural
analogy, Gault argued that the ear and the skin both act as receivers of vibrations originating in external environments. For Gault,
both hearing and touching involved the crucial act of interpreting
and reacting to these received vibrations, a process learned gradually in the early years of an individual's psychological development.

This functional similarity brought him to the conclusion that "hearing . . . is an act of touching."[2] But as the ears had long ago evolved to surpass the skin in the ability to sense vibrations, the skin needed assistance—artificial augmentation—if it was "to be made to substitute for ears."[3] Gault designed the device he dubbed the Teletactor as a means of providing this augmentation. In its protean form, celebrated in a 1926 *Popular Science* article as a "new instrument that enables the deaf to hear with their hands,"[4] the Teletactor consisted of a vibration-producing motor with a single reed protruding from its casing (see Figure 3.1). As the operator

Figure 3.1. A single-unit Teletactor being held by a test subject. The subject's thumb rests on the diaphragm. This image was published in *Journal of the Franklin Institute* 204, no. 3, Robert Gault " 'Hearing' through the Sense Organs of Touch and Vibration," 344. Copyright Elsevier (1927); courtesy of the Franklin Society.

spoke into a telephone receiver, the apparatus translated speech sounds into precisely controlled vibrations sent through the reed. Pressing the sensitive pad of the thumb against the reed allowed the subject to feel its subtle vibrations with a fine sense of discrimination. Gault claimed that the subject, after undergoing an intense training regimen, could learn to interpret and react to the reed's vibrations as if they were the human voice, rather than its electromechanical, vibrotactile transduction. So perfect was this fit between the aural and tactile organs that Gault understood the Teletactor to be "grafting a mechanical ear upon the skin."[5]

Though Gault had abandoned the well-publicized Teletactor project by the 1940s, in subsequent decades, others picked up on its central and coherently articulated aim of transforming touch, through technological augmentation, into what would be repeatedly described as a "communicative sense."[6] Spearheaded by the efforts of University of Virginia experimental psychologist Frank Geldard, this next wave of research took Gault's analogy of touch to hearing as misguided and sought instead to discover a means of coding linguistic signals for touch that was not derived from either the logics of hearing or seeing—a tactile language that took full advantage of what were understood to be touch's unique capacities as an information reception channel. This work self-consciously attempted to depart from Gault's project by taking aim at the special baggage the term "communication" carried in the discipline of psychology. But it essentially followed along the conceptual trajectory that Gault had established, although it employed different technical systems for getting language to pass through the gates of the skin. Psychology took communication as a shorthand for the capacity to transfer knowledge between persons, but psychologists understood this communicative process to be primarily the domain of the eyes and the ears. The naked declaration that touch could become a communicative sense, echoed throughout the decades of research into tactile communication systems, was intentionally pugilistic—a phrasing designed to force the discipline to confront its own biases and assumptions about the tactile sense. These twin aims of engineering effective touch communication apparatuses and reversing psychology's longstanding conceptualization of touch as "nonintellectual" were bound up inextricably with one another; successfully building instruments that facilitated the transmission of messages through the skin would allow these self-appointed pioneers[7] to claim that the wisdom psychologists passed down through the ages had been misguided. The many machines designed with the intent of pushing messages through the skin were understood to be working toward the common goal of salvaging the organ of touch itself from psychology's dustbin.

The first generation of experimental studies on touch, detailed in Interface 2, proceeded under the tacit assumption that furthering knowledge about the tactile sense was a self-justifying project: those who purposed the project showed little concern for articulating the utility of their research. Further, they rarely used their experimental results to intervene in longstanding arguments about the hierarchy of the senses.[8] In contrast, twentieth-century researchers embarked on their projects intent on providing the science of touch with a positive social utility. Different tactile communication projects aimed at different ends—some, like Gault with his Teletactor, sought to help the deaf hear; others, like Geldard, saw the tactile channel as an alternative to the visual and auditory channels that had been "seriously overburdened" by the "insistent demands" placed on them. But their investigations shared a pragmatic orientation, motivated by the belief that touch could have a positive new function in the circulation of information.[9]

The imbrication of the technical and the conceptual in twentieth-century investigations of tactile communication systems yielded four significant and intertwined outcomes. The first was a genealogy of communicative tactility—the articulation of a new discursive history of touch, organized around its capacity to act as a communicative sense. This new genealogy positioned contemporary efforts at engineering touch-based communication systems as the continuation of a conversation that had "waxed and waned over the centuries."[10] Facilitating this shift in the status of the skin senses therefore required a strategic recall and revision of touch's cultural and philosophical positioning—the forging of a disruptive new account of touch that wove together the practical and conceptual. A host of sedimented prejudices against touch had banished the tactual senses from the realm of the intellect; those psychologists attempting to technologically rehabilitate tactility intended to render these prejudices obsolete by showing touch to be capable of taking on new utilitarian functions once they had granted it the proper vocabulary.[11]

The second outcome involved developing new machinic tactile languages, which required engineering psychologists to build and test scores of new electrical, mechanical, and electromechanical tactile communication apparatuses, with each accompanied by its own unique code for transmitting information through the skin. This multiplicity of machines, their inventors claimed, provided tangible evidence of touch's communicative capacity, and as the machines grew in number, they bolstered arguments about the untapped potential of the tactual senses to aid in the transmission of information. During the 1960s and 1970s in particular, as a diverse array of private and public institutions funneled money into this research, the instruments for tactile communication increased rapidly. In a 1977 summary of progress in the field, Geldard identified twenty-five different devices for facilitating skin-based communication (see Figure 3.2). Each apparatus, owing to the idiosyncrasies of its structure and design, brought its own unique coding possibilities that its users were expected to assimilate to. The materiality of the apparatus and the code it employed to communicate through touch were bound inextricably together. While very few of these devices would make it beyond the walls of the lab, their design helped lay a foundation for the engineering of future touch machines.

The third outcome involved the construction of an informaticized model of tactility. A biological correlate of machinic tactile languages, informaticized tactility required the discovery of touch's specific capacities to act as a receiver and processor of machine-generated signals. This research built on, challenged, and furthered extant techniques and protocols for investigating touch, as experimenters sought to uncover the psychophysical limits of touch perception, the underlying anatomical features of the tactile system, and differences between touch's various submodalities. Progress in skin-based communication language systems depended on

Tactile communication systems and devices.

Mechanical	Electromechanical	Electrical
Direct speech mediation	**Skin capacitance effects**	**Direct speech mediation**
Speaking tube	Electrostatic "textured"	Simple dermal electrodes
Electromagnetic receiver	matrix	Vocoder
Teletactor		
Vocoder		
von Békésy cochlear model		
Tadoma ("speech feeling")		
method		
Pictorial display		**Pictorial display**
Tactile TV		Tactile TV
Elektroftalm		Elektroftalm
Optacon		Optacon
Embossed legible print		
Coded language		**Coded language**
Vibratese		Katakana
Body braille		International Morse
Optohapt		Electrocutaneous vibratese
Polytap		
Air-blast symbols		
Visotactor		
Braille and derivatives		
Finger spelling		
String writing		
Tracking and monitoring		**Tracking and monitoring**
Vibrators and air jets on		Single and multiple active
hands or forehead		electrodes on fingers, arms,
		legs, or neck

Figure 3.2. Geldard's typology of tactile communication systems and devices. From Frank Geldard, "Tactile Communication," in *How Animals Communicate,* ed. Thomas A. Sebeok (Bloomington: Indiana University Press, 1977).

constructing the most accurate map of the tactile system that the instruments and methods of the day would allow. Descriptions of tactile communication systems were positioned alongside graphs representing the newly discerned limits, for example, of the skin's ability to recognize high-frequency vibrations at various locations on the body (dubbed "tactograms" by Gault, and framed as a direct analog to audiograms).[12] As with nineteenth-century studies on touch, mapping these updated parameters depended on the progressive refinement of laboratory instruments and measurement techniques. In what would prove to be a significant investigation of vibratory perception, Geldard underscored this point, noting that "recent developments in apparatus and method are important since our future facts are a function of them."[13] The haptic subject, as a figure constantly in dialogue with the demands of new sociotechnical orderings, proved to be crucial here: this subject had to be constantly reaggregated and reshaped in response to shifting technologies and their new deployments. Knowledge about the specific functioning of the various tactile channels was not a static object that could simply be mobilized on an as-needed basis in the design of touch communication systems. Instead, the experimental systems developed around new techniques for transmitting messages through touch revealed critical gaps in the available data about touch's inner workings. Novel tactile communication systems existed in a reflexive relationship with the doctrine of touch: each proposed innovation required researchers to develop new techniques for quantifying and mapping touch's discriminatory capacities.

As a fourth and final outcome, this research produced a self-conscious, teleological understanding that research into touch communication would continue to advance unabated, moving toward any number of the humanistic ends imagined for it. Whether restoring hearing to the deaf, giving vision back to the blind, increasing the pace of information reception, or routing emergency alerts through the "always-on" tactile channel, the thousands of hours poured into experiments with these systems were intended to eventually yield socially useful outcomes. This assumed inevitability organized investigations into touch perception's most seemingly inconsequential and minute characteristics around the unifying macrohistorical goal of making touch machines that would one day achieve a transformative utility, even if the technical limits of existing devices made the near-term diffusion of these machines unlikely and impractical.

Taken together and considered in the broader context of the archaeology of haptic interfacing pursued throughout this book, these developments recall touch's cultivated sensitivity to machine-generated sensation (Interface 1), its subdivision and quantification by experimental psychology (Interface 2), and they foreshadow the eventual emergence of the computer-controlled, dynamically rendered vibrotactile signification systems currently utilized in smartphones, videogame controllers, and wearables (Interfaces 4 and 5). For example, the Taptic Engine (a portmanteau of "tap" and "haptic") used in the Apple Watch and Immersion Corporation's Instinctive Alerts Framework for Wearables both employ complex, coded vibrations to send tactile messages, via the always-on space of the wrist, through the smartwatch. Like the experimental subjects early tactile communication systems were tested on, those who affix these new computers to their bodies are expected to quickly assimilate themselves to tactile languages engineered through practices of iterative design aimed at making machines and users mutually legible.

Furthermore, cutaneous communication apparatuses presented an image of the human sensorium as malleable, composed of isolated organs that each possessed their own set of distinct receptive capacities. Framing the eyes and ears as having a common evolutionary origin in the skin, while not completely dedifferentiating these organs, positioned the skin as a sort of master medium, capable of providing a conduit through which image, sound, and text alike could flow (on being filtered through a suitable machine, of course). Like electricity and language, the skin functioned almost as a universal translator of sense data. In this rendering, tactile sensations were instrumentalized as a means of information transmission— the electromechanical production of tactile sensations was aimed not at storing or replicating or synthesizing tactile experience, but instead at routing language through the skin using touch sensations assigned meanings as part of a given coding system. Treating the cutaneous senses as linguistic conduits meant subordinating the skin's myriad functions to its new role as a receiver of coded messages. Though critics of the monadic methods employed to study touch in the lab, such as the German phenomenological psychologist David Katz, derided the artifice of the experimental conditions required to produce knowledge about the senses,[14] it was precisely the artifice of these machine-generated sensations that allowed touch to be granted its new communicative powers. Whether through targeted bursts of electricity, sequences of finely controlled vibrations at various locations on the body, or puffs of air projected against the skin, engineers selected the tactile stimuli used in touch communication systems precisely because they avoided invoking the messy interweaving of sensations normally involved in organic touch perception.[15]

The pre-twentieth-century ideation of tactile languages presaged the eventual development of electromechanical tactile communication systems, providing a philosophical grounding for the arguments that would eventually be advanced by the psychologists who advocated for the utility of tactile communication. Denis Diderot's "Letter on the Blind" (1749) and Jean-Jacques Rousseau's *Émile* (1762) both suggested that touch could be used as a conduit for language transmission. An exploration of the conceptual grounding for tactile languages therefore serves as the starting point for this chapter. Against this backdrop, I examine three significant paradigms in twentieth-century touch communication research. The first paradigm, the grafting of a mechanical ear upon the skin, involved the aforementioned efforts to make the skin substitute for the ear via the Teletactor, a project first pursued by Gault in the 1920s and taken up again at various points throughout the century. The second paradigm, formulated explicitly by Geldard in the 1950s as a resuscitation of and improvement on Gault's earlier efforts, attempted to divine "the tongue of the skin"[16] on its own terms, without forcing it to conform to the structural logics that governed either hearing or seeing. Along with his former student Carl Sherrick, Geldard gave electromechanical cutaneous communication systems their most sustained and thorough articulation. The third paradigm, an attempt to wed vision to touch, pressed the skin into service as a pathway for optical images. Beginning with Paul Bach-y-Rita's research in the 1960s, the category of devices dubbed tactile vision substitution systems (TVSSs) used the skin as a space that pictures or letters could be drawn on by a matrix of small vibrators in hopes that users could be trained to interpret the vibrations as images. Together, these paradigms embody the humanistic hopes that came to be associated with an increasingly technologized touch during the twentieth century, as the techniques for subdividing and quantifying the tactile senses provided a means for realizing the tactile language system Diderot had envisioned two hundred years earlier.

I employ this somewhat rudimentary typology as both an organizational strategy and as a reflection of the classification schemes
researchers used to structure their inquiries. Though it may imply
clean divisions among the different paradigms, the field as a whole
was not expansive enough to facilitate the pursuit of projects in
total isolation from one another. For example, when investigations
into TVSSs began in the 1960s, the psychologists, engineers, and
neuroscientists involved in this project drew on the psychophysical data generated from earlier research on the capacity for vibratory discrimination at different points on the skin. During its
forty-two-year history, the Cutaneous Communication Lab
Geldard founded provided a hub for projects in each of the three
movements. Gatherings (such as the 1966 International Symposium
on the Skin Senses and the 1973 Conference on Cutaneous Systems
and Devices) assembled investigators with a range of diverging
aims, all dependent on the further accumulation of knowledge
about touch's functions, to consider the potential positive deployments of touch communication technologies.

While researchers in this knowledge network quibbled frequently about which sense touch was best suited to serve as a surrogate for, they rallied around the common belief that touch's
potential as a communicative channel had not yet been fully understood or exploited. No one movement could triumphantly claim
victory, as efforts at substituting touch for the eyes and the ears each
found limited and controversial success. Often, systems that worked
in the highly structured conditions of the lab failed on leaving its
protective walls. The training required to master a given tactile
language frequently proved too intensive and demanding to be
worthwhile. In spite of aspiring to achieve portability, the mechanisms used to project sensations onto the skin were regularly
deemed too cumbersome to be practical in daily use. But regardless of the individual success any one system achieved, the *idea* that
touch could have its own particular, formalized, machine-generated
language gained wide acceptance across a range of specialized
fields. Although no sole signaling system would achieve hegemonic
status or widespread diffusion, by the end of the twentieth century,
the enduring transformation had been accomplished. All that

remained was to concretize a tactile vocabulary, standardize a corresponding transmission system, and train receivers to transcode tactile noise into images, sounds, or linguistic signs.

"Barricaded for Want of Signs": Toward New Languages of Touch

Until the early nineteenth century, speculations on the possibility of creating a formalized language for touch lacked both cohesion and material expression. The firmly established prejudice against touch, inherited from the Platonic ordering of the senses, maintained that touch was a mode of perception suited to neither aesthetic nor rational judgments. As a proximal sense, judgments informed by touch were thought to be governed only by the body and its instinctual desire either to avoid pain or to seek pleasure.[17] As a consequence, sensations obtained through touch were understood to be processed by the body rather than by the mind and as such, could not be given the structure required by a linguistic form. In the argument passed down in one form or another from Plato to Aquinas to Schopenhauer, touch was portrayed as a fundamentally utilitarian mode of sensation, with tactile discriminations insurmountably informed by the overriding imperative to preserve the life of the touching organism. By the time of Gault's intervention in the 1920s, however, these old prejudices were beginning to crumble. Experimental psychologists in the nineteenth century had cleaved pain off from the tactile system, supposedly isolating the physiological mechanisms that generated sensations of pain from those responsible for sensations of temperature, sensation, vibration, and pressure. The notion that touch could be activated without risking the infliction of pain, demonstrated through the design of psychological apparatuses that insulated experimental subjects from harm while exposing them to a variety of tactile stimuli, would prove crucial to later attempts at upending the old argument. But in addition to the nineteenth-century positioning of pain and touch as psychophysically distinct, another substantive contribution came nearly two centuries before Gault's experiments with his Teletactor, when Denis

Diderot—motivated by a belief that the intellect could be spoken to through touch—imagined the possibility of a tactile language. Diderot's 1749 "Letter on the Blind" would inspire subsequent attempts, executed with varying degrees of success, at constructing "tangible alphabets."[18]

In the letter, informed by his ethnographic encounter with a blind man from Puisaux, Diderot criticized the idea that language had to be strictly wed to the eyes and ears. While his perceptual theory embraced a framing of the senses as a means to abstraction, he diverged from earlier accounts by disagreeing with the contention that touch was incapable or unsuited to such abstractions. "The faculty of calling to mind and combining sensations of palpable points" honed by the blind was for Diderot analogous to the sighted man's "faculty of combining and calling to mind visible points." However, abstraction alone was insufficient; the eyes and ears had the advantage of possessing, through a social arrangement, a system of signs that were "common property" and served "for the staple in the exchange of our ideas."[19] By failing to provide touch with its own commonly accepted, shared storehouse of signs, this social arrangement effectively excluded touch. As Diderot explained: "we have made [signs] for our eyes in the alphabet, and for our ears in articulate sounds; but we have none for the sense of touch, although there is a way of speaking to this sense and of obtaining responses." Without such a language to facilitate communication, Diderot feared that the blind, deaf, and mute would "remain in a condition of mental imbecility." Language, for Diderot, served as a medium in which the intellect could grow, but in the absence of such a medium, the intellect withered. The means for constructing such a language seemed readily available: nothing remained "but to fix it, and make its grammar and dictionaries." Locating the skin (along with the eyes and ears) as one of three available entry points that knowledge has to the mind, Diderot claimed that touch's gates had been kept "barricaded for want of signs."[20] To remove this barricade, it was simply necessary to provide a "clear and precise language of touch"[21]—a system of sensations that would be a common property among the entire population.

The blind British geometer and mathematician Nicholas Saunderson's (1682–1739) "palpable arithmetic" and its related apparatuses provided Diderot with his proof of concept. Saunderson's machines, one of which was a matrixed board consisting of differently sized pegs that Saunderson could manipulate with great rapidity and dexterity, allowed him to perform complex calculations and to understand geometrical relations by feeling the relative position and orientation of the pegs with his fingers (see Figure 3.3). But Saunderson's success at developing his own idiosyncratic tactile symbol system, rather than assuaging Diderot, served to fuel his lament: he speculated on what Saunderson would have been able to accomplish if he had been provided with such a system "arranged with signs for touch" at age five, rather than "having to invent it

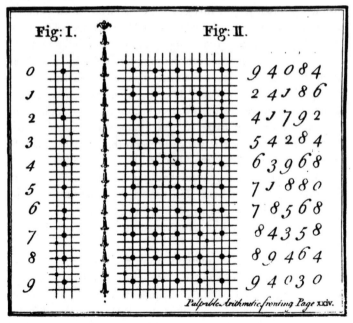

Figure 3.3. Saunderson's palpable arithmetic. From Nicholas Saunderson, *Elements of Algebra* (Cambridge: Cambridge University Press, 1740).

at twenty-five."[22] Anticipating Geldard's later quest, Diderot
sought a fixed and stable pathway for knowledge to pass through
the tactile system that did not attempt to imitate the other senses,[23]
but rather obeyed touch's unique structures and capacities. A
well-crafted system of tactile signs would allow knowledge to
flow into the mind through touch, facilitating the entry of previously excluded populations onto the stage of reason. As a predecessor of both Gault and Geldard, Diderot similarly attempted to
rehabilitate and instrumentalize touch for the purpose of societal
enrichment, projecting transformative hopes on the tactile communication system he imagined.[24]

Valentin Haüy, inspired by the suggestions in Diderot's letter,
took up the project of designing systems that would allow the blind
to read and write by touch. These primarily consisted of raised
wooden Roman letters that could be pressed against paper to
emboss it, and then later could be scanned by the finger of the blind
reader. Haüy's system was one among a number of different methods for raising script, but both the production (making books in
raised script) and reception (reading the raised script with the fingers) quickly proved difficult.[25] Their common problem concerned
the attempt to force the finger to read according to the logic of the
eye; they required the finger to pause ponderously on each letter,
inhibiting the flow of knowledge through the tactile channel. In
spite of these limits, Louis Braille, at age 10, quickly mastered both
reading and writing using the method. Though he had success with
it, Braille recognized its many shortcomings: it was difficult to master, exhausting to use, and virtually worthless outside the walls of
the recently established Royal Institute for the Young Blind, where
it was taught.

Hungry for a better method, Braille encountered Charles
Barbier's *écriture nocturne* (night writing), a raised-dot language
developed as a means of communicating orders to French soldiers in
the cloak of darkness.[26] The raised dots Barbier employed had a
decided physiological advantage over Haüy's raised script; although
Barbier's design predated Weber's research on two-point thresholds, the spacing of Barbier's dots embodied that basic model of
tactile "sensory circles" that Weber would posit in the subsequent

decades. The 2.5-millimeter distance between each dot closely resembled the value Weber and later researchers would obtain through pressing compass-points against the index finger.[27] Night writing, therefore, had succeeded in mimicking the tactile system's structure, but only in part. Barbier used a rectangular matrix consisting of twelve dots arranged in two columns, with six rows per column. Because its height exceeded that of the index finger's pad, it too proved cumbersome, requiring the reader to scan vertically as well as horizontally. In contrast, Braille employed six dots arranged in two columns. This configuration facilitated faster scanning, as the reader's finger only had to be moved horizontally. Braille's system cut the size of the rectangular matrix in half simply by eliminating three rows of dots. Other psychophysical parameters, such as the rate at which the skin recovered from deformation (the dermal equivalent of a visual after-image) and the height required for a dot to be perceptible by the pad of the index finger, were embodied in the material structure of Braille, in spite of the fact that knowledge about such parameters had not been formalized at the time of its invention.[28]

Braille was just one among numerous sensory substation systems devised during the nineteenth century and circulated at conferences throughout Europe, as pedagogues took a pragmatic interest in the idea that deficiency in one of the major senses could be compensated for through touch. As Jan Eric Olsén explains, these explorations were informed by the German notion of *Sinnesvikariat* ("vicariate of the senses"), which held that the senses had the power to substitute for one another as necessity dedicated.[29] The general question of the translatability between the senses had long been a subject of philosophical inquiry, but in the nineteenth century, new discoveries in the physiology of perception, combined with efforts at teaching the blind to read, aimed to provide functional resolutions to these longstanding questions.[30] The theory of *Sinnesvikariat* "embedded the notion of sense analogy in the nervous apparatus of the physiological body,"[31] suggesting that the body had a latent psychophysical capacity to route audible and visual impressions (in reduced and degraded form) through the comparatively more primitive organ of the skin. Though my immediate interest here lies in the systems that transferred either written or spoken language through touch, efforts at educating the blind also involved the development of techniques for producing tactile pictures, raised relief maps, and other tactually perceptible models, all intended to press the fingers into service as a surrogate for the deficient sense of sight.[32]

Along with these sensory substitution techniques, a pair of examples drawn from the pre-twentieth-century history of telegraphy complicated and challenged the notion that touch was incapable of serving as an information-reception sense. The first involved a biomachinic apparatus dubbed the "sympathetic flesh telegraph,"[33] which operated on a principle so fantastical that it is best understood as what Eric Kluitenberg terms "imaginary media." Imaginary media, which serve to "mediate impossible desires," exist at the intersection of articulated social wants and the limits of extant technologies.[34] In the case of the sympathetic flesh telegraph, accounts of which circulated as early as the sixteenth century, this impossible desire concerned the hope of allowing two geographically remote subjects to "communicate all their ideas with the rapidity of volition, no matter how far asunder."[35] As Fahie explained in his 1884 genealogy of electric telegraphy, the skin-to-skin communication system required the communicators to undergo two medical procedures. In the first, a piece of skin would be cut from the hands or arms of each, and then "mutually transplanted while still warm and bleeding." The newly affixed flesh would bond with its host, while "retain[ing] so close a sympathy with its native limb" that the original limb still would be sensitive to any injury inflicted on the flesh now detached from it. In the second procedure, both transplanted pieces of skin would be tattooed with the letters of the alphabet. When the two communicative subjects wished to exchange thoughts, "it was only necessary to prick with a magnetic needle the letters upon the arm composing the message; for whatever letter the one pricked, the same was instantly pained on the arm of the other."[36]

Fahie positioned the flesh telegraph as a subspecies of "sympathetic needle telegraphs," where two compass needles would be placed on distant blocks of stone with the letters of the alphabet carved in a circle. As the sender manipulated the needle on one chunk of stone, the needle on the other would imitate its movements.[37] Like the flesh-based variant, these devices used magnetic needles to write from a distance (akin to the movement of needles on a Cooke and Wheatstone telegraph). But in stone-and-needle communication systems, writing was apprehended through the visual registering of an object's movement; the subject on the receiving end of the message watched the needle move to different positions on the stone, as the sender on the other end slowly spelled out a message. Multiple receivers could witness the transmission of the same message. Presumably, any literate operator could take a turn communicating through the apparatus. In contrast, the skin-based system established an inherently private and inseverable link between communicative subjects, as the reception of messages hinged on a constant exchange of nervous signals between the two remote units of flesh—a discrete channel, always open, that facilitated something akin to telepathy but was routed through a clumsy process of skin-prick writing.

By requiring subjects to decode needle pricks as written language, the sympathetic flesh telegraph both defied and anticipated the structural logic of later skin-based communication systems. The longstanding prejudice against using touch as a message transmission system, as detailed earlier in this chapter, treated pain as an obstacle to tactile communication—a sensation that immediately invokes subrational and instinctive desires, thereby inhibiting the intellectual capacity to process and receive linguistic signs. Skin-graft telegraphy's transmission mechanism, however, instrumentalized intersubjective sensations of pain, using them as a means to indicate the different letters that composed a given message. The painful, pricking sensations involved in message exchange, then, were oriented semantically: communicating the sensation of pain was not an aim of the system but instead an unfortunate and necessary by-product of the writing process. A light touch, or the dispersed pressure of a blunt object, would not produce a signal intense or precise enough to allow for the localization on the confined skinspace that contained the tattooed letters. Pain, instead of clouding the subject's ability to discriminate between stimuli (as in Weber's two-point threshold experiments, where he blunted the compass-points to avoid activating the pain sense), served as a way to ensure that the receiving subject would successfully decode the transmitted message.[38] It was this capacity of pain—capable of interrupting and capturing attention, and incapable of not being noticed—that gave it utility in the skin-graft telegraph.

From a formal and structural perspective, this system employed only the single variable of location, with the position of the needle sticking into the skin used as a binary signal—the skinspace of each individual letter existed either in an "on" state (in pain) or an "off" state (free of pain). Though we might imagine that two communicators would eventually agree to layer other signifiers onto the system—a particularly deep press of the needle, for instance, to communicate anger, or swiping the needle across two adjunct letters to indicate their consecutive usage—Fahie's description gives no indication that such modifications occurred. As a matrix of discrete spaces plotted across a confined region of the skin and then activated by an external stimulus, the sympathetic flesh telegraph's basic design anticipated many of the touch communication systems built in the twentieth century.[39] While the impossibility of this biomechanical communication system makes such comparisons troublesome, the fantasy that animates the sympathetic flesh telegraph suggests a continuity of scientific imagination around touch that reaches back several centuries, as people dreamed of new pathways through which messages could be passed.

The second practice of distance communication through touch—what Edward Knight described as "tangible telegraphs"—involved the use of alternative techniques for receiving messages sent through the Morse electric telegraph system. These tangible telegraphs mobilized the receiver's skin as a means of "communicating intelligence at a distance,"[40] with inventive telegraph operators inserting their bodies directly into telegraphic circuits. To understand how these tangible telegraphs operated, it is helpful to consider electric telegraphy according to the sense modality activated in its reception. The Morse transmission system communicated linguistic elements through a series of long and short pulses of electricity, with the breaks between pulses serving as essential components of a given message. Early in its history, these pulses were transcribed via a striker onto scrolling tape, allowing for a visualization of electrical current intelligible to the trained eye as written language (the "dots and dashes" that signified alphanumeric characters). By 1845, operators had learned to identify or "read" the code by the tapping sound made by the marking lever, and as a consequence, a sounder was added to the striker in order to take advantage of this decoding capacity. Morse began to morph from a code apprehended through vision to one accessible either through seeing or hearing, suggesting that, as Jonathan Sterne argues, practices of receiving electric telegraphy "allowed for an interchangeability among the senses."[41] The tangible reception of telegraphic messages functioned by the same principle of receiving dots and dashes, with telegraphic code written directly onto the body through creative modifications to the Morse apparatus. In one technique of tangible telegraphy, the operator allowed the striker to directly hit their hand, effectively substituting skin for scrolling paper. Another technique did away with the striker altogether, with the operator grasping the telegraph wire directly, so that the electrical pulses would be felt as shocks of varying duration. In both cases, the skin functioned as a receiver of coded messages, with the first operating through the mechanical stimulation of touch and the second working through the electrical excitation of tactile sensations.

Like sound telegraphy, then, tangible telegraphs initially emerged as a sort of ad hoc innovation on the part of telegraph receivers. The operating principle of sensory interchangeability did not come about as a result of any grand philosophical speculation, nor did it mobilize a storehouse of accumulated physiological knowledge about touch (though it is worth noting that E. H. Weber's brother Wilhelm worked with Carl Friedrich Gauss to develop an electric telegraph at Göttingen University in 1833). It was instead demonstrated through practice, as operators using the telegraph machine to write signals onto their skin assimilated to a language of sensations, translating a set of machinic vibrations and shocks into letters and numbers. In one account, necessity prompted the telegrapher to use his skin as a receiving instrument: lacking a suitable telegraph key to affix to the electrical circuit, expert Union telegrapher William Fuller grabbed directly onto the wires. As a military historian explained, the shared accessibility of Morse's variations in duration to the senses allowed him to read incoming messages by shock: "telegraph characters . . . are composed of dots, spaces, and dashes, which are ordinarily read by the ear; but Fuller, having no instrument, could only determine them by electric shocks; i.e., a long shock would indicate a t, longer yet than an l, and yet longer a cipher; a succession of quick ones meant i, s, h, p, and others." Fuller compensated for the somewhat weak strength of the electrical pulses by using his tongue, "which being moist, is perhaps the most sensitive part," to receive shocks from the wire.[42] As with early electrical researchers and autoexperimenting psychophysicists, tangible telegraph operators had to relate to their own sense of touch as the means to accomplish an end, suppressing in the reception process the supposed instinctual desire to avoid displeasing sensations. Particularly in the case of receiving telegraphic electric shocks, the body was inserted directly into the telegraphic circuit as a substitute not for a missing or deficient sense, but instead for a missing instrument used to generate stimuli that would, under optimal conditions, be routed through either the eyes or the ears before being decoded as language.

When reading these tangible telegraphs as continuous with Nollet's experiments networking the tactile systems of monks, it is tempting to suggest that there is something quintessentially tactile about electric telegraphy: perhaps the technology's ontology is not visual, but rather fundamentally tactile. However, Sterne rightly argues that the principle of sensory interchangeability illustrated by sound telegraphy calls our attention to the "*ontological* fallacy"[43] inherent in historiographies that essentialize telegraphy as visual. Rather than seeking to replace the visual ontology of the telegraph with a competing one grounded in its audible variant, he pushes for a treatment of telegraphy that attends instead to the historically contingent characteristics of the technology and its associated practices. Concerning touch, the ability to read electrical messages via shock (or by allowing the mechanism of the striker to repeatedly impact the skin) was idiosyncratic to the structure of the Morse system. The many other competing systems of electromagnetic message transmission experimented with during and before the nineteenth century did not lend themselves easily to reception by touch—needle telegraphs, like Cooke and Wheatstone's, had certain structural characteristics (for example, the use of multiple wires would have required receivers to interpret several different flows of electricity simultaneously) that made them hostile to touch reading. As a reception practice, tangible telegraphy was therefore contingent on the material configuration of the Morse apparatus, its accompanying cipher, and the practical knowledge shared by telegraph operators.

These practices of touch telegraphy call into question axiomatic assumptions about the relationship between media, touch, and communication systems. The notion that electric media at their point of inception were somehow inherently audiovisual is muddied by touch's incorporation alongside seeing and hearing into these electric message transmission systems.[44] Together with Braille, tactile telegraphs presented proof of the concept that touch could be effectively deployed as a means of receiving linguistic messages, prompting the revaluation and revisiting of longstanding debates concerning the connection of the senses to the intellect. Finally, the apprehension of telegraphic messages by the skin—where touch, either by electrical or psychic connection, effectively received signals from a distance—provided an exception to the frequent claim that touch was a sense inhospitable to extension. The thread running through these various upheavals concerns the broader disruption brought on by what Collette Colligan and Margaret Linley frame as the "invention of media" during the nineteenth century, where conceptualizations of the senses and the technologies that facilitated their extension and amplification were reorganized in response to a mixture of new scientific, technical, and physiological knowledge.[45] Our contemporary ontologizing of media as audiovisual, then, is a remnant of the categorization schemes mobilized to help order the chaotic changes that swept across the social body as a consequence of nineteenth-century technological and economic developments.

It is not accidental that these accounts of touch-based telegraphy circulated in nineteenth-century histories and genealogies of the telegraph, as professional engineers and historians alike crafted origin stories, typologies, technical manuals, and conceptual foundations that would make electrical telegraphy intelligible both to the culture at large and to the specialist audiences targeted by their individual works. This understanding of history not as a passive recording of events but instead as something actively recalled and reshaped by the historian suggests that we think of touch telegraphy as an "imaginary in action,"[46] capable of actively impacting the organization and conceptualization of the possible. The piecemeal granting of signs to touch by technical media throughout the nineteenth century prompted later researchers like Gault and Geldard to consider the possibility that the barricades nature had piled in front of touch could crumble with the right combination of physiological knowledge and engineering ingenuity. Intermingling the sciences of psychophysics and electrical engineering would allow the twentieth-century designers of touch communication systems to launch a sustained and organized assault on the received wisdom surrounding touch.

Grafting a Mechanical Ear upon the Skin

In the 1920s, against the backdrop of these prior efforts at communicating through touch, Gault began his experiments attempting to route speech through the fingers using his Teletactor. The foundational assumption underpinning the device's construction—that speech, with the combination of a properly calibrated instrument and a suitably trained receiver, could be rendered intelligible through touch—catapulted Gault into a research program that lasted nearly two decades. Through iterative, experimentally driven refinements, he sought to arrive at an effective pairing of apparatus and teletactile listener. During the project's evolution, the Teletactor underwent a range of mutations, reflecting both new experimental findings discovered through repeated testing and changes in the intended uses for the machine. Originally consisting of only a single vibrating reed tasked with carrying the whole range of speech sounds to the skin, difficulties with distinguishing among these various speech sounds prompted Gault to add four more vibrators to the system. The resulting "Multiunit Teletactor" split sound into five bands, with each band mapped onto one of the digits on the receiver's hand (see Figures 3.4 and 3.5). Another modification used two Multiunit Teletactors, one for each hand, in an effort to construct a tactile equivalent of binaural (stereo) hearing that would allow

Figure 3.4. The hand of a subject grasping the multiple-unit receiver. This image was published in *Journal of the Franklin Institute* 204, no. 3, Robert Gault, "'Hearing' through the Sense Organs of Touch and Vibration," 344. Copyright Elsevier (1927); courtesy of the Franklin Society.

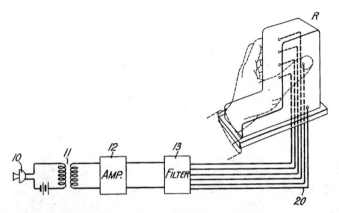

Figure 3.5. Diagram of the Multiunit Teletactor originally designed at Bell Labs. Source: Warren Jones, "US Patent # US 1,733,605—Tactual Interpretation of Vibrations," October 29, 1929.

the teletactile listener to feel sound as it panned from one hand to the other.[47] In spite of these many revisions, the Teletactor failed to gain the widespread adoption Gault had hoped it would, causing him to abandon the project in the late 1930s. It would be resurrected by Norbert Weiner at MIT a decade later and continued in fits and starts throughout the twentieth century, resulting in occasional steps forward. Each apparent surge toward efficacy typically was accompanied by a spate of hopeful attention in the popular press.[48]

While explanations for the failures of these systems abound,[49] my interest lies not in uncovering some definitive answer about touch's capacity to serve as a substitute for the ear but rather in locating the Teletactor as part of a broader effort to make touch serve a new function in the communicative economy. To avoid treating the Teletactor as a stable and finished product, I focus on its location in a technical system that both gave it form and allowed it to stage the development of later touch communication devices. This approach views the technical specificities of a given invention as bound up with its insertion into a configuration of particular historical, scientific, social, economic, and discursive contexts, while also providing a way of laying bare the explicit and implicit influences of those contexts on the technology's development.

It is significant, then, that the roots of the Teletactor project lie in sound reproduction. Gault's initial explorations into tactual communication had been inspired by the curious sensation of a telephone speaker vibrating against his hand. He suspected that, with greater amplification, focus, and control, these vibrations could come to be recognizable—after passing through the hand to the mind—as speech. The first experiments, a "crude" start in a "virgin field,"[50] consisted solely of a long speaking tube, extending from one room to another so as to isolate the speaker from the listener. With the teletactile listener's hand pressed against one end of the tube, the experimenter spoke loudly into the other end. The listener was tasked with distinguishing first between sounds, and then eventually between a set of 38 words. The spatial separation of speaker from teletactile listener, in its attempt to disentangle the organs of touch from those of hearing, recalls Weber's efforts at experimentally isolating touch from vision by blindfolding his subjects during the two-point threshold trials—a separation of the senses accomplished through the material conditions of the experiment. Successful trials prompted Gault to develop the apparatus further by introducing electrical appliances into the system. It was at this stage that the project mobilized not just the "principle of the telephone" that had provided initial inspiration but also a network of telephone engineers and a host of sound reproduction components, appropriated for use in the design of the first Teletactor at Bell Labs.[51] The device combined a radio battery, tube amplifier, vibrating diaphragm, and carbon transmitter to form a relatively primitive mechanism capable of sending speech from the experimenter's mouth to the fingers of several experimental subjects simultaneously, effectively facilitating a wired broadcast model of communication through the skin that resembled early practices of broadcast telephony. Owing to the relative paucity of psychological research and commercial investment into touch communication systems, this practice of borrowing from extant sound reproduction machines when engineering tactile communication machines continued even with the field's maturation in the latter half of the twentieth century. As Cutaneous Communication Laboratory cofounder Carl Sherrick would later observe in tracing his field's history, "the technological evolution of mechanical tactual stimulators can be described as parasitic, depending as it has for much of its existence on advances in acoustical research."[52] While researchers in Sherrick's era would eventually come to identify this dependence as problematically limiting, for Gault and his collaborators, the relationship proved generative, as telephony and radio combined both to suggest an orienting set of principles and to provide a ready-made set of components that could be united to form new communicative apparatuses.

As news of the Teletactor's successful deployments spread in the scientific, medical, and popular press, these analogies to contemporary sound media served the additional function of providing a rhetorical grounding for the device—a way of indexing an unfamiliar product of laboratory research against a set of domesticated communication devices. In his initial presentation of findings to the Franklin Institute in 1927, Gault explained that the telephone's domestication had caused people to forget what a "bizarre thing" it was at its introduction. In the short span of a half century, the telephone had become so thoroughly interwoven into everyday communication that it almost felt "as near to us as hands and feet."[53] Though the Teletactor may have seemed fantastical to his audience, the double analogy provided by the telephone—simultaneously a comparison between the two devices' technical workings and their rate of social adoption—suggested that it would follow a similar trajectory of gradual incorporation into a range of communicative practices. Gault speculated that eventually the Teletactor would become a common feature in classrooms, allowing the deaf to be educated alongside those without hearing impairments (see Figure 3.6).[54] After Gault demonstrated the Teletactor's capacity to intelligibly facilitate the transmission of musical sounds through the fingertips, he envisioned its eventual migration, in miniaturized form, into movie theaters and opera houses.[55] Like the telephone and radio, the device could have utility not only in the

Figure 3.6. Group of deaf subjects at work, with each one holding a Teletactor receiver in his left hand. Each subject simultaneously receives the same impression. This image was published in *Journal of the Franklin Institute* 204, Robert Gault " 'Hearing' through the Sense Organs of Touch and Vibration," 341. Copyright Franklin Society (1927); courtesy of the Franklin Society.

practical art of speech communication but also in the pursuit of aesthetically oriented listening.

This functional analogy between sound reproduction and touch communication was complemented by a second analogy that figured centrally into the device's design and discursive framing: a psychobiological equating of hearing with touching, where the two senses operated by a similar process of receiving vibrations through a sensitive membrane. However, the comparison between the two senses quickly moved beyond the figurative. As suggested by Weber and later by Darwin,[56] hearing emerged in the genetic history of the human organism literally as a more evolved and acute form of touching. What, after all, was the ear, if not a highly developed and extremely acute membrane capable of perceiving vibrations in the air inaccessible to the nonspecialized regions of the skin?[57] Gault presented this as a sequential development traceable to a point

of origin that was a precondition for the later emergence of the specialized sense organs:

> In this adventure we are going backward over the genetic
> history of our organs of sense to the great, undifferentiated
> "mother sense of them all," and we are looking toward a time
> when those who are bereft of the most complicated organ of
> touch—the hearing ear—may fall back upon the very base
> of the sense of reality in the extended array of organs of touch
> in the skin.[58]

Gault was not shy about announcing the far-reaching consequences
of this repositioning; throughout his thirty-plus publications
and lectures on the Teletactor,[59] he continually pressed the case that
the extant models of communication informing psychology needed
to be productively upended. Though they differed from Gault's
preferred approach of routing sound through the fingers, later
investigators of tactile communication drew inspiration from his
attempted revalorization of touch.

Where communication media were concerned, the reduction
of hearing to a form of touch meant that radio and telephone were
not sound reproduction apparatuses analogous to the Teletactor.
Instead, the entire concept of "sound reproduction" was an act of
terminological misdirection: all the various technologies for storing, capturing, and transmitting sound were fundamentally vibrotactile media that produced vibrations specifically targeted to the
ear, not as a distinct organ but as a highly specialized organ of the
vibrotactile senses. In this model, communication media still
functioned to encode biologically derived notions of sensory differentiation. However, the differences they encoded were informed
by a division between hearing and touch that this new account of
touch rendered obsolete.[60] Gault's implicit suggestion meant that
the lines between the senses could be drawn anew, and in doing
so, psychology would break open a storehouse of fresh possibilities
for technologically mediated communication. Further, redrawing
these lines would not only enable the discovery of new and so-
called foreign languages like the vibrotactile one generated by the
Teletactor—it also would require the discipline "to forget a great

deal about the hearing of speech," as psychologists were "biased toward [their] own mountains of data relating to the physics and physiology of hearing."[61]

The task that occupied Gault and his collaborators during the nearly twenty years they spent investigating the possibilities of "'hearing' through touch"[62] involved supplementing these "mountains of data" about hearing with new quantifications of vibrotactile perception, built from repeated experiments with various iterations of the Teletactor. Through rigorously constructed experiments, Gault set himself to work probing the limits of the skin's capacity for distinguishing between the vibrations produced by the instrument. His results called into question the previously demonstrated maximum thresholds of vibrotactile perception, suggesting that the skin's capacity to act as a surrogate for the ear could be enhanced through the design of more complex instruments. In other words, it was not any inherent or hard-coded limits of the human skin that presented an obstacle to communication through touch. Instead it was the lack of a formalized and coherent system of knowledge about how to best pass messages through it that barricaded the tactile channel.

Here it is helpful to understand the Teletactor as an apparatus in the Foucauldian sense of the term. In this conceptualization, apparatuses are techniques that push outward the crucial border between the governable and the ungovernable, and in doing so, produce new and productive forms of subjectivity.[63] But by pushing this border outward, they also reveal sites that lie beyond the reach of strategic governance and management: resistant subjectivities that become objects of power/knowledge. The Teletactor apparatus brought a new dimension of touch—one previously thought to be unreachable and perhaps even nonexistent—within the sphere of technoscientific management. By doing so, it revealed a resource capable of being productively exploited and deployed in the service of communication. To give one specific illustrative example, prior to Gault's trails, 1,552 double vibrations (dv) per second of an armature against the skin was accepted as the highest recorded threshold of perceptible vibrations. If that number represented a hard psychophysical limit of vibratory perception, the

Teletactor would prove ineffective: research carried out that same decade by Gault's collaborators demonstrated that many vowel sounds reached frequencies well above 2,500 dv, with some exceeding 3,500 dv.[64] By showing that the fingertips could reliably distinguish vibrations in excess of 2,000 dv, trials carried out in this experimental system suggested "an undreamt of capacity for refinement of discrimination with the realm of the senses of touch and vibration."[65] With the upper limits of vibratory perception having been shown to correspond to the tonal qualities of speech sounds, the senses of touch and vibration could be used "as the base of supplies for the process of integrating a language sense."[66]

Mapping and testing this capacity for fine discriminations in the fingers provided Gault with a functioning understanding of the "criteria" that could pass recognizably "through the old gates of the skin."[67] To enable teletactile listeners to more easily recognize the criteria that comprised speech, Gault employed an electrical filter that broke the speaker's voice into five separate bands, with each carrying a particular range of frequencies to a specific vibrating reed on a Multiunit Teletactor. By breaking up the signal into these component parts, Gault distributed the load that had initially been placed on the single point of contact among five different contacts, with bassy vibrations below 250 Hertz (Hz) routed to the thumb, 250–500 Hz sent through the index finger, the middle finger responsible for receiving everything from 500 Hz to 1 kHz, the ring finger handling frequencies in the 1–2 kHz range, and the little finger accommodating the high-pitched vibrations above 2 kHz. This system fused knowledge of acoustics, electronics, psychology, physics, and haptics with a nascent theory of neural plasticity, as it tasked the brain with reaggregating intelligible speech out of vibrations that the Teletactor disaggregated.

At the heart of this concerted effort to terraform psychology's disciplinary terrain was the human subject; pushing toward a final and effective version of the Teletactor that would convincingly pass sounds through the fingers required its repeated submission to laboratory trials. This subject had to be managed through the application of experimental protocols and the construction of laboratory equipment capable of factoring out, as much as possible, the variants in subjective experience. Concerning touch, this proved to be no easy task: the measurement of vibration thresholds, the reception of teletactile signals, and the calibration of vibrotactile instruments each depended on experimental subjects maintaining a constant pressure between their fingers and reeds protruding from the Teletactor. The physical fit of bodies into experimental machines, then, posed a challenge that Gault and the investigators who followed in his footsteps would continue to struggle with throughout their many trials. At every stage in the evolution of a given machine, this "lack of uniformity of stimulation"[68] threatened to undermine the reliability and consistency of experimental results.

Training, too, proved to be of paramount concern, presenting perhaps the most substantial obstacle to the Teletactor's uptake outside the confines of the laboratory. The vibrations passed through the machine had to be intelligible—legible—as spoken language to the specific subjects willing to "lend their skins" for Gault's experiments.[69] Repeated trials involved not just acclimating evolving iterations of the Teletactor to the psychophysiological characteristics of an idealized user. They also required the development of standardized training protocols that would allow users to comprehend the vibrotactile sensations felt by their fingers as language. In attempting to master the teletactile language, some of Gault's deaf-mute subjects undertook training programs that exceeded a hundred hours in the laboratory, with further refinements to the training protocols able to gradually shorten the training time necessary to acquire the vibratory language. One helpful modification came from complementing touch with a visual component; after initially separating teletactile speaker from teletactile listener, Gault later allowed listeners to see the speaker, and added a "teletactile auditor" who would transcribe spoken words on a chalkboard as they passed from the speaker's mouth into the listener's fingers. These successive phases of experiments first detached the senses from one another and then strategically wove them back together, so that the relative performance gains that each sense modality added could be quantified and compared. Both learning to "hear" through and speak into the device thus entailed the formulation of complex and routinized training procedures that would allow humans to become legible to one another by careful adjustment to a machinic system.

While the project of grafting a mechanical ear upon the skin would continue throughout the twentieth century, the enduring significance of Gault's work, in excavating the macrohistory of haptic interfacing, lies in the tradition of research inspired not by its success and later continuation, but rather by its failure and (temporary) abandonment. The naked declaration motivating this work—that "there is nothing sacred about signs,"[70] that humans could be trained to arrive through touch at meanings customarily received through the other sense organs—laid a nuanced theoretical scaffolding for subsequent investigators to build on. It also suggested the technical means by which electromechanical tactual stimuli could be made to serve as a medium for the transmission of language.

Divining the Skin's Tongue (or "Vibratese: The Medium Is a Massage")

Though Gault's work on the Teletactor attracted significant attention during the 1920s and 1930s, the apparatus's failure to achieve success outside the lab's simulated conditions condemned it to obscurity, at least for a time. Late in the 1940s, when Geldard began exploring communicative pathways that could provide alternatives to what he described as the oversaturated channels of seeing and hearing, he found Gault's Teletactor to be an inventive first step. However, Gault's goal of helping the deaf to hear caused him to try to make the skin function as if it were an ear. Consequently, Geldard argued that Gault had erred by trying to force the skin to assimilate to a language incompatible with the tactual organ's constitutive structures: owing to its "intrinsic inability to handle moment-to-moment vibrations of the finest sort,"[71] the skin proved to be incompatible with the high-frequency vibrations produced by the Teletactor. Through all the efforts undertaken by Gault to route sound through the fingertips, "no one paused to ask the skin what language it could compass. No one considered what . . . the tongue of the skin might be."[72] The validity of the lofty claims Gault advanced about touch's capacity to serve as a communicative sense hinged on the Teletactor's success; to Geldard, its failure represented a "major disappointment,"[73] because it dissuaded further investigation of skin-based language transmission. With a missionary's zeal, he took it upon himself to resuscitate the project. Unlike Gault, Geldard sought to develop an application that would have utility not just to those with hearing impairments but also to individuals challenged by the overcrowding of the visual and aural channels in everyday life more generally. His passionate advocacy for the utility of touch communication systems entailed a simultaneous critique of conditions in a mediatic modernity, where "our eyes and our ears are assaulted so continuously, such frequent and insistent demands are placed on them, that the visual and auditory channels are seriously overburdened at times."[74] The "nearly universal dependence on vision and audition in all important affairs of human existence"[75] proved problematic for functionalist rather than philosophical reasons: the channels of vision and audition were simply becoming too crowded, suggesting that forward social progress depended on discovering additional sensory pathways to exploit.

Capitalizing on his growing stature in the field, Geldard campaigned to win converts to his cause. In a 1956 address to the Experimental Psychology division of the American Psychological Association, titled "Adventures in Tactile Literacy," he issued a call to arms: "the human integument" he claimed, "has been the object of precious little research effort on the part of psychologists."[76] Highlighting his field's sustained inattention to the skin was a rhetorical strategy; identifying the crisis brought on by the "seriously overburdened" visual and auditory channels, Geldard urged his colleagues to follow him in considering the possibilities for relieving the stresses placed on seeing and hearing by routing information through an alternative sensory pathway. To show how flexibly the body could be used as an information-reception instrument, he arrived at touch as the most pragmatic complement to vision and hearing only after a careful process of eliminating competing alternatives. Recalling the tradition of tangible telegraphy, Geldard

explained the different pathways Morse could traverse to reach the
mind:

> It is possible to transmit intelligence successfully by any of
> the sense avenues, including the ponderous chemical senses,
> for instance, by suitable utilization of International Morse
> Code. A skilled receiver can get meanings just as promptly
> and accurately by feeling dots and dashes as by listening to
> them. It would be possible to tap out Morse with spaced suf-
> fusions of salt on the tongue, with differently sized packages
> of radiant heat on the forehead, or even by a series of injec-
> tions of acids at some not too inconvenient spot on the
> integument.[77]

As quickly as he introduced these various possibilities, Geldard dis-
missed them as highly inefficient modes of transmitting information. In International Morse, he estimated the transmission time
of the message "Now is the time for all good men to come to the
aid of the party" (the same sentence Gault had used in his early
trials with the Teletactor), at "a little over a half hour, for spaced
suffusions of salt on the tongue; about an hour or so, for packages
of radiant heat on the forehead (assuming the aid of a cooling
system to keep down the blisters); and the better part of a day, for
acid injections in the skin, also assuming the assistance of a suitable
counter-irritant to hold the pH within bounds."[78] In conceptualizing this human–machine messaging system, Geldard disregarded
the sensations of pain that would be produced by the transmission process; the management systems he proposed were only
intended to ensure that the interfaced skin did not become so
deformed that it would no longer be capable of accurately receiving signals. Were these skinspaces able to more quickly receive
messages, they might have resulted in a man–machine interface
where all systems were active at once: heat on the forehead, salt on
the tongue, and an acid injector on each arm, each wired to a telegraphic transmitter and with each communicative node receiving
a different component of a given message.

But Geldard's transmission system was destined to remain the
figment of his cyborgian imagination: as he concluded, the "chem-
ical senses . . . by accident of slow stimulus transport, are so pedes-

trian as not to be serious contenders in the world of communication."[79] Through either the mechanical or electrical transfer of
energy to the skin, touch proved uniquely suited to serve as an
alternative pathway for communicating intelligence. And while
these examples showed Morse to be a viable code for a range of
skin-based communication, he believed that in its simplicity, Morse
failed to capitalize on the skin's full and unique potential. He
instead wanted to design a system that was not just a tactile analog
to Morse but would surpass it in the speed with which it communicated messages.

Vibratese, Geldard's proposed language for speaking in "the
tongue of the skin," was transmitted through a mechanical apparatus calibrated to empirically derived tactile capacities for noticing and not noticing changes in stimuli.[80] This apparatus consisted
of five vibrating motors, distributed at different points on the torso.
Geldard and his collaborators engineered each motor to generate
three steps of intensity (low, medium, and high) in three different
intervals (short, medium, and long). Employing this combination
of location, intensity, and duration allowed Geldard to produce
forty-five distinct tactile sensations. Of those forty-five sensations,
twenty-six were assigned to letters, ten to numbers, four were
taken up by commonly used words ("of," "and," "the," and "in"),
another was used to indicate a space between characters, and four
were left unassigned for later use (see Figure 3.7 for details). Where
Morse functioned by duration alone (dot, dash, or space) and sent
through signals a single point of contact, Vibratese combined
multiple points of contact with meaningful shifts in stimulus
intensity to increase transmission speeds.

After training their subjects to decode Vibratese, Geldard and
his students measured the subjects' message reception and error
rates, and triumphantly declared the system a success. They claimed
victory on two fronts: in addition to having the advantage of routing information through an unoccupied sensory channel, their
results showed that Vibratese allowed for more efficient message
transmission than the Morse system. An expert in International
Morse Code could receive twenty-four words per minute by ear;
with Vibratese, Geldard asserted that a properly trained receiver

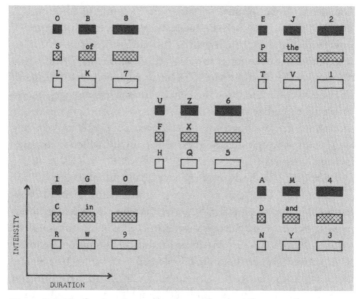

Figure 3.7. The final coding scheme used for the Vibratese characters employed three steps each of location, duration, and intensity. White symbols are the least intense, checkered are more intense, and the cross-hatched symbols are the most intense vibrations. From William Howell, "Training on Vibratory Communication System" (master's thesis, University of Virginia, 1956), 21. Diagram by William Howell; courtesy of Steven Howell.

would obtain speeds of up to sixty-seven words per minute.[81] While he had articulated a wide range of potential uses for new touch communication systems at the start of his presentation, this particular application evidenced both Geldard's background in combat aviation[82] and his continued affiliation with a host of military research organizations. Along with much of Geldard's later research into vibratory perception, the project that Vibratese grew out of was underwritten by funding from the U.S. Office of Naval Research, as the recognition that psychology played a vital role in national defense prompted a series of new grants in the area of "human engineering."[83] Like Barbier's *Ecriture Nocturne,* institutional investment aimed at producing improved systems for relaying messages in combat conditions fueled the mid-twentieth-century interest in touch communication. As the dominant mode of combat communication, Morse transmission rates provided the bar against which Vibratese would be measured. Geldard there-fore devoted extensive time to quantifying the performance gains that could be won with a shift to "the vibratory language."[84]

A third front in the battle with Morse concerned the question of training Vibratese receivers: no matter how quickly the system could work under optimal conditions, if new receivers could not be rapidly brought to proficiency in the vibratory language, it would prove impractical for use in the field. Here again, Geldard's military background informed his practice, as he drew lessons from the well-tested methods used to train radio operators in Morse during World War II. Using similar techniques allowed him to quickly raise receivers to competence in decoding Vibratese: after roughly twelve hours of practice ("a couple of days' work in code school"), recognition rates for single Vibratese characters approached 100 percent, and "the symbol-signal connections" were solidified sufficiently to allow the learning of words and short messages.[85] As Katherine Hayles highlights in her discussion of World War II–era military telegraph training manuals, interest in the reception capacities of human operators indicates the extent to which communication codes depended on an embodied notion of signal coding; learning to read by telegraph required submission to a highly regimented process of embodied sensory training, with bodies expressing the hard limits that the senses placed on infor-mation transmission.[86] The Vibratese project aspired to transcend the old limits on the human receiver's capacity to process infor-mation by opening up a new and more efficient communicative pathway.

The primacy of these considerations in the search for the skin's tongue suggests that the project was as much guided by specific social and political considerations as it was by Geldard's purported drive to uncover the "bare facts"[87] underpinning the cutaneous system's operation. An ideal "teletactile listener"—a deaf subject prepared to devote dozens if not hundreds of hours to learning to hear through the fingers—informed the design parameters of

Gault's Teletactor. But the features of the Vibratese apparatus were constructed with an altogether different implied ideal receiver: the equivalent of a radio telegraph operator, the "Vibratese feeler" would need to be able to train rapidly for quick deployment in stressful combat conditions. Paul Virilio's thesis concerning the relationship between warfare and the senses directs us to consider "the history of battle" as "primarily the history of radically changing fields of perception."[88] From this perspective, the psychotechnics of the sense organs can be understood as another front in the theater of war. Those soldiers who were to be outfitted with the Vibratese apparatus would, Geldard's project implied, come to relate to the rapid, pulsing vibrations moving about their torsos as the data of combat, to be processed and faithfully transcribed at a rate imparted through rigorous, standardized training procedures. Both the Teletactile listener and the Vibratese feeler were to acquire their vibro-tactile literacy in institutional contexts, but in the latter case, the trainee's capacity to quickly acclimate to a new language was framed as central to the maintenance of U.S. military dominance.

Much of Geldard's presentation focused on demonstrating Vibratese's technical operation, with the contents of messages transmitted through the new machine seen as irrelevant. But he also wanted to advance a conceptual argument that echoed the one Gault had offered decades before, using the messages sent through Vibratese to help press his case. In filmed demonstrations of Vibratese prepared for the 1956 lecture, Geldard transmitted passages drawn from a pair of psychology's foundational figures through the apparatus as a means of demonstrating the efficiency and efficacy with which his new language could communicate complex ideas. The first passage, from Johannes Müller, compared the spatial properties of the senses of sight and touch. The second, a famous passage from William James's *Principles of Psychology,* addressed the relationship between bodily states and emotion.[89] Geldard's decision to use these messages from what he understood to be his discipline's nineteenth-century founding was not insignificant. Like Gault, he imagined himself to be reversing centuries of dogma about touch's capacity to receive coded messages. Now that its own intellectual history could be transmitted through the skin, psychology—informed by a tradition that treated touch as nonintellectual—had to confront and reevaluate the foundational axioms that caused it to regard touch as a sense incapable of language acquisition. The length of each passage served the dual purpose of staging the conceptual upheaval Vibratese implied while also showcasing its advantage over Morse. The passages were entered, via the standard typewriter that functioned as the sending portion of the Vibratese apparatus, and then decoded by the subject connected to the five vibrating motors that made up the system's receiving end. In a clever trick of audiovisual rhetoric, the James passage was then displayed on the screen while it played via a tape recorder in International Morse at the rate of twenty words per minute, with the Morse audio quickly lagging behind the image of the passage's text. This presentation technique effectively drove home the point that communication through a touch-based coding system possessed a decided performance advantage over its aural analog.

Where the Teletactor project failed in the short term both as an applied technology and (in Geldard's view) as an object lesson, Vibratese succeeded, not so much at gaining widespread adoption or in dislodging entirely the assumed incompatibility between touch and language, but rather in spawning a sustained and widespread investigation into touch's potential as a communicative channel. The notion that there existed languages unique to touch, declared by Geldard's use of the phrase "tactile literacy," suggested a path forward for the research paradigm he attempted to establish. Geldard had made clear in the 1956 address that Vibratese should not be treated as the endpoint of psychologists' investigations, insisting that the design he showcased be treated provisionally as one of many possible outcomes. The "tongue of the skin" was not, then, the final product of a discrete research project. Instead, it was intended as an epistemic orientation for future investigations. At a general level, it implied a preferred set of psychophysical methods for stimulating, testing, quantifying, and transcribing cutaneous sensations. More specifically, as Geldard conceptualized it, the tongue of the skin involved a direct rejection of the sensory substitution model

Gault had embraced, along with a host of preferences for how best to go about speaking to the integument. Location, duration, and intensity—the three basic variables used in the Vibratese language—Geldard considered to be the most fundamentally and demonstrably useful "building blocks," or "collections of stimulus properties"[90] available for the coding of cutaneous communication, regardless of the signifiers that would be attached to the signifying sensations.[91] Further, owing mainly to the "omnipresent pain" that had been the "great stumbling block" for systems that attempted to communicate language through electrocutaneous cues, he indicated an emphatic preference for the use of vibratory rather than electrical stimuli.[92]

In spite of these dogmatically expressed preferences, Geldard envisioned an expansive, interdisciplinary field devoted to launching a sustained "attack"[93] on the problem of touch communication, where no approach would be considered heretical (provided, of course, its success could be validated through laboratory trial). Drawn from a diverse array of fields, with each investigator seeking some strategic deployment of the accumulating psychophysical knowledge about touch, this imagined network of research would develop novel methods and socially transformative devices for the circulation of language through the skin. Geldard's Cutaneous Communication Lab (initially dubbed the "Princeton Cutaneous Research Project"), cofounded with Carl Sherrick when they moved to Princeton University in 1962, played a central role in realizing this vision, owing in part to the significant funding the lab's directors (Geldard at its inception, then Sherrick, and finally Roger Cholewiak) were able to secure during the lab's forty-two years of operation.[94] Over the next decade following Geldard's 1956 address, the field matured rapidly, with a host of projects springing up in various labs around the United States, each devoted to the same overarching aims of assessing the current state of psychophysical knowledge about touch and inquiring into the practical possibilities for positively deploying that knowledge in the design of touch communication systems. In 1960, more than a dozen of these investigators (including Geldard and Sherrick) gathered at the U.S. Army Medical Research Laboratory in Fort Knox,

Kentucky, for the Symposium on Cutaneous Sensitivity, where they began what would be a continuing partnership, fueled by substantial military investments and support from a variety of academic institutions.[95] Researchers in this emerging knowledge network engaged in the sort of practical sharing of experimental results necessary for forward movement in any scientific enterprise, exchanging apparatus schematics, experimental findings, testing protocols, and coding schema. Their common interest in what was frequently articulated as a neglected and underexploited avenue for communication also served as a bonding agent, feeding forward a shared sense that they were working at the frontier of an unmapped territory. Successive conferences, such as the International Symposium on the Skin Senses in 1966 and the Conference on Cutaneous Systems and Devices in 1973, reinforced this idea that the intellectual project they pursued positioned them at the margins of psychological investigations into the senses. At the close of the 1973 conference, Sherrick described the study of touch as "confined to the intellectual ghetto" and criticized his colleagues in the field of psychology at large for the "chauvinistic condescension" inherent in the phrase the "minor senses," where touch, along with taste and smell, had been grouped.[96] Even fifty years after the extensive battery of experiments carried out by Gault and his collaborators, touch still inhabited a "vast wasteland" in psychology research that lay between the poles of seeing and hearing.[97]

Although he trumpeted Vibratese's success, Geldard remained keenly and intimately aware of the practical limitations that would inhibit its widespread adoption. Securing the array of vibrators to the chest left the subject "literally trapped in a forest of concrete-rooted supports, flexible goosenecks, and long, flat springs"[98] that were needed to maintain the isolation of the vibrators from one another and to fix the pressure they placed on the subject while inactive (see Figure 3.8). In spite of its aspirations to utility, it could have no life beyond the walls of the lab—and especially no success in the chaotic conditions of the battlefield—without drastic refinements to its design that would allow the system to be made portable and receive messages wirelessly. In spite of these limits, the device succeeded in what was arguably its primary purpose: to

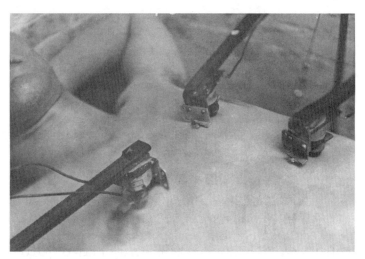

Figure 3.8. Three of the five vibrators in the Vibratese apparatus positioned on the chest. From William Howell, "Training on a Vibratory Communication System" (master's thesis, University of Virginia, 1956), 20; courtesy of Steven Howell.

attract followers and devotees to Geldard's mission of remaking touch as a communicative sense.

The image of the Vibratese arms closed around a subject fixed in place recalls the Apparatus for Simultaneous Touches (see Interface 2)—an immobilizing frame that experimenters enclosed their subjects within so that they could lay bare touch's capacity for the synchronous perception of multiple contact points. Read in the context of Geldard's intellectual biography, the similarity between the two machines is more than simply an accident of appearances, becoming both an illustration of the process by which scientific projects (and the experimental systems that constitute them) are transmitted from one generation of actors to the next and a way of showing the discontinuities that emerge in subsequent iterations of experimental machines. I want to establish a genealogical continuity here between the nineteenth-century apparatus and its twentieth-century successor: Krohn and Bolton designed and tested their apparatus in the psychology lab at Clark

University, which had only recently been established there by G. Stanley Hall. When Hall, who studied in Wilhelm Wundt's Leipzig lab before taking residence first at James's lab in Harvard and later founding his lab at Clark, published his essay "The New Psychology" in 1901, it was accompanied by an image of Krohn and Bolton's apparatus, along with a history of experimental psychology that positioned Weber and his two-point threshold experiments at the origin of the new science.[99] And Hall himself, during his time at Johns Hopkins, had collaborated with Herbert Henry Donaldson in developing new apparatuses to investigate pressure, temperature, and pain spots on the skin. Geldard arrived at Clark for his graduate study in the final years of Hall's career, but during that short time, he served as Hall's reader (necessitated by the aging professor's failing vision) and also worked as a research assistant for Joseph Jastrow, who had developed a range of widely adopted aesthesiometric compasses used in testing two-point thresholds (see Figure 2.13 for one example). The discontinuity between the machines, emblematic of the shift Geldard effected in the aims of scientific research on touch, concerns the different ends to which they aspired. As I suggested in Interface 2, the use of touch as a communicative sense was almost immanent in the nineteenth-century machines built to measure its discriminatory capacities. Geldard followed implications that seemed to be inherent in the materiality of the machines themselves, taking what had been machines used in the production of knowledge *about* touch and transforming them into machines used in the communication of knowledge *through* touch. The human skin, particularly its "vast expanses" that remained "practically unused, except for regulation of water and heat economies of the organism," presented to Geldard "a great many sites of potential stimulation" that could be "capitalized on" by a well-designed message transmission apparatus.[100]

While many other tactile communication systems developed in the field's early years were, like the Vibratese apparatus, highly impractical for routine use, pragmatic social utility was nevertheless central to their inception and design. Taking cues from Geldard's 1956 address, subsequent researchers (many working on grants from the U.S. military) described creative uses for these emerging communication media. Most of these imagined deployments revolved around the demonstrated capacity of tactile messages to directly, immediately, and forcibly capture the attention of their intended recipients. In this technocommunicative ideation, touch possessed a set of essential structural characteristics (its "tongue," as Geldard repeatedly put it) that both differentiated it from the senses of seeing and hearing and implied a set of preferred uses for the newly opened tactile channel. Always on, ever vigilant, but infrequently occupied, touch promised to remain open and attentive, even during sleep; as Beverly von Haller Gilmer phrased it, "the skin as a sensory channel is . . . rarely ever busy," and "vibratory stimuli cannot be 'shut out.'"[101] According to this normative modeling, the tactile channel could compete successfully with the visual and aural channels, whose attention in the Cold War era was under siege by other electrical communication technologies. In a report prepared for the Civil Aeromedical Research Institute, Glenn Hawkes located the attention-capturing capacity of contemporary audiovisual media centrally when describing the value of communication through touch: even if a group of soldiers "happened to be watching television or listening to the radio,"[102] a tactile alert could immediately interrupt that activity to warn them of some impending danger. While the other senses might wander, touch remained in a state hypervigilance, constantly monitoring its surface for any irregularities or interruptions. This condition implied a particular design and structure for the tactile stimuli used to communicate messages: their artifice and strangeness were features rather than bugs, allowing them to be distinguished immediately from the regular bumps and impacts touch receives from the environment. The condition of being always-on suggested an optimal placement on some "unused" portion of the skin unaccustomed to receiving cues from the environment. The skin's vast expanse became an untapped and underexploited resource in the communicative economy, one that could be capitalized on through the design and iterative refinement of signaling systems and languages.

Although this research may have developed in a military context, infused with the hopes of weaving touch into a combat communication infrastructure, designers imagined civil applications for the technology, too. As von Haller Gilmer explained: "it is possible through a cutaneous communication system to provide information of such diversification as stock market quotations, weather forecasts, and gun laying data."[103] In other words, the applications for touch communication were thought of as value-neutral language systems unmarked by the context in which they developed. However, when von Haller Gilmer detailed possible uses for tactile communication systems, they remained infused with a decidedly militaristic flavor, in spite of his efforts to shake loose those associations. They expressed a biopolitical fantasy of managing the attention systems of communicative subjects in a range of military contexts: civic defense personnel living under the imminent threat of nuclear warfare, pilots attempting to navigate in situations that confused the eyes and ears, and soldiers operating in remote environments where orders needed to be relayed without optical or acoustical indicators (see Figure 3.9).[104]

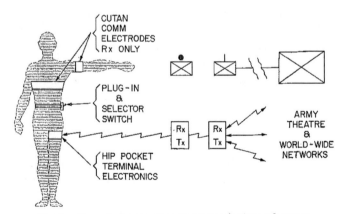

Fig. 1. Concept of use: cutaneous communications sub-system.

Figure 3.9. Cutaneous communication subsystem. From John Hennessy, "Cutaneous Sensitivity Communications," *Human Factors* 8, no. 5 (1966): 463–69. Copyright 1966 by John Hennessy. Reprinted by permission of SAGE Publications.

Following this pragmatic imperative, investigators searched for more effective and efficient tactile communication systems. Each novel mode of stimulation prompted new systems of experiments, configured with the intent of revealing the unique possibilities each machine possessed for encoding messages, improving transmission speeds, and capturing the attention of message recipients. Next to vibration, the most mature field of investigation involved the induction of tactile sensations through the application of controlled pulses of electrical current. While Geldard expressed a clear dislike for this technique, partially because of how easily electrical stimulation could conjure the ever-present specter of pain, others identified electricity as having distinct advantages over vibratory stimulation in its ability to more accurately and rapidly communicate information, and they set themselves to work designing electrocutaneous communication systems. As proponents such as Hawkes, von Haller Gilmer, John Hennessy, and Robert Gibson claimed, electrical systems offered two primary advantages over their vibratory counterparts. First, they allowed for more precise control over the duration and intensity of the stimuli. With no motor to account for, signals could be started and stopped more rapidly, limited only by the capacity of the human receiver to distinguish successfully between pulses. Second, these electrocutaneous stimulators avoided yielding sound as an unintentional by-product, allowing a more private and less socially disruptive pathway for the transmission of tactual messages. Vibrotactile systems purported to allow for messages to be passed discretely through touch, but a practiced interloper could intercept vibratory messages by listening for the motor's sound as it sent messages to the receiver's skin.[105] In contrast, electropulse signals were free from what Hennessy termed "auditory contamination," allowing for the further isolation of a touch freed from the interference of the other sensory channels.[106]

Their efforts at building electrocutaneous communication systems required them to specify, as precisely as their instruments would allow, that elusive threshold between touch and pain—the point where electricity ceased being signal and instead became noise.[107] These investigations quickly revealed that some portions of the skin would be unusable in an electrotactile communication system, as even the slightest current activated pain alone. Other areas of the skin, while initially capable of receiving electrical stimuli without invoking pain, became irritated after only a few pulses, rendering them unsuitable to receive messages. Gradually, these repeated trials shocking experimental subjects yielded a map of the skinspaces most capable of providing a pathway for electrotactile messages, which were plentiful enough, researchers concluded, to facilitate a range of possible coding schemes. Weber's old two-point threshold experiments were systematically conducted again, but this time, a pair of electrodes replaced the compass points, and a computer rather than the experimenter modulated the application of stimuli.[108] The stimuli were moved about the body, the applied amounts of current were varied, and trials were repeated on a range of test subjects, all with the intent of arriving at the optimal combination of discriminatory acuity, stimulus intensity, and site dormancy. The flurry of activity devoted to discovering novel ways to pass messages through the old gates of the skin left no method untested, and no portion of the skin—except for those comparatively small patches of the integument that covered the genitals—was exempted from being poked and electrified.[109]

As with the design of vibrotactile languages, engineering electrocutaneous communication systems necessitated the search for specific technical knowledge about the body's capacity for electrotactile discrimination, fueled by refinements in the ability to control the application of electrical current to the skin. Research in this tradition recalls the model of epistemic, nontraumatic shock mobilized around electrical machines during the eighteenth and nineteenth centuries (see Interface 1). Calibrating the stimuli to the receiver's skin again required the experimenter to arrive at the same sort of golden mean—between absence and presence—that proved so vexing for electrotherapists. Too weak a stimulus would go unnoticed, while too strong a stimulus would inflict pain on the receiver, and even, in extreme cases, cause damage to the nerves so severely that they would take hours to recover their sensitivity (see Figure 3.10 for a sampling of electrodes). Gibson, in his countless trials mapping electrocutaneous sensitivity, was able to arrive

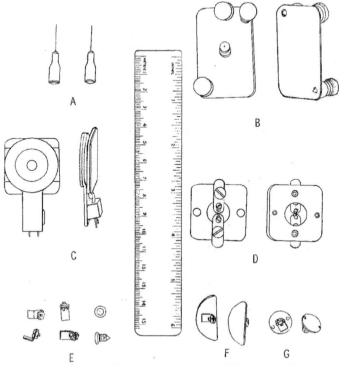

Fig. 2. Representative electrodes employed in the cutaneous sensitivity research project: (A) needle; (B) pin-modified EKG; (C) flat concentric; (D) pin; (E) "pucker"; (F) hemispheric; (G) round.

Figure 3.10. Representative electrodes used to investigate cutaneous sensitivity. From John Hennessy, "Cutaneous Sensitivity Communications," *Human Factors* 8, no. 5 (1966): 463–69. Copyright 1966 by John Hennessy. Reprinted by permission of SAGE Publications.

at this mean when he adjusted the pulse from one of his electrodes so that it felt like "a light tap from the blunt end of a fountain pen."[110] Fittingly, the feeling of a writing instrument pressing into the body provided the phenomenological ground for Gibson's description of the properly calibrated electropulses, as he redeployed the skin as a surface that could be legibly written on by computer-modulated flows of electricity. By the early 1970s, both humans and machines seemed up to the productive task of communicating coded messages through targeted electrical shocks.

In spite of their pragmatic aspirations and engineering successes, proponents of electrotactile communication, like their counterparts working on vibrotactile systems, failed to move their machines beyond the confines of the lab. Though electrical apparatuses were less cumbersome and more precise than their vibratory counterparts, they remained haunted by the unshakable specter of pain, as electricity constantly threatened not only to temporarily numb but also to permanently injure the nerves it targeted. Von Haller Gilmer, one of the most outspoken proponents of skin-based electropulse communication, cautioned that even "those skin areas where pain-free pulses can be sensed day after day can easily be made painful by increasing one of the dimensions of intensity, duration, or frequency."[111] In addition to the psychophysical component of the challenge posed by the slippage from electrical touch to electrical pain, electricity had taken on some additional cultural baggage since the nineteenth century, with electroconvulsive therapy—which had only recently been adopted in the United States—looming large in the cultural imagination, and the electric chair reaching its apogee around the time of these investigations. For the moment, in spite of these efforts, the widespread adoption of machines that would transmit messages directly to the skin via electrical current was to remain a fantasy, albeit one that would continue to capture the imaginations of researchers in subsequent generations.

The Skin as Optic Nerve

With investigations into TVSSs in the 1960s, the notion that the skin could serve as a vicariate for one of the major senses—the driving idea behind Gault's Teletactor—gained a new expression, organized similarly around the idea of a structural affinity between touch and the sense it was asked to emulate. Inspired by the enthusiasm for skin-based communication Geldard had rekindled in his 1956 address, psychologists and engineers set themselves to work at two interwoven tasks: designing machines that could transcode images for the skin, and uncovering the pertinent psychophysical parameters that would allow their subjects to interpret artificial tactual stimuli as optical impressions.[112] As with the previous projects taken up by Gault, Geldard, and others, success depended both on calibrating the apparatus to the hard limits of human tactile senses and on the ability to rapidly train individual subjects to decode machinic stimuli. Given a properly calibrated apparatus and enough training time, researchers suspected that their blind subjects could learn to interpret patterns of stimuli projected onto the skin as images, effectively enabling the act of "seeing through the skin."[113] The goals of these various projects were both practical and theoretical, aimed simultaneously at providing lightweight, portable, and economical machines that could provide the blind with an awareness of the three-dimensional world, and at studying the mechanisms involved in perceptual learning more generally. Thus, TVSSs functioned both as attempts to realize what Paul Bach-y-Rita described in 1972 as the "age-old dream" of "the replacement of lost sensory systems" and as epistemic things that could provide insight on the brain's neuroplasticity.[114]

In this brief discussion of TVSSs, I use Bach-y-Rita's Tactile Television system (developed at the Smith-Kettlewell Institute of Visual Sciences beginning in the 1960s) as a representative case study. While other experimental machines developed contemporaneously with Tactile Television—including the OPTACON (*Opt*ical-*Ta*ctile-*Con*verter) and Elektroftalm[115]—similarly took aim at relaying images through the skin, Bach-y-Rita remained the most ardent devotee to the TVSS project until his passing in 2006, as he continually attempted to raise awareness about the interchangeability of the senses purportedly demonstrated by the Tactile Television system.[116] These efforts differed in their specific configuration and mode of tactual stimulation (some used electricity, others vibrators, and one early machine even employed targeted jets of air), but they shared the common underlying principle of attempting to translate visual impressions for successful transmission through the skin. Taken together, and in conjunction with the other tactile communication systems discussed in this chapter, they illustrate the breadth and depth of the sustained "attack" on touch executed by psychologists during the second half of the twentieth century.

Driven by the question "are eyes necessary for vision or ears for audition?"[117] Bach-y-Rita believed he could take advantage of the brain's plastic capacity to succeed where other sensory substitution systems had failed. "In the past," he noted, "many efforts at providing information to the blind have been based upon hopelessly old-fashioned ideas about the way the perceptual system works."[118] In addition to this reconceptualization of the brain as a malleable organ, Tactile Television capitalized on new methods for mapping and quantifying the tactile channel to meet the "prosthetic challenge . . . of using the tactile sensory system to carry optical information from an artificial receptor to the brain."[119] While Gault and Geldard had each presented models of touch as an information reception channel, subsequent work in psychology fused nascent findings in information theory together with psychophysics.[120] As suggested by G. A. Miller, a "bit" of information, from the sensory standpoint, was the amount of information necessary for a given sense organ to make an absolute (correct) judgment about the stimulus. From this perspective, the senses were coequal but differentiated by the character of information they were able to receive, as well as by the rate at which they could receive it. Applying information theory to the variegated and dispersed organs of touch entailed a massive battery of tests intended to map the information reception speeds of various points on the skin. For example, at different points in the palm, reception speeds varied from 2.7 bits per second to 3 bits per second; any increase beyond that rate caused errors in the correct identification of stimuli.[121] Owing to the heterogeneity and complexity of the skin's structure, this project proved to be immense, accompanied by new and more detailed graphic representations of touch, new investigations into the anatomical structures of the skin (primarily concerning the end organs and receptors responsible for the reception of touch data), and the development of new testing machines and protocols. Furthermore, it generated new controversies, typically in the form of sharp disagreements about experimental methods and their corresponding results.[122] But these minutiae were not as important for

Bach-y-Rita as the more general utility information theory had for understanding tactile signal reception. The affinity between the two suggested a particular organizational structure for the TVSS he built.

In one of its earliest iterations, Tactile Television employed a user-controlled television camera connected to a twenty-by-twenty matrix of vibrators (four hundred in total) spaced twelve millimeters apart, mounted to a dentist's chair (see Figures 3.11 and 3.12).[123] When the user sat in the chair, the ten-square-inch matrix of vibrators pressed firmly against their back. As the user moved the camera across its field of vision, the vibrators activated to indicate the

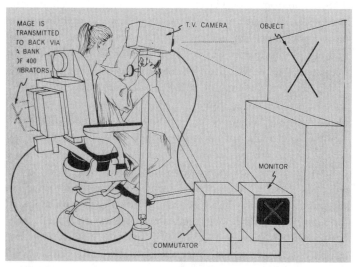

Figure 3.11. Schematic representation of the Tactile Television system. A subject directs the TV camera to an object of regard, which is converted to a pulse-sampled video image by the commutator, which in turn electronically switches each of the vibrating stimulators in the tactile array to present a 20-line (400-point) mechanical image on the skin of the subject's back. From Carter Collins and Frank Saunders, "Tactile Television: Electrocutaneous Perception of Pictorial Images," in Paul Bach-y-Rita, ed., *Seeing with the Skin: Development of a Tactile Television System* (San Francisco, Calif.: Smith-Kettlewell Eye Institute, 1970). Figure courtesy of the Smith-Kettlewell Eye Research Institute.

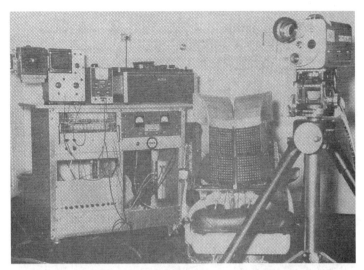

Figure 3.12. The hardware (camera, chair, oscilloscope, and control unit) that makes up the Tactile Television vision substitution system. The digitally sampled television camera with zoom lens is seen at right; the electronic commutator and control electronics with the monitor oscilloscope and videotape recorder are on the left. In the center, the 400-point two-dimensional tactile simulator matrix array is shown mounted in the back of a dental chair for projecting mechanical television images onto the skin of the back of blind subjects. From Carter Collins and Frank Saunders, "Tactile Television: Electrocutaneous Perception of Pictorial Images," in Paul Bach-y-Rita, ed., *Seeing with the Skin: Development of a Tactile Television System* (San Francisco, Calif.: Smith-Kettlewell Eye Institute, 1970). Photograph courtesy of the Smith-Kettlewell Eye Research Institute.

presence of letters or objects, dynamically drawing shapes on the user's skin through what Bach-y-Rita referred to as "tactile image projection." The camera relayed images to two components of the system simultaneously: an oscilloscope, viewable by the experimenter as a matrix of lit dots, and the vibratory display, "viewable" by the subject seated in the dental chair as rumbling across their skin (see Figures 3.12 and 3.13). Compared to the video image, the images on the oscilloscope and vibratory array were relatively low resolution, capable of only hinting at the objects captured by

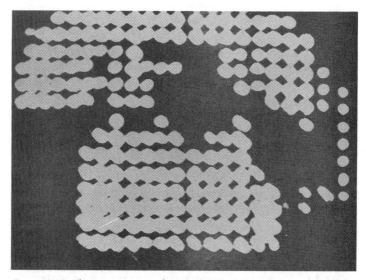

Figure 3.13. The appearance of a telephone, as captured by the Tactile Television camera and represented on the oscilloscope monitor. Bright dots correspond with active vibrators in the matrix. From Carter Collins and Frank Saunders, "Tactile Television: Electrocutaneous Perception of Pictorial Images," in Paul Bach-y-Rita, ed., *Seeing with the Skin: Development of a Tactile Television System* (San Francisco, Calif.: Smith-Kettlewell Eye Institute, 1970). Photograph courtesy of the Smith-Kettlewell Eye Research Institute.

the camera. But the oscilloscope's scant array of pixels—mapped onto the tactile pixels (later termed *taxels*) pressed into the subject's back—visualized the challenge that faced the subject, as they were asked repeatedly to identify objects, shapes, and letters using the scant data provided by the vibrators. The tactile display used in these trials was effectively monochromatic, rendering pixels as either on or off, with no gradation in the intensity of the stimulus (a later version of Tactile Television used stepped electrical stimulation to allow the presentation of a tactile equivalent of grayscale). In spite of these limits, Bach-y-Rita found that after only ten hours of training, subjects (seven congenitally blind college students) had gained the ability to "discriminate geometric forms; read block capital letters at 5–10 sec per word; identify

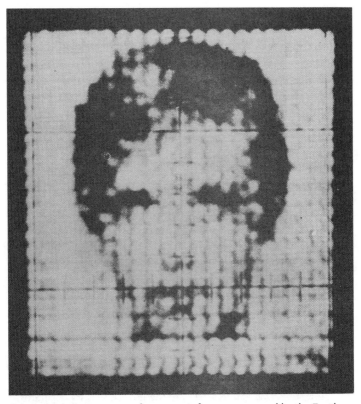

Figure 3.14. Appearance of a woman's face, as captured by the Tactile Television camera and represented on the oscilloscope monitor. The woman's face shows the effects of temporal integration (longer exposure). The resultant blurring (which also occurs on the skin) causes the points to coalesce. From Carter Collins and Frank Saunders, "Tactile Television: Electrocutaneous Perception of Pictorial Images," in Paul Bach-y-Rita, ed., *Seeing with the Skin: Development of a Tactile Television System* (San Francisco, Calif.: Smith-Kettlewell Eye Institute, 1970). Photograph courtesy of the Smith-Kettlewell Eye Research Institute.

more than 25 common objects at 1–5 sec latency; identify and describe the arrangement of four or more such objects placed randomly on a tabletop; and accurately identify the movements, postures, and distinctive characteristics of other persons."[124] Like the Teletactor and Vibratese before it, Tactile Television appeared to be a success, although Bach-y-Rita remained keenly aware that this success was a construct of the lab's carefully controlled testing conditions.[125]

Bach-y-Rita attributed the users' newfound ability to see through the skin to their training in actively moving the camera across its field of view. He noted that the perceptual organ always consisted of both a receptor organ and a motor mechanism for guiding it; in the case of the eye, vision involved not only light stimulating the dense network of nerves in the retina but also the exertions required to move the eye muscles. In addition to substituting the nerves of the eye for those in the ten-by-ten square of skin on the user's back, Tactile Television swapped the eye muscles for those in the user's hand and arm, which governed the movements of the camera as "an artificial receptor surface" akin to the eye.[126] In other words, actively panning and zooming the camera helped users learn to see through Tactile Television to such an extent that Bach-y-Rita noted a correlation between their manual dexterity and their ease of recognizing objects through the machine. Practiced users gradually experienced the effacing of the apparatus, with objects the vibrators presented to the skin appearing to exist not on the immediate surface of the skin, but rather "located in the three-dimensional space before them."[127] In spite of its inescapable artifice, the camera became "an extension of the normal sensory apparatus," folded seamlessly into the perceptual schema.[128] The system accomplished a complex reformatting of vision, dispersing a set of functions normally localized to the nerves in the eye throughout the cyborgian assemblage of camera, commutator, vibratory array, hand, arm, and back. The televisual image, downscaled to a resolution appropriate for passage through the cutaneous nerves, was mapped onto a skinspace that had itself been thoroughly transformed by its rendering in the labs of sensory psychologists.

As part of the sustained and ongoing attack on the old gates of the skin, Tactile Television aimed at providing touch with a new utility in the communicative economy. Whether by substitution, supplementation, or augmentation (and Bach-y-Rita suggested that these all be considered part of the same intertwined project), psychologists and engineers mobilized the laboratory protocols that constituted the science of haptics to fuel an attempted transformation and revaluation of the tactual senses. This biopolitical rethinking of touch—as an information reception channel that possessed calculable discriminatory capacities, capable of being folded productively into extant communicative networks—promised to salvage touch from its previous state of neglect.[129]

Before concluding this section, I summon McLuhan's ghost for a brief digression on Tactile Television. Media theorists will recognize an obvious parallel between Bach-y-Rita's system for relaying images through the skin and McLuhan's contemporaneous suggestion that television functioned, phenomenologically, as a tactile medium, "an extension of the sense of touch."[130] McLuhan's account of televisual tactility depended on a notion of tactile vision, where the eye, presented with certain types of images, would take on a tactile function, as the image cued sensations of touch rather than of sight.[131] The low resolution of the televisual image—a characteristic specific to television in McLuhan's era, rather than an enduring ontological feature—had precisely this effect, purportedly invoking the tactile sense through the eye. When considered in the context of McLuhan's broader political project, this was not merely a formal quality of the medium. Television, for McLuhan, served a utopian and ideological function: owing to its ability to activate this synesthetic mode of perception in the viewer, it held the potential to undo the earlier specialization and fragmentation of the senses accomplished by print media. In its synesthetic effects, television "reverses this literate process of the analytic fragmentation of sensory life."[132] But while both authors sought a dedifferentiation of the senses through perceptual technologies, the models of Tactile Television they leaned on diverged dramatically. For Bach-y-Rita, it was touch's specificity and calculability—obtained through its subjugation to the lab experiment—that allowed him to map the image onto the subject's skin in such a way that it would productively achieve this pragmatic synesthetic function. For McLuhan, it was touch's multiplicity—its easy conflation and collapsing of all the sense modalities—that provided its liberatory potential. Touch, in McLuhan's corpus, was nothing tactile: it constantly slipped from specifying a literal act of touching to being instrumentalized as a metaphor for his utopic desire to undo the damaging specialization of the senses brought on by language. He frequently described touch as "total, synesthetic, involving all the senses," akin to the Aristotelian *sensus communus* (common sense) that functions to translate between the senses.[133] Furthermore, the two models differed materially: Bach-y-Rita's Tactile Television apparatus enabled a televisual form of seeing through the skin, mapping the image onto touch; McLuhan's theorization of television suggested a televisual form of touching through the eye, mapping touch through the image.

McLuhan's presence here serves as a way of indexing just how thoroughly tactility had been worked over by experimental psychologists at the historical moment when McLuhan advanced his perplexing, counterintuitive theory of tactile television. The model of touch as calculable, mathematicized, and instrumentalized as a networkable communicative modality contrasts sharply with the fuzzy visualist, quasiphysiological account of touch that McLuhan borrowed from early twentieth-century aesthetic discourse, and it illustrates the different imaginaries mobilized contemporaneously around the common technical object of television. Touch underwent a radical reconfiguration in the labs of engineering psychologists, prompted in no small part by its forceful contacts with technical media like the telegraph, telephone, and television. However, this reconfiguration escaped the notice of media theorists. Consequently, the field fed forward stale ideas and sedimented assumptions about touch's purported inability to be folded into communicative networks, proceeding with a model where media act on touch only synesthetically—only through the imprecise activation of a tactual *faculty* that circumvented those nerves in the skin so crucial to the operation of tactile communication apparatuses.

Transmitting to an Empty Channel

Throughout the twentieth century, engineering psychologists undertook a sustained attack on touch. Intent on transforming it into a communicative sense, they sought to perfect what Carl Sherrick colorfully called "the art of tactile communication."[134] In the process, they crafted a new narrative—embodied in the various machines, protocols, and coding systems built in their labs—that portrayed touch as an empty channel, waiting to be flooded with information. As an untapped and unexploited resource, touch became part of a rationalized, calculated, and carefully managed communicative economy of sensations. However, in spite of the many victories claimed in the lab, the fantasy expressed through their research remains unfulfilled. Though the story as I have told it describes touch as being neatly folded into the communicative economy, another narrative could just as easily suggest that the twentieth-century efforts to engineer touch communication machines tell the tale of touch's capacity to *resist* being folded into that economy by emphasizing the repeated failure of touch communication systems to achieve the widespread success and adoption their proselytizers hoped they would. In light of that counternarrative, it might be tempting to ask whether the old view that Diderot, Gault, Geldard, and all their fellow travelers labored so doggedly to unseat had been correct all along: perhaps it is simply the case that touch, while not wholly unsuited to serve as a language reception sense, possesses certain structural features that make it inhospitable to language. From this perspective, the story has played out precisely as we should expect: touch is theoretically capable of serving as a language reception sense, or a vector through which images and sounds can pass; in the overdetermined conditions of the lab, it can obtain precisely these functions. But once touch leaves the lab, once it passes from the realm of the possible to the domain of the actual, whatever new powers it had won fade rapidly.[135]

However compelling—and even accurate—this counternarrative may be, I have resisted embracing it for several reasons. First, I want to understand these efforts as part of a broader macrohistorical system of attempts to productively bring touch under the dominion of modern technoscience. The specifics of these attempts matter immensely, as do the motivations that drove and continue to drive them, for they reveal much about the hopes and desires expressed through touch machines. At various times in the twenti-eth century, vibrotactile and electrotactile apparatuses were imag-ined as machines that could ameliorate sensory impairments, allow-ing the blind and deaf to more fully realize their socially useful potential as information processors;[136] they were thought of as ways to ease the "overload of the ordinary channels of communica-tion";[137] and they were considered as a potential means of establish-ing an always on, uninterruptible connection between individuals and alert systems networked to global flows of data. Taken together, these projects engaged in a collective discursive repositioning and rearticulation of touch that ascribed to it a whole new range of use values, themselves reflective of broader biopolitical needs that touch was expected to serve, once it had been transformed into an agent of the intellect. Examining minutiae of this archive provides a win-dow into the complex technoscientific imaginary mobilized around touch.

Second, this effort to make touch function as a receiver of coded messages is still ongoing, and it seems to be enjoying some minor victories, particularly in the area of mobile communications. Vibrating alerts transmitted from cellphones signal to their users that they are receiving some sort of incoming communication. For some time now, it has been common for these patterns of vibra-tion to differ, depending on the nature of the incoming commu-nication: as with the assignment of different sounds to different events, the vibration pattern activated by an incoming call will be easily distinguishable from that of an incoming SMS message. Users often learn these languages of touch without ever being aware that they have done so. This theme will be taken up in greater detail in Interface 5. For now, it is enough to point out that today we are still faced with the problem Diderot identified in his 1749 letter: we have yet to fix this language of touch, to "make its grammars and its dictionaries."[138] Instead, various technology firms currently compete for dominance in this area, with no one system yet achiev-ing the sort of stability or dominance required for its tactile signs to become "common property."[139] This commercial investment in vibrotactile communication renders obsolete Sherrick's claim about the parasitic dependence of advances in tactual stimulator technology on advances in acoustical research. A drastic spike in market demand for the vibrating motors used in touchscreens and other mobile computing technologies has fueled dedicated research into new components, materials, and signaling systems.[140]

Finally, the methodological orientation I have maintained throughout this book embraces a perspective that treats individual experiments, machines, and inventors not in isolation, but as parts of intertwined, institutionally grounded networks of human and nonhuman actors. From this perspective, failure frequently proves generative: it yields new data useful to other actors in the network, and it produces machines that may lie in stasis for decades before gaining renewed utility as a consequence of technological advances or shifting social needs. Even the most spectacular technological failures still function to redraw the ever-shifting border between the known and the unknown, between the governable and the ungovernable. When Gault began his investigations into the possibilities for touch communication, he lamented the lack of attention it had received in comparison to the senses of seeing and hearing. Sherrick, echoing Gault's observation half a century earlier, described "the world of haptic experience" as "one of shadows and echoes" that remained "littered with the partly explored phenomena and underdone analogies that result from the ad hoc hypotheses and horseback expeditions of previous generations of researchers."[141] Though touch was not governed as precisely or completely as these researchers dreamed it could be, owing to the new machines, funding streams, and measuration protocols, its terrain was at least mapped with much greater care and precision than was thought possible at the start of the century.

While all of these efforts were aimed at communicating and coding messages through touch, they intentionally avoided the transmission of tactile sensations as an end in and of itself. This began to change in the late 1950s, as a new tradition emerged in the labs of military researchers that treated touch machines not as a means of carrying coded messages or as a way to reroute data from the other senses, but rather as a way to transmit, store, synthesize, and encode the feeling of virtual or remote objects. The designers of these new machines aspired to transform the information gleaned about objects through the touch act—characteristics about the tactile materiality of objects, such as their weight, texture, shape, and temperature—into a form that could be sent through electronic networks. This next stage invoked a new logic, one where touch could become like the senses of seeing and hearing, not when it acquired a similar set of established scientific facts or when language could flow through it, but instead only once it had obtained its own set of representational media. Marvin Minsky, advocating a move away from the sensory substitution tradition in favor of a focus on remote manipulation, noted in his foundational 1980 essay "Telepresence": "it seems quite ironic to me that we already have a device that can translate print into feel, but that we have nothing that can translate *feel* into feel."[142]

Interface 4

Human–Machine Tactile Communication

Man–machine tactile communication . . . emerges not as
a supplement for computer-generated visual displays but
primarily as an entirely-new man-machine communication
medium of vast importance for its own unique abilities to
represent surfaces and objects. The tactile channel is a
competitor to the visual channel, and this situation is
something new to the field of computer science.
—A. Michael Noll, "Man–Machine Tactile Communication"

In a paper presented at the 1965 Congress of the International Federation for Information Processing, Ivan Sutherland detailed his
vision for an "ultimate display"—a computer-controlled sensory
environment capable of providing full immersion in a "mathematical wonderland constructed in computer memory."[1] In its ideal
form, the display would "serve as many senses as possible,"[2] including computerized displays of data perceptible by smell and taste.
But it was touch that occupied the bulk of Sutherland's attention in
his short essay, as he described a new type of "kinesthetic display"
that would be able to render the mathematical world of the computer to the operator's body using force cues to indicate the presence of virtual objects. As Sutherland explained, "the ultimate
display would, of course, be a room within which the computer

can control the existence of matter."[3] Objects displayed in the room would appear, to the sense of touch, as if they were real: "a chair displayed in such a room would be good enough to sit in," while handcuffs displayed in the room would be "confining."[4] More striking than Sutherland's desire that the ultimate display be able to bind and restrain its user was his wish to have the machine assume total control over the user's life; in arguing that the kinesthetic simulation should have a perfect fidelity to that which it represented, Sutherland suggested that "a bullet displayed in such a room would be fatal."[5]

Sutherland's idea for an electromechanical force-rendering mechanism capable of giving tangible materiality to computer-generated environments had been prefigured by research during the 1950s and 1960s on robotic remote manipulation systems. Funded by the U.S. military, engineers at the Argonne National Laboratory and, later, at General Electric (GE)[6] sought to extend the human body's capacity for action into distant environments using devices that mimicked the motions and actions of the human hands. In attempting to hone the functioning of these devices, they discovered that without feedback from the remote robot hands to those of the human operator, the devices functioned poorly. To address this shortcoming, they added force-reflection mechanisms that would allow the operator to feel an approximation of the forces encountered by the remote robot as it manipulated objects. Sutherland suggested that, rather than manipulating and sensing remote objects, the mechanisms controlling the production of force feedback sensations could be governed instead by a computer and then synchronized with computer-generated images and sounds to create the impression of interacting physically with objects. A growing awareness, prompted by the absence of kinesthetic feedback from early remote manipulation devices, that touch feedback sensations were crucial to the completion of everyday tasks animated research on force feedback systems. The project of electromechanically extending the body and its senses across space thus brought with it a new appreciation of touch, couched in increasingly technicist terms, along with a set of mechanisms for facilitating the transmission of touch feedback from remote object to operator. By the time Sutherland delivered his address in 1965, the extensive research into force feedback had laid both the conceptual and technical groundwork for touch's further reshaping in the design labs of computer scientists.

Beginning in the late 1960s, in what I designate as the experimental era of touch feedback computer interfaces, the computer scientists inspired by Sutherland's essay attempted to engineer a "computer display for the sense of feel,"[7] though they took careful steps to keep their devices from inflicting the damaging effects Sutherland hoped to achieve with his ultimate display. In 1967, at the University of North Carolina at Chapel Hill, Frederick Brooks and J. J. Batter began work on what they playfully titled Project GROPE.[8] Funded by the U.S. Atomic Energy Commission, Batter and Brooks used GROPE-1, the first in a series of sequentially numbered GROPE interfaces built from 1968 to 1990, to give computer-generated objects the illusion of having a physical materiality.[9] Working out of Bell Labs in New Jersey, A. Michael Noll took up a similar project also in the late 1960s, designing a force-rendering system that used a single point of contact between user and virtual environment to render shapes for the sense of touch. A University of California–San Diego research team[10] diagrammed plans for a pair of displays—"Touchy Feely" and "Touchy Twisty"—that would allow the computer to render position, torque, force, and orientation data of virtual objects, projected onto the user by a point of contact they dubbed the "Touchstone."[11] These early devices served more as proofs of concept than as revolutionizing inventions, prompting their designers to detail the potential utility of future, presumably more robust and accurate, force displays. The handful of experimental era publications on force feedback provided designers a space to articulate the strategic aims of their devices. Through their practice, engineers sought to simultaneously disrupt the conventional visualist configuration of computer interfacing and to bestow new powers on the sense of touch via computer augmentation. "Man-machine tactile communication," as Noll dubbed it,[12] provided the means to contest vision's dominance in the area of computer displays; the project of resisting ocularcentrism in the new medium of the computer became bound up with the successfully designing force feedback mechanisms. The desire to write touch into the realm of the image animated engineers' labor, as they sought to rebalance the technologized sensorium by building new computer interfaces.

In the early 1990s, as the field began to attract increased attention from those working outside its borders, touch computing researchers partnered with psychophysicists and cognitive psychologists who specialized in the study of haptic perception. The adoption of the designation "haptic interface" to describe the devices that resulted from research into touch feedback computing came about as a consequence of this interplay between disciplines. I therefore suggest that the terminological shift from force display to haptic interface indicated the importing of the technoscientific model of touch developed in the nineteenth century to computing. This model brought with it an epistemic framework for arresting, apprehending, and quantifying tactile processes that proved necessary to refining the design of touch feedback interfaces. At an institutional level, the project of designing haptic interfaces gave rise to the "new discipline" of computer haptics. Framed as "analogous to computer graphics,"[13] this new discipline provided a home for the growing number of researchers, scattered across a range of fields, who wished to contribute to the project of engineering touch computing interfaces. By the late 1990s, the bourgeoning interest in computer haptics fueled the formation of new research labs, professional organizations, conferences, and journals. Incorporating the term "haptics" into the names of these publications, gatherings, and associations quickly became standard practice, as haptics provided a shorthand for the progressive maturation and formalization of a new research paradigm while simultaneously indicating the field's wholehearted embrace of the technoscientific model of touch inherited from the nineteenth century. Computer haptics also brought with it a new institutional investment in uncovering, with ever-increasing precision, the parameters of haptic perception. Refining the techniques initially used in the nineteenth century to produce the mass haptic subject became central to forward progress in this new discipline. Innovations in haptic interface design depended on a cross-disciplinary network of researchers

dedicated to the constant solicitation, quantification, and aggregation of new technical knowledge about the microprocesses of haptic
perception.

Inspired by these interdisciplinary partnerships, haptic interface
designers throughout the 1990s rapidly invented new devices during what would later be dubbed the "epoch of haptic interface."[14]
These devices not only evidenced the possibilities of reconstruct-
ing touch using computing technologies but also highlighted the
challenges designers faced as they attempted to make touch analogous to the already-mediated senses of seeing and hearing. Rather
than synthesizing a natural and holistic touch, each interface
engaged in a piecemeal and selective reconstruction of the haptic
system by producing sensations targeted at a specific subcomponent
of touch. I illustrate this process of selective construction through a
comparison of several representative devices that each embodied
different paradigms of touch feedback interfacing. The devices
extended the human haptic system into virtual worlds selectively,
reflecting designers' strategic aims as they sought to productively
deploy digitized touch in a range of industrial, scientific, and medical contexts.

Throughout the chapter, while building on the themes of the
instrumentalization, rationalization, and disciplining of touch
raised throughout *Archaeologies of Touch,* I also highlight the operation of two interwoven narrative strands that gave structure and
form to the design process. The first concerns what I describe as
the *logic of analog medialization*—a narrative framing of changes in
the technologized sensorium that suggested that touch could
become like the technologically augmented senses of seeing and
hearing through the acquisition of its own mediatic apparatuses.
From the earliest writings on touch feedback computing, engineers
described force displays as analogous to the audio and visual displays
that transmitted data from the computer to the organs of seeing
and hearing. Here, I echo Lev Manovich's recent claim that the
early pioneers of computing can be read simultaneously as computer engineers and as media theoreticians. He further suggests
that they were also, at times, media historians.[15] As such, these
pioneers considered the media-historical implications of their

research, positioning their efforts in a broader continuum of inventions that sought the gradual and iterative extension of the human senses via electromechanical machines. Analogies between the operation of visual and haptic computer displays frequently reinforced this purported equivalence through the use of visual metaphors to explain the function of haptics. For example, the phrase "tactile pixel," or *taxel,* refers either to a discretely controlled element of a tactile display that can combine with other taxels to present an image for the sense of touch, or to properties assigned to a computational object that give it the illusion of a tangible materiality when rendered by a touch feedback device.[16] The technique of "haptography" later developed by Katherine Kuchenbecker, too, employs a visual logic to set an aspirational model for haptic interfacing. In an attempt to emulate painting's eventual displacement by photography as the predominant means of precisely and reliably capturing lifelike images, Kuchenbecker designed a complex system for capturing data about the haptic properties of computer-rendered objects that would be capable of producing a "haptic realism" akin to the camera's visual realism.[17] The outcomes of these projects are less important to my argument here than the explicit discursive framing of these efforts as attempts to emulate the operative technical logic of visual and audio media.

The second narrative strand involves the notion of a *master device*—an imagined endpoint of research on haptic interfacing, where a single apparatus would faithfully replicate the entire range of human haptic experience. The constant invocation of Sutherland's ultimate display implied a belief that the various mechanisms designers used to synthesize haptic experience would gradually unify, yielding a machine capable of assuming full control over the haptic system of its user. The final hegemonic design could then be added to the audiovisual components of the computing apparatus to form an audiovisual–haptic mediation machine. The narrative of the master device combines with the logic of analog medialization to provide a teleological orientation for design practice, where all haptic interface research pushes toward a predetermined and inevitable end. In later decades, proponents described the search for this master

device as the quest for a "Holy Grail" of haptics technology,[18] imply-
ing the eventual emergence of a unifying, standardized mode of
interacting haptically with computer-generated worlds.

Woven together, these two narrative strands provided a sort of
cohesive internal logic to the technical evolution of haptic interfaces that guided specific decisions made in the design process.
However coherent and self-evident this logic may have seemed to
the designers who embraced it, the analysis I undertake in this
chapter shows it to be impinged on by a range of external factors
that ensured this logic would be fraught with contradictions.
Enmeshed in a nexus of technical systems, the technogenesis of
haptic interfacing proceeded according not to an autonomous and
self-organizing logic, but was instead shaped by "the play of constraints imposed by the interdependencies between technical elements and those intrinsic to the system[s]."[19] Together, considerations
of computing power, the manufacturing costs of devices, the psychobiological limits of the human haptic system, and potential market
demand for new interfaces steered the development of each individual touch machine designed within the new discipline.

By providing a window into the complex dynamics at play in
the crafting of this new research paradigm, I want to portray both
touch and the computerized sensorium more generally as objects
contested and actively reshaped through the design process. The
technoscientific retrieval of touch by these early interface designers
involved not simply mobilizing a static conceptualization of tactility, but rather, entailed a wholesale reimagination and rearticulation of touch's social, psychological, and economic utility. Haptic
interface design gradually reordered touch both as a problem to be
solved through the discovery of new knowledge about human and
machine haptics, and as the solution to a set of problems associated
with visual displays. Viewed through the lens of a media archaeological analysis, the technical lineage of haptic interfaces becomes
not simply a progressive parade of devices advancing toward a predetermined end, but also an expression of designers' desire to grant
touch a new set of powers that would make it adequate to an
emerging economy of computerized information circulation.

Remote Manipulations

The early kinesthetic displays designed during the experimental era borrowed technically, materially, and conceptually from research carried out in the related areas of remote manipulation and exoskeletal robotics. In the first decades of the Cold War, those research streams received substantial injections of funds from the U.S. military, owing in large part to the need—prompted by the Manhattan Project—for new systems that would allow for the safe and dexterous handling of toxic materials (see Figure 4.1). As Blake Hannaford explains in his history of telerobotics, the goal of protecting workers from radiation was at odds with the need to manipulate radioactive materials into exact shapes.[20] In the initial set of systems, described by Hannaford as "essentially tongs that grew longer and longer,"[21] the lack of tactile and kinesthetic feedback made precise and controlled movements challenging. To address this problem, Raymond Goertz, an engineer for the Atomic Energy Commission working at Argonne National Laboratory, developed a set of mechanical arms that transmitted feedback to the tactile channel, via stainless steel cables and pulleys, from remotely manipulated objects to the operator. To allow the robot arms greater freedom of movement and to facilitate the manipulation of objects at longer distances, Goertz cut the steel cables running from slave to master, replacing them with a system for translating the mechanical forces applied by the operator into electrical signals that could be communicated via alterations in current to the motors in the remote arms.

Though this translation allowed the operator's body to be projected (however selectively) through space, in its initial form, the system was unidirectional—it did not project the space encountered by the remote manipulator back to the operator. The move to electrical control over the arm therefore entailed a loss of crucial information that had been provided, albeit imperfectly, by the earlier mechanical system. When working with hazardous and costly nuclear materials, such a loss of sensation meant the operator was more prone to make errors in manipulation that could prove costly and dangerous. Attempting to address this problem,

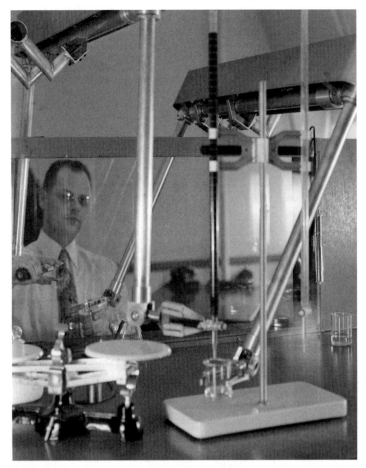

Figure 4.1. Raymond Goertz using an early mechanical manipulator.
Photo courtesy of Argonne National Laboratory.

Goertz introduced in 1953 what he termed "force-reflection" into
the robot arm that allowed the slave (robot arm) to communicate
information about the physical state of the manipulated object back
to the master (operator).[22] He sought to transmit the force experienced by the robot arm at a distance to the human operator
through the use of electrical current, thereby establishing a mimetic

relationship between the remote arm and the operator. Goertz accomplished this using the voltage of the remote arm as an indicator of the force it required to move objects; by applying proportional resistance to the unit moved by the operator, this system allowed the operator to feel what the remote arm felt. When the slave arm grasped a heavy object, the increased voltage required to lift it was transmitted to the operator's unit on the other end, where synchronized motors displayed the same force (see Figures 4.2 and 4.3). Dubbed the Argonne Remote Manipulator (or ARM),[23] the device enabled the operator to feel at a distance through the use of electrical signals, effectively facilitating the extension of the human nervomotor system beyond the limits of the operator's body.

Ralph Mosher, working for GE's Research and Development Center in the 1960s, incorporated similar force feedback mechanisms in designing a series of exoskeletal apparatuses that he intended to serve as "man-amplifiers." Inside a "Cybernetic Anthropomorphous Machine" (or CAM),[24] a human operator could "perform superhuman tasks, while remaining unharmed by the effects of his labor or dangerous working conditions."[25] In promotional documents for Mosher's CAMs, GE engineers positioned the machines as superior to those controlled by "computers and pre-programmed planning,"[26] because they offloaded perceptual and cognitive labor onto human operators, and by doing so, they left the robotic system with only the task of amplifying the physical rather than mental aspects of human labor. CAMs, in short, embodied GE's maxim "machines should work . . . people should think." As Mosher explained, the incorporation of kinesthetic and touch feedback systems was essential to accomplishing this cybernetic fusion of human and machine: a successful CAM "must be equipped with a kinesthetic sense corresponding to the human one."[27] An autonomous robot, absent "human sensing" (shorthand in Mosher's writings for the whole range of touch sensations), would make dangerous and damaging errors when performing even the most rudimentary of tasks. In a series of simplistic cartoons produced for Mosher's article, a hapless robot, "lacking human sensing," snaps a chair in two, rips a door from its hinges, and crumples a pipe (see Figure 4.4). Human figures juxta-

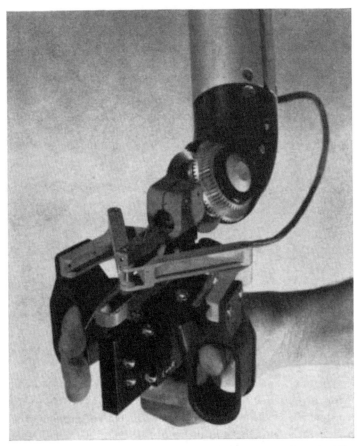

Figure 4.2. The master grip of an Argonne Remote Manipulator (ARM). The ARM provided force feedback to the fingers and thumb. The loose fit of the manipulator meant a loss of feedback response. From Ralph Mosher, "Industrial Manipulators," *Scientific American* 211, no. 4 (1964): 93.

posed with the robot performed each task with pleasurable ease, owing to the feedback mechanisms present throughout the human nervous system.

Building on the principles of Goertz's ARM, Mosher and GE trumpeted force feedback as the solution to this problem—the key

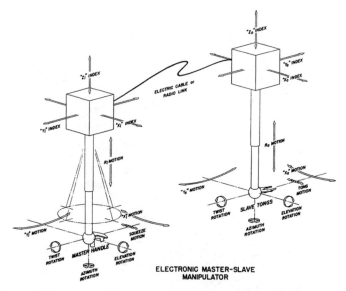

Figure 4.3. Diagram of Raymond Goertz's force-reflection system. From Raymond Goertz, J. R. Burnett, and F. Bevilacqua, "Servos for Remote Manipulation" (Lemont, Ill.: U.S. Atomic Energy Commission, 1953), 17.

to "making the machine a true extension of the man"[28] was to build an apparatus conditioned to the operator, where the forces encountered by the machine could be accurately and precisely projected back onto the human controller. The incorporation of touch feedback allowed the operator to feel as if the remote hands were actually part of his body; as explained in a promotional pamphlet for CAMs:

> machines of today can feed back sensory information to the operators thus retaining his "sense of feel" for the task being performed. By responding to this information he is able to comply with any interferences or restrain[t]s which occur and act as if he were performing the task with his own hands.[29]

With enough training, the link between operator and machine could become so tight that it would allow dexterous handling of

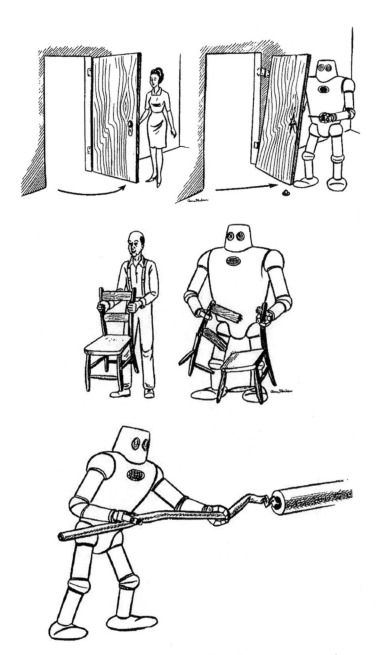

Figure 4.4. Mosher's hapless robot, lacking a haptic system, struggles with the performance of simple tasks, such as opening a door and lifting a chair. From Ralph Mosher, "From Handyman to Hardiman," in *Automotive Engineering Congress 1967* (New York: Society of Automotive Engineers, 1967), 5. Figures courtesy of miSci, Museum of Innovation & Science.

even the most delicate materials. One frequently circulated image (Figure 4.5) showed robotic tongs delicately gripping a light bulb, tacitly suggesting that the force reflection and spatial correspondence capacities of the CAM were accurate enough to allow the operator to apply the correct amount of pressure necessary to grab the glass bulb, while not providing so much force that the metal arm would crush the bulb.[30] As with learning to properly feel electricity, the disciplining required to register the differences between tactual stimuli in psychophysical studies, and the repeated drilling needed to master tactile languages, cultivating a skilled

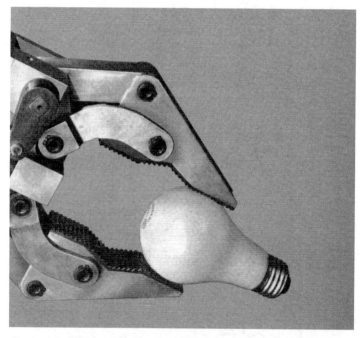

Figure 4.5. GE's force feedback–enabled robot arm delicately grips a GE light bulb. From General Electric Specialty Materials Handling Products Operation, "Cybernetic Anthropomorphous Machine Systems" (Schenectady, N.Y.: General Electric, 1968), 7. Photograph courtesy of miSci, Museum of Innovation & Science.

touching subject was essential to the successful deployment of these machines. However, GE's promotional materials often obscured the cognitive demands CAMs placed on their operators. For example, the "walking truck" prototype required its operator to assimilate to a complex set of controls to precisely move—and feel—the machine's four legs (see Figure 4.6). The company repeatedly

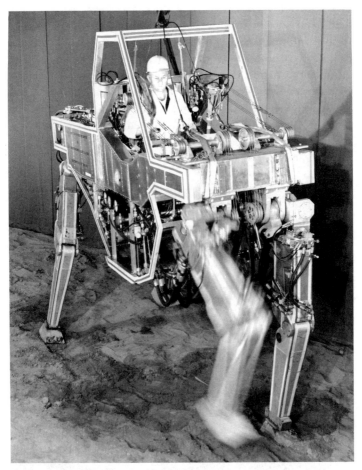

Figure 4.6. Mosher operating GE's walking truck prototype (1969). Photograph courtesy of miSci, Museum of Innovation & Science.

framed force feedback as a means of desubjectifying—a mechanism that would allow the operator to act as if the remote hands were his own—but in practice, the artifice of the jolts, jerks, and vibrations the operator was tasked with interpreting presented a nearly insurmountable challenge (Figure 4.6).

Investigations into force–reflection systems, funded by significant investments from the Advanced Research Projects Agency, the U.S. Office of Naval Research, and the U.S. Army, provided a conceptual framework for the emergence of force feedback computer displays in the subsequent decades. For both Goertz and Mosher, touch's initial absence in robotics systems provided an occasion to examine and articulate its neglected centrality, at a pragmatic level, to human action. The design of force feedback systems ordered touch as an engineering problem, capable of being addressed by building robotic apparatuses that mimicked, however imperfectly, the physical structures responsible for producing both movement and the resulting touch sensations (see Figure 4.7). Extending and

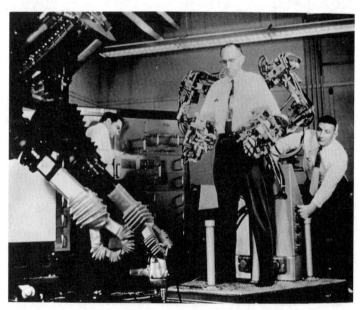

Figure 4.7. Mosher preparing the Handyman for use (1959).
Photograph courtesy of miSci, Museum of Innovation & Science.

amplifying the human body, which were necessary to make labor adequate to "the Space Age need for machines to do work in environments uninhabitable by man,"[31] prompted new and progressively more detailed examinations of human kinematics and the physiological mechanisms responsible for producing sensations of force (as in James Egleson's diagram of a human arm opening a door; see Figure 4.8). Improvements in remote manipulation systems and CAMs were measured by the degrees of freedom available to the machine compared to those of the human operator.[32] Interest in the physiological processes responsible for producing the perception of weight—a question Weber had taken up when he began his research on touch in the 1820s—acquired a new, utilitarian relevance, as engineers tried to bring the weight-registering mechanisms of remote systems into harmony with human discriminatory capacities.[33] Pioneers in remote manipulation proposed terms like the "'really there' index" to "evaluate the operator's direct sensing of the remote task and his identification with the remote hands as his own."[34] Subsequent research into haptic human–computer interfaces built on this conceptual foundation, steered by the imperative to allow humans to act not in distant and dangerous environments, but instead in spaces that existed only in the memories of computers.

At the material level, Goertz's ARM provided the nascent haptic interface for designers attempting to make good on Sutherland's vision with a tool for manual interaction in computer-generated worlds. In one of these early efforts, J. J. Capowski, a doctoral student at the University of North Carolina at Chapel Hill who was affiliated with Project GROPE, utilized a pair of Model E-3 ARMs donated to the university as a means of inputting three-dimensional positional information to a computer.[35] Capowksi's system, combined with a vector graphics display, allowed a single point of contact controlled by the ARM to move through three-dimensional visual space. As with the graphical component of the display, force feedback sensations were also controlled by computer software, but instead of transmitting forces encountered by the slave unit, the ARM manipulated by the user displayed force sensations calculated by an algorithm responsible for simulating the physical properties of virtual objects. I consider the full ramifications of this

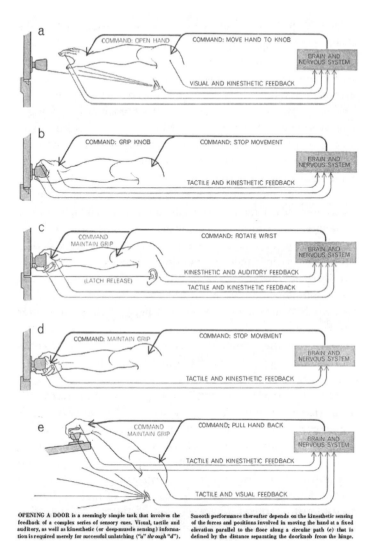

OPENING A DOOR is a seemingly simple task that involves the feedback of a complex series of sensory cues. Visual, tactile and auditory, as well as kinesthetic (or deep-muscle sensing) information is required merely for successful unlatching ("a" through "d"). Smooth performance thereafter depends on the kinesthetic sensing of the forces and positions involved in moving the hand at a fixed elevation parallel to the floor along a circular path (e) that is defined by the distance separating the doorknob from the hinge.

Figure 4.8. The sensorimotor processes involved in opening a door, divided up into discrete actions that could be replicated by a remotely manipulated robot. From Ralph Mosher, "Industrial Manipulators," *Scientific American* 211, no. 4 (1964): 91. Drawing by James Egleson.

new display paradigm in the next section. Here I highlight the way that engineering practice helped establish a direct technical lineage from remote manipulation to force displays, as computer scientists followed Sutherland's suggestion to appropriate remote manipulators for use as computing displays. While force displays became a discrete area of investigation throughout the 1970s and into the 1980s, engineers continued to borrow technology from older and considerably better-funded research into remote manipulation, leading Project GROPE investigator Ming Ouh-Young to conclude in a 1990 review of the field that "advances in teleoperators help greatly the engineering of a virtual-world force display."[36]

The Experimental Era: Interfacing against Ocularcentrism

While Sutherland's imagined "ultimate display" provided a fantastical endpoint for future computer scientists to work toward, his essay was strikingly short both on the conceptual implications of force displays and on the technical innovations that would be required to build the display function he envisioned. Those who set themselves to the task of making high-fidelity force displays a reality envisioned a new, computer-augmented tactility capable of reshaping both touch itself and the configuration of the electronically mediated sensorium. The engineers working on touch computing, projecting this interface technology several generations into the future, described potential applications that ranged from assisting the visually impaired to the feeling of textiles from a distance. They lauded the tactile channel for its superiority over the visual channel at collecting and processing particular types of information about the external world, lamented vision's dominance in the computer display paradigm, and praised touch's capacity to reveal characteristics of objects inaccessible to vision. In short, they treated force displays as an opportunity to positively deploy touch, and through this positive deployment, they sought to transform touch into a sense that could have use value in the emerging economy of computerized information manipulation. Tactile human–machine communication provided a means to rebalance a technologized sensorium increasingly thrown off-kilter by the rise to dominance of visual media. The computer display—still in a relatively protean state in the late 1960s and early 1970s—became a space where vision's inevitability could be contested and resisted.[37]

As with other countervisual strategies, experimental era writings on force displays valorized touch for its purported capacity to resist and overturn the deleterious effects of ocularcentrist mediatic orderings of the senses. For many of ocularcentrism's critics, disentangling the elevation of vision from the imaging apparatuses that aided in its ascent often proved impossible; critiques of visuality comingled with polemics condemning the mechanisms and logics responsible for image making.[38] But in several instances, though media technology caused the initial wound, it also provided the bandage. For example, McLuhan lauded television for fostering a "maximal interplay of the senses"[39] that was at its core primarily tactile rather than audiovisual. Laura Marks located a "haptic visuality" operating in a particular type of cinematic image—a mode of tactile seeing capable of serving as a "feminist visual *strategy*" that could supplant "phallocentric models of vision" founded on a distancing of perceiver and perceived.[40] Both authors praised the tactile subjectivity activated by a particular type of mediatic image (in McLuhan's case, televisual; in Marks's, cinematic) for its capacity to summon a neglected and forgotten orientation to the world. In these accounts, touch became synonymous with a corporeal immediacy—with the return of the body to the act of mediatic seeing.

Unsurprisingly, such countermodern and feminist themes were absent from the early scientific literature on force displays, and it is not my intent to suggest too close an alignment between the explicitly political positions staked by media theorists and the strategic, purportedly depoliticized accounts of touch offered by the engineers of kinesthetic displays. Nevertheless, Noll and Atkinson et al. imagined touch interfacing as a means of disrupting established perceptual conventions of interacting with computers. As society became increasingly dependent on computing, such a disruption would have substantial quasi-utopian consequences. These computing pioneers took for granted the imperative to fold the body into the mediatic ordering of the senses through the use of sensory prosthetics. Informed by the assumed inevitability of human–computer interfacing, they set to work designing new instruments that could help rebalance the technologized sensorium. By fusing the body more fully with machines, they hoped to grant new powers to a sense of touch neglected by imaging technologies.

Of the handful of engineers concerned with touch computing during its experimental era, Noll provided the most forceful and direct critique of visual interfaces. Before beginning his doctoral research into touch interfaces, he had worked as an acoustics researcher for Bell Telephone Laboratories.[41] Since the 1920s, when Gault founded the Vibro-Tactile Research Laboratory under the Bell umbrella, it had provided a home for technoscientific inquiries into touch's potential to serve as a communicative channel.[42] Throughout his dissertation, Noll demonstrated a keen awareness of his device's own historicity, positioning human–machine tactile communication in a lineage of other human–machine communication apparatuses. One of only a small handful of engineers working at the nexus of the visual, audio, and tactile modalities of human–computer interaction, Noll was uniquely qualified to recognize the technical challenges presented by each mode of information display. While tracing the genesis of computer displays brought him to acknowledge with seeming reluctance the value of visually presented information, a deep dissatisfaction with the state of the field motivated his research into touch communication. As he explained, "computer graphics has been given too much emphasis in its role as a form of man-machine communication,"[43] crowding out the search for other, perhaps more efficient, means of relaying information from the computer to its human operator. The growing interest among his contemporaries in real-time interactive stereoscopic displays provided a focal point for his critique of computer visualization. Users attempting to act on computer-generated images rendered by stereoscopic displays often had trouble "latching onto a line or object in three dimensions."[44] Force cues, he speculated, could aid users in their interactions with these stereoscopic images. By varying the resistance of the input device as it was manipulated, users could effectively "feel"[45] objects in three-dimensional space. The tactile channel, previously dormant in interactive stereoscopy, could come to life with the addition of force signals to computer-generated images. Though he imagined touch feedback computing as a display paradigm with far-reaching conceptual implications, Noll clearly understood his research into human–machine tactile communication "as a practical solution to a problem encountered in real-time interactive stereoscopy."[46]

Situated as both a pragmatic and epistemic response to the limits of data visualization technology, the simple force feedback cues employed by Noll's device allowed its users to perceive qualities of objects inaccessible through graphical displays alone. The users of Noll's interface manipulated a joystick, held with a ball mounted at its top, in three dimensions (see Figure 4.9). As the device captured the stick's movements, it displayed them visually via stereoscopic display. Contact with onscreen objects caused the custom-built software to apply force to the stick-mounted ball; three 8-bit numbers, one each for the x, y, and z axes, controlled the intensity of the applied force by regulating the voltage of three motors housed in the device. In testing intended to evaluate the device's utility, producing sensations for users allowed them, with the aid of stereoscopic displays, to identify the contours of different computer-generated shapes. However, Noll observed that many users would abandon the visual display altogether, opting instead to try to identify the shapes solely by touch. This inspired him to carry out a new battery of tests designed to measure his subjects' accuracy in shape perception without the use of visual information and without being told the range of shapes that would be presented to them. Most subjects were able to correctly identify cubes and spheres without the aid of the visual stereoscopic display, suggesting that the device had exceeded its intended functionality.

This practical finding had substantial conceptual implications. Though initially framed as a mere supplement to visual displays, capable of complementing and enhancing their functionality, these results prompted Noll to speculate that a computerized touch divorced from its dependence on visual cues had the capacity not only to allow for more efficient interactions between humans and

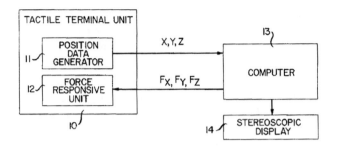

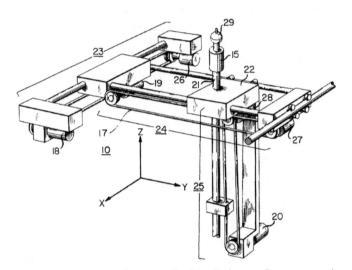

Figure 4.9. Patent image for A. Michael Noll's force-reflecting joystick. A. Michael Noll, "United States Patent #US 3919691 A — Tactile man–machine communication system," November 11, 1975.

computers,[47] but also to activate a mode of knowing that challenged, rather than merely confirmed, visual impressions. As a "competitor to the visual channel,"[48] the tactile channel had the potential to upend vision's dominance in computer interfacing by mobilizing a sensory epistemology with its own unique set of biases and predilections. "This situation," Noll claimed, presented "something new to the field of computer science." By the close of his thesis, he had revised his initial claim that tactile communication was merely a "supplement for computer-generated visual displays," concluding instead that it "emerges . . . primarily as an entirely-new man-machine communication medium or channel of vast importance for its own unique abilities to represent surfaces or objects."[49] Noll envisioned psychologists, empowered by a set of instruments that enabled new modes of touching, investigating long-held assumptions about the relationship between optical and tactual knowledge; "deliberately induced clashes and offsets between the tactile and visual communication channels" could upend older paradigms and generate useful knowledge about the way the brain resolved conflicting sensory impressions.[50] As with Volta's electromotive apparatus nearly two hundred years earlier, new strategic applications of electricity to the body held the potential to generate new knowledge about the operations of the senses and interactions among the sense modalities.

Working contemporaneously with Noll, Batter and Brooks similarly understood their GROPE-1 and its later iterations as the means to enable a better pairing between human and computer. Noting that "a fundamental problem in the design of man-computer systems is coupling the man and computer so that information transfer best utilizes the sensors, effectors, and natural action-patterns of the man," they recognized contemporary input and output mechanisms as inherently limited, due to their failure to make use of "the reflexes and natural nervous system functions of the man."[51] It was possible, according to Brooks, to "build yet more powerful tools by using more senses,"[52] and by doing so, to overcome the "relatively poor coupling mechanisms [of] keying and reading alphanumeric representations."[53] Thus for both Noll and the Chapel Hill investigators, touch feedback computing provided a strategic response to the limits of visualization technologies, aimed at a tighter fusing of human and computing machine.

The specific computerized worlds these new touch machines allowed their users to enter helped justify, legitimate, and valorize researchers' efforts. In both Touchy Twisty and Project GROPE, designers sought to make the interactions between molecules tangible and feelable through the kinesthetic display of force. Here, too, touch's capacity to apprehend physical qualities inaccessible to vision proved to be of paramount importance. The new technologies of what Atkinson et al. dubbed "computing with feeling" had the potential to instantiate a mode of knowing the physical world that simultaneously challenged and complemented visual impressions, as forces previously unfelt by human hands came within the tactile field of computer users:

> Chemists can already, with graphics displays, visually interact with molecular models. This, however, is not enough. Chemists also need to be able to manipulate these models, to touch them in order to more deeply comprehend their structure, to reach in and alter them, feeling the forces and torques involved, feeling the interactions between different parts of a molecule and between different molecules.[54]

Atkinson et al. understood touch machines as activating a fundamental corporeal knowledge of the physical world not present in graphical display systems alone. This new knowledge of microscale objects and forces could undermine the data provided through a visual inspection unaided by touch. In the decades that followed, proponents of touch feedback computing continually returned to this theme, consistently casting touch as a force that could, by conjuring a counterhegemonic sensory epistemology, work against the ocularcentrism of dominant mediation technologies in general and of computer graphics displays in particular.

Touch's ability to mount this challenge to the sensory ordering of extant interfacing technologies depended not on its simple folding into the computing apparatus, but on a more complex transformation in its constitutive abilities that experimental era researchers believed would accompany the reengineering process. Given suitable machines for facilitating its extension and amplification, a touch granted new powers would have the potential to usher in sweeping utopian changes in fields from biology and medicine to art and aesthetics.[55] Atkinson et al. presented a vision of touch liberated from its prior constraints:

> Touch is now only a proximate sense; we can perceive objects or forms by touch only if we are directly in contact with them. Thus we are restricted to sensing objects which share the general size range of our bodies, which exist in time long enough for humans to perceive them, and which, if we are wise, are non-injurious. Touch communication mediated by touch interface and computer can remove all these limitations, expanding our touch horizons beyond present limits.[56]

With the extension of touch across space facilitated by new tactile interfaces, the authors suggested that human bodies would no longer need to travel; the distant material world could be rendered as computer code and then reconstructed by touch interface, thereby eliminating the need to move bodies through space. Prefigured by research into remote manipulation systems in prior decades, they imagined touch communication technologies as providing a more efficient allocation of human and nonhuman resources: "why transport people, with the resultant waste of energy and human time; why not transport signals instead?"[57] Noll also believed that tactile human–machine communication held the potential to bring about something akin to "teleportation" with "the senses of vision, hearing and touch . . . extended over great distances."[58] The "separation of communication from transportation"[59] James Carey described as being initiated by nineteenth-century telegraphy extends in this model beyond data for the eyes and ears to encompass the human kinesthetic system. As I showed in Interface 1, the conceptualization of the nervous system as a series of electrical signals produced by the application of eighteenth-century electrical machines to the body always contained the promise of reducing all human sensations to the equivalent of pulses on a telegraph wire. But prior to the technoscientific reengineering and reimagining of touch prompted by Sutherland's essay, kinesthetic sensations had not generally been treated as part of this electronic liquidation of the sensorium.

Following this same logic, future touch machines also promised to allow the archiving of things and experiences as computer code, thereby permitting the searching, sorting, manipulation, and recall of touch experiences. As Atkinson and his colleagues explained, "computer memory can store touch experiences, catalog them, and make them available whenever wished."[60] Touch could not only reach out across space but could also reach back through time, to feel objects and bodies that no longer existed outside their abstraction as lines of computer code. Batter and Brooks also understood that touch interfacing allowed the experiencing of new sensations, as that which had previously been inaccessible to touch now came within its reach: "force fields that are microscopic, astronomical, or otherwise not reproducible on Earth."[61] Once these machines were sutured onto it, touch would acquire powers previously reserved for the senses of seeing and hearing, with the new instruments of touching framed analogously to prior perceptual augmentations. Or, as the University of California–San Diego team stated, "human–machine touch communication can serve as a touch telescope or microscope."[62]

Though these engineers may have sought to liberate touch from its dependence on vision—to allow for the technical expression of something akin to Larissa Hjorth's "logic of the haptic"[63]—their inventions and imaginations remained bounded within a set of three orienting assumptions associated with positivist empiricism and its related imaging technologies. The first assumption concerned the question of sensory subordination and the extent to which touch machines succeeded in undoing or reversing the subordination of tactility to the image. While Noll and Atkinson et al. understood kinesthetic displays as independent of computer graphics, for Brooks and the other Chapel Hill researchers, the image ultimately served as an anchor for kinesthetic cues. Informed by experiments where test subjects were asked to map different force fields using the GROPE-1 interface in conjunction with a visual display, Brooks speculated that "kinesthetic displays will be useful chiefly, and perhaps only, as adjuncts to visual displays."[64] Unlike Noll, who used his apparatus to call into question the assumed dominance of visual impressions over data from the other sensory channels, Brooks understood the primacy of the visual as a design principle. Guided by the maxim "when concurrent visual and kinesthetic cues conflict, the visual ones dominate," Brooks inscribed touch with a subordinate role in apprehending computer-generated objects.[65]

In his later critique of the rhetoric mobilized around virtual reality technologies, Ken Hillis charged that virtual reality's promise of a "corroboration among the senses" masks the extent to which "VR [virtual reality] privileges sight." Despite claims to the contrary, "touch or tactility in a VE [virtual environment] remains a very visible tactility" where "touch is made a proof of what has first been seen."[66]

The second assumption follows from Hillis's suggestion about the primacy—the first-ness—of the image in the design of virtual environments. Engineering touch sensations in the experimental era depended on the capacity to visually represent the force cues that would be generated by touch machines as lines of computer code—sequences of numbers that told the interface how much electrical current to pass through a motor in order to give the corresponding images "the semblance of a resistant materiality."[67] To be rendered for the user's kinesthetic system, touch sensations first had to be made visible on a computer screen, where they could be coordinated and synchronized with the visual and audio aspects of the display. Later "haptic effects" software represented touch sensations not as lines of computer code but instead through graphical depictions of vibration patterns or force effects that could be manipulated and reshaped, bringing about a corresponding set of alterations in users' tactile encounters with these devices.

Firmly entrenched in a positivist tradition that treated instruments as neutral conduits through which the natural world passed untainted, experimental era researchers labored under the third assumption: the belief that prosthetics, by aiding and augmenting the human senses, could compensate for any infirmities of the human's natural perceptual apparatus. However much the computer scientists and engineers writing in the experimental era wished to liberate touch from its subservience to external forces, the emergence of this ostensibly new machine-augmented haptic epistemology unfolded in a scientific context where instruments were thought to be, like the sense organs themselves, neutral conduits through which the natural world could be accessed. This enframing reflects a more general midcentury attitude embodied in the engineering of sensing technologies. As Caroline Jones explains: "modeled on the functionality of humans, machines stood for the positivist objectivity to which the modern human increasingly aspired."[68] Experimental era force displays established a tradition carried through to the future engineering of touch machines that linked these new devices to prior technologies for augmenting the human senses. Atkinson et al.'s newfound ability to reach in, alter, and feel molecules took for granted the microscope's status as a truth-production instrument.[69] Further, the assumed continuity across microscopy, computer-assisted vision, and human–machine tactile communication not only collapsed the differences between sense modalities but also obscured the differences between the earlier project of magnifying the natural world and the later one of making it pass through a computational filter.

In spite of these conceptual boundary conditions, experimental era researchers succeeded in establishing an autonomous new research trajectory around tactile human–machine communication that would be continually pushed forward by computer scientists in the subsequent decades. Latour's suggestion that "a new visual culture redefines both what it is to see, and what there is to see"[70] can be ported here to touch; tactile human–machine communication aimed simultaneously at a reinvention of touch and an attendant transformation in the objects that could be brought within the tactile fields of those using touch machines. However, though experimental era researchers may have sought to accrue new powers to touch through their efforts, they generally did not reflect critically on touch's constitution. That is to say, they took the psychophysical and psychobiological characteristics of touch to be relatively static. Extending touch outward primarily involved designing and refining the robotic devices that gave touch its new powers, writing software to control the robots' movements, and testing the efficacy of their inventions. While they occasionally dipped into the literature that explored the mechanisms responsible for touch perception[71] and acknowledged that touch machines functioned by deceiving the sense of touch, they did not investigate, as Geldard and his collaborators had in building their devices, the specific inputs necessary for that deception. In contrast, future progress in the engineering of touch machines involved turning the empiricist lens inward on the haptic system itself, so that the design of illusion-producing machines could build on a richer knowledge of touch's inner operations.

From Force Display to Haptic Interface

During roughly the first twenty years of research into touch feedback computing, investigators had interchangeably employed an array of terms to refer to their work. Kinesthetic display, tactile computer graphics, force feedback, tactile man–machine communication, physically tangible graphics, and force display were used synonymously to refer to a host of different techniques for rendering computer-generated information to the sense of touch. It was not until the late 1980s that the term "haptic" was introduced into the lexicon of interface design. Thereafter, the phrase "haptic interface" became a designation not only for the products of future research but also one projected back onto the research carried out in the preceding two decades. Computer scientists and interface designers rapidly adopted the new term, using it in titling journal articles, research centers, conferences, edited volumes, and devices.

At first glance, such a move might seem to be a question of mere semantics, worthy of nothing more than a footnote. However, considered from a genealogical perspective, the shift in terminology from "force display" to "haptic interfacing" acquires an increased significance. It reflects a crucial point of contact between computer science and cognitive psychology, where computer scientists found that their ability to produce effective systems for transmitting information though touch was constrained by a lack of knowledge about the biological and psychological aspects of touch perception. Cognitive scientists who studied haptic perception recognized this gap as well and attempted to bring their findings to bear on the emerging and intertwined fields of robotics and information display.[72] Susan Lederman, a cognitive psychologist who had published extensively on haptic perception, was perhaps the most outspoken in arguing that engineers could benefit from exchanges with those in her field. Lederman suggested that, by not accounting for the psy-

chophysical mechanisms responsible for haptic perception, interface designers and roboticists were limited in their abilities to faithfully replicate and synthesize the human haptic system.[73] Together
with fellow cognitive psychologists Roberta Klatzky and Jack Loo-
mis, Lederman frequently provided quick primers on "the scientific
approach to studying biological sensing"[74] intended both to help
orient neophytes to her field and to illustrate the practical benefits
that would accrue to computer scientists and roboticists willing to
reach beyond their disciplinary boundaries. A detailed knowledge
of the biological "subsystems" responsible for haptic perception,
such as "the mechanoreceptive transduction units" that receive
"signals through the mechanical medium of the skin,"[75] would
enable roboticists building virtual and remote sensing systems to
conform their designs to the parameters of those human operators
the systems were expected to merge seamlessly with. Lederman
answered the question of "what is touch good for?"[76] by pointing to
the growing body of well-developed research that provided empirical evidence of touch's use-value.[77] She also specifically encouraged
the adoption of haptics as the preferred terminology for describing
investigations in this burgeoning field, keenly aware that it implicated these studies in a tradition that included the research David
Katz and J. J. Gibson had carried out earlier in the century.

This process of importing haptics to the practice of designing
force displays had four interrelated consequences for the nascent
field. First, it provided a unified designation for the emerging
research carried out in touch computing, effectively becoming an
umbrella for researchers to gather under. Second, it brought to
information display a body of extant findings about the psychophysics of touch perception, recalling Titchener's 1900 definition
of haptics as the "doctrine of touch."[78] Similarly, the rise of haptics in the late nineteenth century had been accompanied by a set
of scientific techniques, also part of the same "doctrine," for further refining empirical knowledge about haptic perception. This
extant collection of investigative methods and instruments provided future interface designers with the means to isolate and
study specific components of the haptic system on an as-needed
basis. Third, importing haptics to the design of force display devices

established symbiotic partnerships, both at the individual and institutional levels, between cognitive psychologists and computer scientists that would serve as the groundwork for haptic interface research in the coming decades.[79] Fourth, this generative encounter between computer science and psychology provided interface designers with an appreciation of touch's immense complexity that proved simultaneously intriguing and daunting. As they probed the vast literature on the psychophysics of the haptic system, these designers quickly realized the substantial difficulties they would face in attempting to replicate its functions.

With research into force displays gaining steam in the 1990s, during what Hiroo Iwata later termed the "epoch of haptic interface,"[80] the field devoted to the study of haptic perception became increasingly intertwined with the engineering practice of building haptic illusion machines. This arrangement quickly proved beneficial to experimental psychologists interested in touch. As Geldard observed a half-century earlier, in psychology, the study of touch had taken a back seat to studies of seeing and hearing (a lament echoed frequently by Lederman), and the merging of haptics with information display provided psychophysical research on touch with a pertinence and practical utility that had previously seemed persistently beyond its grasp.[81] In computer science, the expanding body of works that constituted the canon of force display swelled to encompass foundational texts in the science of touch; suturing their new field to a more established and tested one gave haptic interface design a legitimacy it lacked on its own.[82] Throughout the 1990s, engineers set to work revising their genealogies of haptic information displays. By the decade's close, they located their research in a broader continuum that situated Ernst Heinrich Weber's publications on the skin senses as "the first significant scientific work on touch."[83] In the wake of this partnering, the terms "haptics" and "haptic interface" gradually came to be used synonymously, as the science of touch and the devices that electromechanically extended it were increasingly conflated in both the popular press and technical discourse.[84] In their 1997 review of progress in haptic interface design, Srinivasan and Basdogan offered the following definition, suggestive of the two fields' progressive intermingling:

> Haptics refers to manual interactions with environments, such as exploration for extraction of information about the environment or manipulation for modifying the environment. These interactions may be accomplished by human or machine hands and the environments can be either real or virtual.[85]

Moreover, "haptics" possessed an exotic connotation absent from the more familiar "touch." As roboticist Thomas Sheridan noted at the Workshop on Human–Machine Haptics in 1997, "haptics is not a term well-known to the public."[86] The cybercultural moment of the 1990s, when neologisms that each seemed to portend a coming technocultural revolution spontaneously generated at the blurred line between science and literature, amplified the mystique that enveloped haptics. Although the term had been used in specialist scientific and aesthetic discourses for roughly a century, it nevertheless retained a neologistic feel that allowed it to blend seamlessly into increasingly common conversations about virtual reality. Nowhere was this more evident than in Howard Rheingold's best-selling 1991 book *Virtual Reality,* which featured accounts of the author's hands-on experiences with haptic interfaces alongside interviews with their enterprising designers. He borrowed the title of the book's first chapter, "Grasping Reality through Illusion," from a 1988 article by Brooks, and the "conversion experience" that transformed Rheingold into a proselytizer for virtual reality came about as a result of "directly maneuvering two molecules" with his hands using the GROPE-III (a modified ARM) at Brooks's Chapel Hill lab.[87] As Rheingold described it, the virtual reality paradigm rested on a promise to bring the body more fully into computer-generated worlds through the digital synthesis of a wider range of senses. What demarcated virtual reality from prior mediation schemes was its capacity to generate illusions that could trick the tactual senses, along with the senses of seeing and hearing. Frequently throughout *Virtual Reality,* Rheingold and his informants mobilized this logic of analog medialization, where touch would become like the mediated senses of vision and hearing through the acquisition of electromechanical prostheses.[88] However, as a designator, "touch" failed to adequately capture the scope of the transformation in the sensorium instantiated by virtual reality. In contrast, "haptics" suggested and embraced touch's transformation through the application of scientific methods; it marked off the novel model of touch—as an information-reception system capable of being deceived through the proper application of stimuli—from previous instantiations. The substitution of "haptics" for "touch" can be read as part of a defamiliarization strategy, intended to signify a rupture between the sensory orders of the past and those of the coming age being realized in the labs of virtual reality designers.

Within this newly adopted institutional and epistemic frame of the haptic, researchers proceeded to segment the study of touch into increasingly specific categories intended to accelerate the pace of knowledge production. Mandayam Srinivasan, who founded the Laboratory for Human and Machine Haptics at MIT in 1990 (less formally known as the "Touch Lab"), divided the research carried out in his lab into the categories human haptics, human–machine haptics, and machine haptics,[89] an explicit recognition of the interdependencies between the investigative tracts.[90] A subfield of human–machine haptics, Srinivasan and Basdogan defined the "new discipline [of] *Computer Haptics* [as] concerned with generating and rendering haptic stimuli to the human user, just as computer graphics deals with generating and rendering visual images."[91] The logic of analog medialization here implied a particular, deterministic path forward for the new enterprise. Noting that "rapid improvements in computer graphics hardware" had "enabled the generation of photorealistic effects, animation, and real-time interactive simulations," they expected "similar rapid progress to occur in computer haptics."[92] But with these new divisions, the logic of analog medialization took a curious, though familiar turn, as the expected rapid progress forward in computer haptics necessary for the field to generate useful and commercially viable products depended on intensifying the study of haptic perception itself. The search for the basic parameters of haptic perception—the discovery of the "human abilities and limitations" that set "the performance specifications of haptic devices"[93]—became a fundamental precondition for the field of computer haptics' further development. If haptic interfaces depended on the capacity to trick touch into believing in the presence of what was absent (that is, on the creation of "haptic illusions"), then haptic interface design depended, like the design of prior media that encoded data for the eyes and ears, on discovering the specific conditions under which touch could be deceived. To produce haptic media that could function like the optical and acoustic ones they were modeled after, haptic interface engineers had to retrace the same steps taken in the design of those prior sensory deception apparatuses. Transforming touch into an analog of vision and audition meant evolving it not only through the comparatively simple suturing of prostheses onto the body, but also by mobilizing a technoscientific knowledge production regime aimed at making haptics, as the science of touch, adequate to the "urgent . . . need for more effective interactions between human and computers."[94]

Thus the transition from force display to haptic interface mobilized not just a language of scientific touch but also an accompanying host of experimental methods for arresting, apprehending, and quantifying touch's psychobiological processes. In embracing the legacy of touch's nineteenth-century reshaping, these computer scientists devoted themselves to an ongoing search for the fundamental laws that underpinned and governed haptic perception, committing to a reaggregation of the haptic subject that would continue for as long as interface designers quested after new and more accurate simulation systems. One hundred years earlier, experimental psychologists used the term "haptics" to indicate touch's passage, via the psychophysicist's lab, into a scientific modernity from which it had previously been denied entry. The proselytizers of this new science declared that touch had acquired a "doctrine" akin to those of sight and hearing. In the 1990s, the new discipline of computer haptics called on this doctrine in a self-conscious effort to push touch, however belatedly, into a mediatic modernity defined by the capacity of electromechanical stimuli to delude the sense organs.[95]

Where experimental era research into force displays proceeded with little reference to the vast psychophysical literature on touch, the terminological shift to haptic interface in the 1990s signaled the importing of a particular conceptualization of touch that quickly came to occupy a hegemonic status in the field. One of the first monographs dedicated solely to force feedback—Grigore Burdea's 1996 *Force and Touch Feedback for Virtual Reality*—opened with a detailed rehearsal of the physiology of touch, where he described the neural mechanisms and skin receptors involved in various tactile processes (including an analysis of just-noticed difference capacities for the tactile system).[96] By providing an origin story for designers, the founders of the science of touch became the forefathers of haptic interface design. Engineers began referring directly to the experiments carried out by Weber, Charles Sherrington, Katz, Max von Frey, and other central figures, positioning these experiments as essential precursors to their own research. In Heideggerian terms, haptic interface designers treated this body of preexisting knowledge about touch, and more broadly, the human haptic system in general, as the "standing-reserve" of haptic interfacing.[97] The move from force display to haptic interfacing mobilized the haptic's earlier technoscientific ordering of touch as an exploitable resource. Touch's prior reconstruction in the experimental scientist's lab allowed it, a century later, to be deployed productively in the needs of a science of information circulation. As with Jonathan Sterne's listening subject, Anson Rabinbach's human motor, and Crary's attentive subject, identifying "the relevant psychophysical parameters of the human operator"[98] became a matter of extreme urgency. Quantifying performance, measuring task completion times, and minimizing error rates emerged as prime directives in an instrumentally oriented study of perceptual processes intended to lay the groundwork for the design of new sensorimotor apparatuses. Future research and engineering practice therefore functioned to deeply inscribe this hegemonic notion of touch, along with its attendant subdivisions of haptic perception into its component sensations, in the material configuration of touch feedback interfaces, all along assuming that touch passed unaltered through the technoscientific filter of the haptic.

The Technogenesis of Touch during the "Epoch of Haptic Interface"

In the 1990s, fueled in part by advances in computer processing power and financial investments in emerging virtual reality technology, research into computer haptics produced a host of new devices. During this "epoch of haptic interface,"[99] engineers experimented with different configurations of physical contact between bodies and computer-generated environments. Guided by the drive to discover ever more efficient haptic interfacing schematics—itself part of an overarching push to facilitate heightened engagement with computers and the virtual worlds they animated—these engineers devised batteries of tests to quantify any gains in performance won by their inventions. Informed by close readings of the growing body of literature on haptic perception and by the partnerships with cognitive scientists discussed in the preceding section, interface designers were poised to develop devices that could bring about the transformative changes imagined by their experimental era predecessors. As a result of their efforts, the touch machines engineered and tested in research labs began gradually migrating into various commercial sectors, finding practical application in the areas of computer-assisted design, surgical training and medical simulation, museum display, assistive technology for the visually impaired, military simulation,[100] videogaming,[101] and teledildonics.[102] Like other fields of computing, and cybernetics more generally, engineering touch feedback as a technical object involved the enclosure of humans within the broader social feedback loop of the design process, where user experience was solicited through the careful construction of iterative testing protocols.

In attempting to facilitate modes of interacting with computer-generated worlds that would purportedly be more natural, designers embraced the fragmentary and subdivided model of touch designated by haptics. Doing so proved necessary for touch's successful extension into virtual worlds. To arrest and quantify the organic process of touch perception, psychophysicists in the late nineteenth and early twentieth centuries developed instruments for stimulating specific submodalities of touch. These different instruments provided crucial knowledge about the skin's sense of spatial resolution, the nerve structures responsible for the perception of temperature, the mechanisms that produce sensations of pressure, the receptors that yield sensations of pain, and the skin structures that enable feelings of vibration. Informed by and building on this paradigm, designers had at their disposal a whole range of techniques for artificially inducing tactile simulations, as well as a set of measurement practices that would allow them to assess the extent to which a given device succeeded at generating a convincing set of haptic illusions.

Part of the labor of haptic interface design, then, involved deciding which of the various haptic sensations the interface should be capable of inducing. Though such decisions were, in principle, guided by the overarching aim of producing a haptic simulation faithful to the reality it synthesized, in practice, designers faced three sets of constraints that made generating a one-to-one experience of haptic space not only impractical but also undesirable. Out of this "play of constraints"[103] emerged a whole range of competing designs, each of which fell short of producing a device that accurately synthesized the full range of human haptic experience. The limits of microprocessor power, system latency speeds, and other technical factors made up the first set of constraints, as the extant state of computing technologies imposed boundaries on the range of haptic sensations that these systems were capable of producing. As with the computational capture, storage, transmission, and playback of data for the eyes and ears, the accuracy of sensations rendered by the machine depended on engineering algorithms capable of working efficiently within these constraints.[104] The second set of constraints, which bled into the first, involved the neurophysiology of haptic perception itself. The complexity of the physical mechanisms responsible for receiving, coordinating, and assembling touch sensations made constructing devices capable of precisely stimulating them difficult, and closer inspections of the human haptic system served to heighten awareness of challenges its structure posed. For example, the close clustering together of mechanoreceptors[105] in the skin made the targeted stimulation of individual receptors difficult and placed burdens on the attentive capacities of the users tasked with differentiating between stimuli. Taken together, these psychobiological constraints at times seemed to lend credence to the enduring suggestion that, through some essential character, "touching resists virtualization,"[106] as frustrated designers pressed the haptic system to disclose its secrets in increasing detail. Finally, the third set of constraints concerned the need to find practical, commercially viable applications for the products of haptic interface research; considerations of usability, of marketability, of ergonomics, and of large- and small-scale production costs gave structure to the specific interfaces designers built. Robust systems capable of providing highly accurate and detailed haptic renderings proved costly to manufacture, while interfaces with lower production costs typically provided less-than-convincing haptic illusions.[107] The strategic reconstruction of touch that interface designers undertook thus required them to navigate three interrelated economies simultaneously: economies of computational information processing and transfer, economies of the neuromotor system as revealed through lab study, and economies of capital investment and market demand.

The various haptic interfaces built both during and subsequent to the epoch of haptic interface can, therefore, be read as contingent outcomes of design processes, formulated as a means of navigating the technical, biological, and economic constraints that engineers faced. Each interface embodied a disposition toward one model of technologized touch over competing sets of models, configuring the range of tactile sensations the interface was capable (and, by extension, incapable) of synthesizing. These interfaces also worked to further inscribe the subdivided model of touch at both the software and hardware levels, with each of the various subcomponents of haptic perception—including the mechanisms responsible for sensing texture, movement, weight, pressure, and temperature—acquiring its own unique pairing of haptic algorithm and interface. The divisions between haptic modalities were specified in the engineering process and encoded at the material level in the configurations of both haptic interfaces and the algorithms that governed their movements.

I illustrate this dynamic technogenesis first by showing how designers built machines that intentionally avoided the infliction of pain, and second, by contrasting several representative devices produced during the epoch of haptic interface.[108] Examining the specific considerations that informed design processes shows haptic interfacing to be a strategic and ideologically loaded orchestration of machinic sensations. It is the outcome of an arduous technoeconomic selection process where designers chose which haptic sensations ought to be reproduced by the interface, and then assembled these sensations together in a necessarily incomplete and unfaithful simulation of touch. Haptic interfacing is not, then, simply the value-neutral reconstruction of a strictly psychobiological system. Nor is it reducible to a reading that takes the body more generally as always already a meeting point of the transcendental/biological and the historically specific/technical, as Mark Hansen argues in his analysis of mixed reality systems.[109] Of all the design imperatives that guided engineers, none was adhered to more rigidly than the dictate to avoid inducing pain or discomfort in the user. In selecting among the various haptic sensations their devices would induce, they actively excluded a range of sensations and haptic effects that were crucial in Sutherland's initial formulation of the ultimate display. Haptic interfaces effectively functioned to shield users from precisely the sorts of dangerous and life-threatening sensations Sutherland hoped the display would one day produce. Where handcuffs in Sutherland's display would be confining, and the bullets it displayed fatal, the twin imperatives of pain avoidance and comfort maximization steered design during the epoch of haptic interface. Because pain and discomfort were thought to cloud the user's capacity for accurately assessing information taken in by touch, the infliction of these sensations was to be minimized. Outlining a normative set of best practices for haptic interface design, Srinivasan and Basdogan suggested that "making the human user comfortable when wearing or manipulating a haptic interface is of paramount importance, since pain, or even discomfort, supersedes all other sensations."[110] This design principle was embedded in the material configurations of the various interfaces built during the epoch. Interfaces therefore served as a type of sensory armoring, preventing their users from experiencing the painful aspects of virtual worlds while enabling them to experience those sensations thought to be pleasurable or useful. Weber and the psychophysicists who followed in his footsteps designed their instruments with a similar constraint. Except in rare instances where the study of pain was an aim of the experiment, they too sought to avoid the infliction of pain, similarly assuming that pain overrode the discriminatory capacities of touch's other submodalities.[111]

While pain avoidance and discomfort minimization were intertwined in the design literature, it is worth separating them here. The assumption that discomfort and pain were coequal in their psychobiological capacity to "supersede" all other sensations would not have been one shared by earlier experimenters, or in culture more generally only a century earlier. As John Crowley argues in *Invention of Comfort,* the notion of comfort emerged as primarily a flexible social category, rather than a hard-coded biological one, linked to bourgeois norms of bodily comportment.[112] More specific to the experimental tradition, autoexperimental investigations into sensory perception frequently required experimenters to subject themselves to trials that seem to our modern sensibilities to involve an impossible amount of discomfort. In spite of those conditions, through disciplined attention, they were able to accurately discriminate between fine differences in stimuli. My point here is to suggest that the imperative of comfort maximization was informed as much by economic considerations as it was by psychobiological ones. Comfortable systems were simply more likely to be used, and therefore, more likely to succeed in the emerging commercial market for haptic interfaces.

Following the twin dictates of pain avoidance and comfort maximization ensured that the haptic simulation written onto users' bodies would always be unfaithful to that which it represented, suggesting an incommensurability between audiovisual and haptic realism. While the teleology of the master device ostensibly implied that haptic interfaces would strive for a full replication and synthesis of the haptic system, in practice, designers took steps to avoid reaching this end. For example, the VR Thermal Kit (Figure 4.10) used the temperature sense to create the illusion of moving through a three-dimensional sense. Employing three computer-controlled fans positioned in a triangle shape around the computer monitor, the VR Thermal Kit blew heated or cooled air onto the user. In conjunction with an element for heating and cooling placed directly against the skin, the VR Thermal Kit allowed the users to feel, for instance, the warmth coming from a virtual fireplace and to coordinate their haptic impressions of the computer-generated world with those derived from the graphical display.[113] But the VR Thermal Kit could not, by design, raise or lower temperatures so much that it would prove uncomfortable or damaging to its user; walking directly into the fireplace would not result in

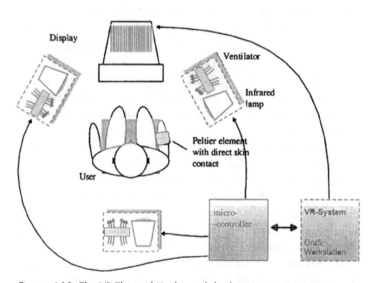

Figure 4.10. The VR Thermal Kit thermal display system. From Jose Dionisio, Volker Henrich, Udo Jakob, Alexander Rettig, & Rolf Ziegler, "The Virtual Touch: Haptic Interfaces in Virtual Environments," *Computers & Graphics* 21, no. 4 (1997): 459–68. Copyright Elsevier (1997).

the heating element becoming so hot that it would burn the skin. As long as the dictate to keep interfaces from inflicting harm was obeyed, haptic simulation systems would remain necessarily unfaithful to the virtual worlds they allowed users to enter.[114]

While there was broad agreement that interfaces should exclude the infliction of pain, on which sensations to include, however, no such consensus existed. The types of haptic experiences interfaces produced varied wildly, with each design paradigm combining, coordinating, and cuing the subcomponents in different ways. As one designer observed, "real-world haptic interaction typically involves many independent, or nearly independent, forces acting together."[115] The task of producing convincing haptic illusions entailed masking a given device's representational limits. By weaving together the correct subset of haptic sensations, designers could distract attention from those sensations that were absent. Arriving at this optimal subset required an iterative trial-and-error engineering process in which user and haptic simulation machine were incrementally brought into harmony with one another.

Here, the PHANToM provides a paradigmatic case study to illustrate the way the economic, the biological, and the technical became imbricated through the design process. Engineered at MIT's Artificial Intelligence Laboratory in 1993 by Thomas Massie and Ken Salisbury and funded by a grant from the Department of Defense, the PHANToM employed a single point of contact between user and virtual environment to render sensations of force and contact for its users. In the initial design, this point of contact took the shape of a "thimble-gimble"[116] that the user inserted the tip of their index finger into. Attached to a robot arm that used three motors to both track its movements and apply force sensations to the user, the thimble-gimble could simulate the weight, texture, and stiffness of objects in three-dimensional virtual space. The robot arm's range of motion effectively comprised a defined haptic field for its user to operate within. In doing so, it synthesized a limited subset of haptic sensations that reflected Massie and Salisbury's assumptions about which touch cues would be useful to the device's intended audience (see Figure 4.11). In a 1994 presentation where they justified the specifics of the PHANToM's configuration,

Figure 4.11. Original version of the PHANToM. From Thomas Massie and Kenneth Salisbury, "The PHANToM Haptic Interface: A Device for Probing Virtual Objects," *Proceedings of the ASME: Dynamic Systems and Control Division 55* (New York: American Society of Mechanical Engineers, 1994): 300. Photograph courtesy of Kenneth Salisbury.

they explained that "force and motion are the most important haptic cues . . . even without detailed cutaneous information (as with a gloved hand or tool), the forces and motions imparted on/by our limbs contributes significant information about the spatial map of our environment."[117] Thus the PHANToM's inability to generate cutaneous sensations embodied a strategic approach to the artificial synthesis of touch, as this subset of the haptic sensations was given a lower priority than the feelings obtained from kinesthetic perception.

The reduction of the hand to a single point of contact also reflected a prioritization scheme, as Massie and Salisbury based their decision on a logic of efficient programming and engineering; rather than having to track and render multiple points of contact between user and virtual world, this single-point system lowered the demands placed on the computer's hardware and software while keeping manufacturing costs down and placing less of a cognitive load on users. The PHANToM entailed "an effort . . . to achieve an effective, affordable, force-reflecting haptic interface with existing technologies."[118] The engineering processes ordered the haptic system as an economy of sensations that could be efficiently managed by the interface. For instance, Massie and Salisbury reported that most users could be "convinced that a virtual surface with a stiffness of at least 20 Nt/cm [newtons per centimeter] represents a solid immovable wall."[119] Using an algorithm that controlled the electrical current supplied to the device's motors, the PHANToM could apply up to 35 Nt/cm, and by doing so, convince its users that they had come into contact with a solid object. The discriminatory capacities of the average user, determined through repeated psychophysical testing, provided the device's designers with a functional knowledge about touch that allowed them to adjust the machine's technical specifications to the intended user. The capacity for haptic discrimination (for example, the threshold at which the user experiences a sensation as indicating the finger's contact with a solid surface) was mapped onto the design of the interface hardware and embedded in the coding of the software that controlled it. Through repeated experimentation, carried out over successive iterations in the design process, these thresholds were *produced,* and this body of knowledge about the specific capacities of the various haptic subsystems gained positive value in helping engineers craft efficient mechanisms for circulating data through touch. Srinivasan's schematic diagram of the interaction between human and machine haptics (see Figure 4.12) did not merely represent the components of two systems that met at the site of haptic interface. It also provided a blueprint for the networking of scientific actors that would be required in order to transform touch into a sense that could have its own media of storage, transmission, and retrieval.

In contrast to the PHANToM, the CyberGrasp (see Figures 4.13 and 4.14) captured the movements of the whole hand and then

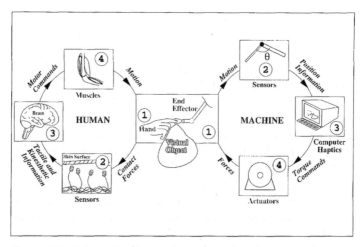

Figure 4.12. Schematic of human haptics, machine haptics, and their interaction. Copyright 1996 Mandayam A. Srinivasan, The Touch Lab, MIT.

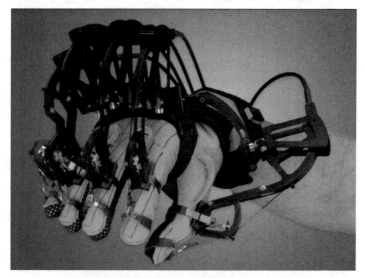

Figure 4.13. The CyberGrasp and CyberForce haptic interfaces. Pictures provided by CyberGlove Systems LLC.

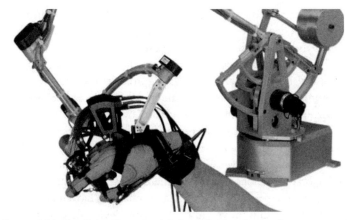

Figure 4.14. The CyberGrasp and CyberForce haptic interfaces.
Pictures provided by CyberGlove Systems LLC.

rendered force sensations that could be targeted to the hand's individual digits. Developed by Virtual Technologies Inc. with funding from the U.S. Office of Naval Research, the CyberGrasp
merged a glove that sensed the position of the user's fingers with a
cable-driven exoskeleton that applied force to the individuated
fingers. When squeezing a ball in virtual space, the user would feel
the contact between their fingers and the ball, as the exoskeleton
applied force separately to each digit. The CyberGrasp rendered
virtual objects it encountered in computational space as force data
displayed to the whole hand, whereas the PHANToM projected
force onto the user only through a single point of contact between
the hand and the virtual environment. The CyberGrasp therefore
proved adept both capturing the hand's position in virtual space
(twenty-two sensors distributed throughout the glove provided the
computer with a detailed image of the user's movements) and at
facilitating the perception of virtual objects' shapes and density (the
exoskeleton rendered 12 newtons of force per finger). Unlike the
PHANToM, however, it could not simulate the weight of these
objects. Furthermore, the device's ability to capture the movements
of the digits in space and to specifically target force sensations to
each digit individually brought with it a cumbersome materiality.

Weighing in at 400 grams and connected by six cables to the computer that controlled its operations, long sessions interfacing with the CyberGrasp often fatigued the user.[120]

Because the devices each facilitated the perception of different types of haptic sensations, each required its own idiosyncratic methods for capturing, storing, and retrieving data. The model of a virtual object intended for rendering by the PHANToM could not simply be translated for reproduction by the CyberGrasp; the former pushed a complex set of force sensations through the single point of contact between user and object, while the latter distributed force sensations across five different points of contact, each of which required a unique bit of data to control its movements. Haptic rendering thus proved to be "a highly device-dependent process,"[121] where the technical specifications of the device set the parameters for the types of haptic information about the virtual world that could be stored and transmitted. The virtual environments imprinted in the computer's memory therefore bore the stamp of the nineteenth-century subdivision of touch into component sensations.

In spite of their differences, the interfacing paradigms embodied by both the PHANToM and the CyberGrasp prioritized the hands as the locus of haptic experience. In contrast, other systems took feedback focused on the hands to be limiting and responded by pushing "towards full-body haptic feedback."[122] Similar to the VR Thermal Kit, full-body systems like the TactaVest sought to create a sense of haptic space by projecting sensations onto the torso. However, even these "full-body" systems, in spite of their aspirations to totally envelop users in computer-generated haptic fields, fell short of this aim, involving instead strategic decisions about which body parts would be targeted by feedback systems. Typically, "body" functioned as a convenient shorthand for the torso, although some systems merged feedback mechanisms targeted at the torso with those that stimulated the hands and feet.[123] In addition to selecting which of the subcomponents of the haptic system a given interface would reproduce, designers also chose which skinspaces to project machine-generated sensations onto.

Considered against this backdrop, haptic interface design can be read as a normative process of selecting which objects, forces, and contacts could and should be rendered as sensations, and which were to be left intangible to computer users. Haptic interfaces function as sites of strategic inclusion and exclusion, embodying preferences about which constitutive elements of touch ought to be reproduced by a given interface. The engineering of haptics technology involves a wholesale reconstruction of touch that must necessarily be incomplete. The touch that emerges out of this engineering process can only ever be a set of fragments, a sampling of sensations stitched convincingly together according to the dictates of the design process. I suggest, therefore, that "touch" provides too general a designation for this reconstituted haptics. It is not enough to say that haptic interfaces extend touch itself into virtual worlds. Instead, we must be attentive to the specific parameters and biases of each particular device, to the specific values and strategies embodied and encoded in the various design paradigms. This allows us to confront, rather than obscure, touch's thorough technogenetic transformation during the epoch of haptic interface.

The Logic and Illogic of Analog Medialization

The human–machine tactile communication paradigm, understood by its early proponents as holding the potential to bring about drastic upheavals in a wide range of human activities (including consumption, social interaction, artistic production, and labor) remained an unrealized promise as the epoch of haptic interface drew to a close. The logic of analog medialization informing the search for the master device proved problematic. As the mechanisms responsible for touch perception were distributed throughout the body, touch's technological synthesis could not be neatly localized to a single organ or set of organs. In its biological structure and psychophysical functioning, touch seemed a poor analog to the senses of seeing and hearing. As Iwata explains:

> Compared to ordinary visual and auditory sensations, hap-
> tics is difficult to synthesize. Visual and auditory sensations
> are gathered by specialized organs, the eyes and ears. On the
> other hand, a sensation of force can occur at any part of the
> human body, and is therefore inseparable from actual physi-
> cal contact. These characteristics lead to many difficulties
> when developing a haptic interface.[124]

Because of this incomplete and selective reproduction, these devices often highlighted their own limits rather than showcasing their strengths, hinting at the difficulties facing those engineers who sought a holistic reconstruction of touch akin to the one Suther-land envisioned. Intended to make interactions with virtual environments feel more natural than those that depended on vision alone, these haptic interfaces often had the opposite effect, calling attention instead to the artifice of virtual experiences.

Out of the chaotic epoch of haptic interface, then, no single device emerged victorious. None could lay claim to the title of "master device" or "ultimate display," and the "Holy Grail" of haptics remained perpetually beyond the reach of researchers. With the turn away from fully immersive virtual reality interfaces in the early 2000s, the brief public fascination with haptic interfaces began to fade. However, virtual reality's decline left in its wake a discrete and fully formed field of investigation and design, complete with a network of researchers and engineers, conferences, laboratories, corporations, academic departments, and journals each dedicated to furthering the project of evolving touch through the design, testing, and commercial deployment of haptic interfaces. The mass haptic subject—produced by collecting, quantifying, and aggregating knowledge about the microprocesses of haptic perception in structured laboratory experiments—became a site of new financial investment. On a practical level, this institutionalization of computer haptics entailed the concretization of best practices in haptic interface design; the development of materials (textbooks, courses, equipment, etc.) used to train new generations of researchers; and the formulation of narratives circulated throughout this discursive network to justify, valorize, and legitimize the field.

Furthermore, while full-body, exoskeletal, and single-point interfaces did not prove commercially viable on a mass scale, selected applications of haptics research gradually began to be woven into everyday media and communications devices, as haptic interfaces seeped out from the specialized scientific, industrial, military, and medical contexts in which they had originally taken shape. Almost without notice, and certainly with little critical commentary from media theorists or historians, experiencing machine-generated haptic sensations became commonplace in interface culture. Initially this took the form of simple vibrating alerts generated by pagers, cellphones, and videogame controllers, but these signaling mechanisms and the patterns they generated grew steadily more complex. In the early 2000s, haptic interfaces, as a subset of virtual reality, seemed to be "failed" technologies just as they were beginning to enjoy their most widespread successes. Subsequently, a new narrative strand emerged around haptics, woven together with the narratives of analog medialization and the master device. As the popular press regularly forgot and rediscovered the technoutopian potential of haptic interfaces, journalists and designers framed them as technologies on the cusp of arrival—always just over an approaching horizon, always promising to bring about vast cultural, economic, and aesthetic transformations. Haptic interfaces therefore became suspended in a state of perpetual immanence, all while they continued their gradual march toward ubiquity.

As I have shown throughout this chapter, the technogenesis of computerized tactility obeyed no interior, structuring, evolutionary logic. Instead, its development was contingent on the strategies and aims of actors working in specific institutional contexts. The project of engineering human–machine tactile communication simultaneously invoked technical, biological, social, aesthetic, political, epistemic, and market considerations. In Interface 5, I follow this relationship between touch and computing technology into the twenty-first century, when the cultural protocols associated with technologized touch began to shift radically. The rise of mobile touchscreen interfaces (notably in the Nintendo DS and the Apple iPhone) brought new valorizations of touch's role not just in human–computer interaction, but also in culture more generally, as advertisers encouraged consumers to see touch as a neglected experiential modality that could be rediscovered by touching newly sensitive screens. At the same time, the material dedifferentiation of interfaces accomplished by the touchscreen's flat glass prompted an appreciation for the tangibility of media forms and the other objects subsumed within the touchscreen computer. The proliferation of touch interfacing depended on convincing potential users of the improvements it would bring to their interactions with computers—on the cultivation of a haptic subjectivity that would prompt users to identify haptics as a vital and necessary component of the human–computer interfacing schematic. Where haptics in the twentieth century was confronted primarily as an engineering problem, in the twenty-first century, it became a marketing problem, one that could be solved through actively crafting a desire for the technologies of mediated touch.

Interface 5

The Cultural Construction
of Technologized Touch

Touching is not good. Or so we're told. Please do not touch . . .
yourself, your nose, wet paint, that zit, grandma's best china.
You name it, you can't touch it. We think that's wrong. Why
shouldn't you touch what you want? What if you could touch
the games you play? What if you could make something
jump or shoot or run just by touching it? Let's face it,
touching the game means controlling the game. And when
we say control, we mean precision control. One right touch
and you're master of the universe. One wrong touch and
you're toast. Forget everything you've ever been told and
repeat after us. Touching is good. *Touching is good.*
—Print ad for the Nintendo DS portable gaming system

As Nintendo prepared for the 2004 release of its DS portable gaming system, the company faced a crucial challenge: how should it
go about advertising and marketing a device whose primary distinctive feature was a touch-sensitive screen used to control and
manipulate gamic actions? How could the company acclimate players to actively touching, stroking, and swiping screens formerly
encountered as nonresponsive? The solution provided by marketing firm Leo Burnett took the form of a $40 million multiplat-
form advertising campaign organized around the slogan "Touching
Is Good" (see Figure 5.1). In addition to a blitz of television and

265

Touching is not good.
Or so we're told. Please do not touch...
yourself, your nose, wet paint, that zit,
grandma's best china. You name it,
you can't touch it. We think that's wrong.
Why shouldn't you touch what you want?
What if you could touch the games you
play? What if you could make something
jump or shoot or run just by touching it?
Let's face it, touching the game means
controlling the game. And when we say
control, we mean precision control.
One right touch and you're master of the
universe. One wrong touch and you're toast.
Forget everything you've ever been told
and repeat after us. Touching is good.
Touching is good.

NINTENDO DS.

touch control • wireless chat • voice control • wireless play

Figure 5.1. Print ad for Nintendo DS. Copyright Nintendo
Entertainment, 2004.

print ads, Nintendo initiated a guerrilla marketing campaign in which they distributed mannequin hands—imprinted with the DS's logo—to their fans, encouraging them to make their own Touching Is Good–themed videos to share via the campaign's online hub at touchingisgood.com. Echoing the narratives mobilized by haptics researchers in the preceding decades, the campaign positioned touch as a neglected and marginalized experiential modality that could be rediscovered through a pleasurable tactile interfacing with computers. The DS ads involved a strategic recall of touch that helped redefine and reconceptualize its value in contemporary interface culture.

Just as the DS foreshadowed the release of other touchscreen-based portable digital media technologies (such as smartphones, tablets, e-Readers, and laptops), the Touching Is Good ads staged future advertising campaigns that would mobilize touch, alongside the finger as its iconographic figuration, in marketing strategies for marketing mobile computing devices. Apple's "Touching Is Believing" ads for its iPod Touch and iPhone devices, Hewlett-Packard's "Touch the Future" theme for its touchscreen laptops, and Barnes & Noble's "Touch the Future of Reading" for the touchscreen version of its Nook e-Reader each employed similar juxtapositionings of text and image to highlight the tactile sensitivity of the represented device's screen. By 2010, print and Web ads that featured a finger reaching out to touch an interactive screen, accompanied by some variant of a playful touch-centered tagline, became a commonplace feature in both popular press outlets and their online counterparts. Taken together, these ads served the common function of visualizing an invisible trait of touchscreens, and in doing so, helped condition consumers to the new bodily habits required by screenic touch.

While the rise of touchscreen interfaces provided touch with a new primacy in digital interfacing, the gradual displacement of buttons, keys, and knobs by the flat glass of the sensitive screen did little to alleviate the situation that had prompted the formation of computer haptics in the 1990s—computing, much to the disappointment of haptics proponents, still depended heavily on routing messages through the visual and aural channels. To some haptics researchers, the touchscreen even seemed to be a backward step. Immersion Corporation's Christophe Ramstein summarized this perspective bluntly in 2007, when he suggested that "from a tactile standpoint, the touchscreen is literally dead."[1] According to this view, the touchscreen, in flattening and purportedly dematerializing the human–computer interface,[2] paradoxically resulted in a decreased level of tactile engagement with data, rather than bestowing the utopian and humanizing benefits promised by touchscreen advertisements.

This chapter therefore tells two intertwined stories about touch. The first is of the new qualities, characteristics, and values it acquired with its depiction in ads for touchscreen interfaces. In these narratives, touch's utility, epistemic function, and positioning in the broader ecology of mediated perception each underwent a drastic revision, as the category of touch itself was adjusted to meet changing conditions of information manipulation. The second is the story of the exertions by haptics marketers to articulate the need for touch feedback through carefully crafted documents explaining the urgent problems caused by touch's conspicuous absence in computing. The proliferation of haptics came to depend increasingly on the successful appeal to and cultivation of a *haptic subjectivity* in consumers, who were asked to identify themselves as deficient information-manipulating subjects in need of the rehabilitation haptic feedback promised to provide. As with prior media technologies, the utility of haptic interfaces did not exist inherently in the technology but instead had to be actively constructed, and the ascent of potential users had to be won through compelling rhetorical appeals. At stake in these appeals was both the success of the individual firms whose survival hinged on the widespread uptake of haptics, and, more generally, the viability of haptic interfacing as a paradigm of human–computer interaction. This constant construction of new subjectivities through apparatuses, as Maurizio Lazzarato explains, serves an essential role in the operation of capitalism, where apparatuses function not merely as technical things, but as "social instruments for decision-making, management, reaction, technocracy, and bureaucracy."[3] Identifying with this haptic subjectivity meant acquiring a particular instrumentalized attitude toward one's own sense of touch. Subjects had to learn to see a more perfect bonding with computing machines as a desirable outcome enabled by their assimilation to the artificial languages and rhythms of haptic feedback.

By pivoting slightly away from the lab in the first part of this chapter, I am not suggesting that its importance diminished following the epoch of haptic interface. Instead, this move recognizes that what went on in haptics labs became progressively bound up with the need to market and commercialize the products of lab research. In short, while during the twentieth century haptics had been primarily confronted as a design problem, in the twenty-first century haptics quickly came to be a marketing problem. My orientation to the mutual reciprocity of design and marketing draws inspiration from Kittler's claim that our sense perceptions are the "dependent variable" in a "compromise between engineers and salespeople." For Kittler, the qualities of images and sounds—"how poor the sound from a TV set can be, how fuzzy movie images can be, or how much a beloved voice on the telephone can be filtered"[4]—arose out of this compromise. With haptic interfaces, that compromise impacts the tangibility and hapticity of media worlds: the degree of accuracy with which an interface renders the resistance provided by a sheet of virtual fabric as the user tries to pierce it with a virtual needle, for example, depends on a whole range of technical affordances baked into the device. These include the quality and strength of the motors used to generate resistance, the speed and accuracy with which the software samples and captures the user's movements, and the refresh rate of force sensations sent back to the device by the haptic effects software. But as with the fuzziness of the movie image, the technical affordances of human–machine haptics are shaped by market factors weighed alongside engineering ones. Using a higher quality motor may prohibitively increase the manufacturing cost of the haptic interface device, higher rates of position sampling and force rendering might prove too taxing for a standard consumer-grade processor to handle competently, and obtaining the license to use a more accurate feedback mechanism may prove too costly. The task then falls to the marketer to persuade users that what they are experiencing is a convincing enough haptic simulation, in spite of its shortfalls in fidelity or resolution. These marketers, in other words, serve as key agents in the construction of what qualifies—and fails to qualify—as technologized touch. In much of this chapter, my focus is on that point of intersection between marketing and design, where touch interface and haptics technology were made legible to their prospective audiences.

Articulating the value of haptics, then, has involved managing public perceptions of haptic interfacing technologies and attempting to control the life that haptics lives in the popular technocultural imagination. For much of their recent history, haptic interfaces have been suspended in a state of perpetual immanence, appearing always on the cusp of an emergence that promises social transformation and revolution. Journalists gleefully chronicled apparent breakthroughs in research (such as the 2002 "transatlantic touch" demonstration that took place simultaneously at MIT's Touch Lab and University College in London),[5] consistently describing haptics as "next-generation" technology. These stories parroted and perpetuated the logic of analog medialization—first media for seeing and hearing, and now, for touching!—mobilized in early force feedback research. From roughly 2000 on, haptic interfaces were rediscovered with a clockwork regularity, always encountered as exotic objects, and always inspiring a technofetishistic hope in the capacity of computational prostheses to erase and overcome the body's hard physical limits. Individual devices, no matter how circumscribed their intended applications, bore the burden of making good on every promise cooked into haptics since the technology first captured the public imagination in the 1990s.

But inevitably, tomorrow never came. During the epoch of haptic interface, the market for haptic feedback interfaces had remained relatively limited, confined primarily to specialized, high-end applications, such as scientific data visualization, military simulation systems, and medical uses in laparoscopic and remote surgery. Most consumer-grade devices came in the form of force feedback "haptic peripherals" intended for in-home videogaming (force-reflecting steering wheels for racing games, force feedback joysticks, and rumble-enabled videogame controllers).[6]

Although these mass-marketed gadgets helped begin the long process of domesticating haptics, they were generally portrayed—and
even denigrated—as relatively rudimentary instantiations of the
technology. With the move into the new millennium, however,
haptics as a discrete area of research developed so rapidly that a full
cataloging of publications and research areas is virtually impossible.[7] The products of lab research crept outward into a whole range
of social spaces. For example, in 2001, Immersion Corporation
partnered with computer peripheral manufacturer Logitech to
release the iFeel Mouse—a standard, two-button mouse that added
vibrational cues to let users know when they had moved their cur-
sor over specified desktop icons and hyperlinks. With the iFeel
Pixel™ tactile effects software Immersion developed, coders and
end users alike could create customized patterns of vibration,
bringing tactility to the experience of the desktop graphical–user
interface and making navigation easier for those with vision impairments.[8] Haptic interfaces were incorporated into museum displays,
fueled by a new focus on the tactility of artwork and aesthetic
experience that digital interfaces seemed capable of bolstering.[9]
Drawing inspiration from the tactile communication systems
described in Interface 3, pagers and mobile phones commonly
employed simple vibrational patterns to pass alerts and messages to
their users, resulting in a hypervigilant attentiveness to any vibrations that interrupted the tactile field (accompanied by the curious
sensation that came to be problematically labeled "human phantom vibration syndrome").[10] After the release of Nintendo's
Rumble Pack for its N64 system and Sony's DualShock controller for its PlayStation console in 1997, vibration feedback quickly
became a standard feature in console gaming. Microsoft adopted
an identical rumble configuration with the release of its Xbox console in 2001, and by 2007, there were in excess of 500 million
vibration-enabled controllers in circulation worldwide. In spite of
the near ubiquity of haptics applications, popular press narratives
continued to treat haptics as exotic, foreign, and futurological, with
interface designers confirming this framing by routinely describing
such commonplace applications as "rudimentary."[11] Repeated calls
to fix and standardize languages of machine-generated vibrations,

similar in spirit to the plea Diderot made in the eighteenth century, obscured the extent to which such standards—or protocols—were already being offered and adopted, both at the material level in haptics software and hardware and at the cultural level in the negotiation of meanings and practices associated with technologized touch. The actuality of haptics always and necessarily fell short of its overhyped potentiality.

Against the backdrop of these developments, the common refrain that the sense of touch is neglected in Western culture, echoed throughout range of disciplines that includes psychology, aesthetics, computer science, communication, and anthropology, seems to be obsolete. Summarizing a host of recent cultural and technological developments, including changing media interface technologies, a renewed interest in massage therapy, and the aforementioned resuscitation of concern over touch in aesthetics, Robert Jütte entertained the notion that we have entered "a haptic age" (originally suggested by a German journalist in 1999) defined by an increased focus on touch as an epistemic and experiential modality.[12] Taking into account the fluctuating interest in touch throughout the twentieth century, which included McLuhan's emphasis on mediatic tactility in the 1960s, Jütte was reluctant to endorse the journalist's epochal pronouncement and opted instead to summarize recent cultural trends and steps forward in human–computer interface technology that hint at—but stop short of being determinative of—touch's elevation in the hierarchy of the senses. Although I echo Jütte's restraint for making essentialist and premature pronouncements about the character of our present technohistorical moment, throughout this chapter, I show how touch's interface with computational technologies has already brought about substantial revaluations and transformations in its constitution and cultural status. Regardless of touch's future expressions through media technology, this thorough territorialization and remapping of touch accomplished by computer haptics will endure as an artifact of touch's material and discursive history. The case studies in this chapter each push toward the goal of charting the multifaceted new vision of touch expressed in the design and marketing processes for various haptics applications, as touch was made adequate to the new demands imposed on it by the imperatives of shifting information-circulation practices. Touch, in short, is no longer neglected, but the conditions of its attending to—of its strategic retrieval—are governed by a fundamentally mediatic logic that can only come into view through the empirical investigation of its various technological reformattings.

Touching Is Good

Owing to the rapid pace at which touchscreen interfaces came to first complement and then displace button-based modes of mobile human–computer interaction, it is rather easy to forget or obscure the process by which the new mode of interfacing became naturalized and seemingly instinctual. For Gitelman, the sort of cultural amnesia that routinely follows the uptake of a new communication medium makes pressing the need to study media at the historical moment when they were new, before all the protocols that support and underpin their usage become naturalized and taken for granted. Once sedimented, these protocols provide the often unstated background consensus shaping the accepted meanings and habits associated with a given media technology. The narratives crafted in ads for mobile touchscreens proved crucial to wiping away the old protocols of information manipulation, depicting past modes of interacting with computers as obsolete. The embrace of touchscreen interaction would facilitate entry into a new era of interfacing defined by the circumvention of buttons, knobs, and keys. More specifically, the keyboard/mouse combination served as the a priori of interfacing—a taken-for-granted interaction schematic that seemed to have emerged fully fledged in some ancient age of computing, in need of discarding rather than iterative improvement.

The Touching Is Good campaign established the basic trajectory that later ads would follow, engaging in reflexive revaluation of touch's status in the broader ecology of mediated perception. The ads in Nintendo's 2004 campaign recapitulated a social history where touch was gradually neglected and forgotten, only to be remembered and reawakened by the beckoning of now-sensitive screens. This cultural imaginary of touch as sense modality repressed and marginalized in contemporary culture generally, and in modern media systems more specifically, allowed Nintendo to frame the DS as more than simply a new game interface, but as a product that would facilitate a new tactual mode of being in the digital world. Touch could be rediscovered—a lost sensory modality recaptured and remembered—through the consumption of a new digital media interface. Conjuring touch as a prerational, presocial, and prelinguistic mode of encountering the world, unencumbered by the burdens civilization imposed, the DS simultaneously gestured backward and forward in media historical time. Naturalizing touchscreen interfacing hinged on touch's ideation as the most natural—and least culturally dependent—mode of encountering and knowing the world.

Touching Is Good continued a conversation around mediatic touch initiated by AT&T's McLuhan-inspired "Reach Out and Touch Someone" campaign in 1979. Consisting of print and television ads, Reach Out and Touch Someone featured heartwarming scenes of loved ones whose physical and emotional isolation from one another could be shattered by the miraculous technology of distant speaking. Reaching out and touching someone—telephonic touch—was accomplished not by the hand, but rather, by the voice. Through the campaign, AT&T repeatedly attempted to retrain telephone users to think not about the negative emotional aspects of long-distance telephony—such as the high cost of long-distance calls, and the anxiety that late-night calls often evoked—and instead focus on the positive affective state that could be activated by a call from a cherished friend or family member.[13] In this depiction, touch provided a metaphor for emotional contact and connection, with the telephone represented as the means of facilitating the collapse of affective distance between remote speakers. The model of touch advanced in the AT&T campaign, however, was nothing tactile: it highlighted instead the frequent oscillations in accounts of mediatic touch between touch as something that designates an act of material contact and touch as a synesthetic figuration that erases the difference between (or translates among) sense modalities.[14] While the tactile fusing of hand and communicative apparatus featured regularly in these ads acts as a figuration of the annihilated emotional gap between remote communicative subjects, the extension of the finger through the telephone network was only ever representative of the voice's transmission across electric networks. When electronic media extended touch, it was always only ever by passing it through the filter of another sense modality. For McLuhan, this was a feature rather than a bug, as it indicated touch's power to serve as a mediator of the other senses. And, crucially, touch shared this function with electricity: both had an equal capacity for synesthetic translation and sensory dedifferentiation that McLuhan identified as potentially liberating and disruptive. The specialization and fragmentation of the senses initiated by language intensified with the rise of print; electronic media held the potential to undo the damaging effects of this disassembling.

Though Touching Is Good sought a similar upheaval in the ordering of the senses, it did so by insisting on touch's specialization rather than its dedifferentiation. Emphasizing the literal point of contact between finger and screen, the campaign equated touch's newfound ability to directly manipulate screen images with a more general rediscovery of a sense modality rhetorically positioned as lost and neglected. Mediatic touch here provided a means of mastery over virtual worlds, where touchscreen interfacing allowed for a precise type of manipulation absent from previous interface schematics. The relationship between the user and the interface took center stage in these ads, showing visual markers of the user's contact with the newly sensitive screen. In Reach Out and Touch Someone, AT&T had focused on the emotive expressions of the remote subjects touching through the telephone; in the Touching Is Good ads, the bodies of users were frequently missing, with fingerprints on the screen as the only traces of their contact or with the hand disembodied from the user and the index finger melded into the game world (see Figure 5.2). Touch was something that bodies did, but the body that touched the screen was rendered as absent in these ads. Mobilizing a long history of iconographic figurations that localized touch to the fingertip, Nintendo reinforced the notion that the nimble index finger embodied the essential totality of touch. Technologized touch—the touching that is good—reduced the whole of the tactile system to the single point of contact between finger and screen.

Nintendo's appeal derived its urgency not from the positive claim that touching is good, but instead from the negative set of conditions that made touch a sense in need of rejuvenation and rediscovery via the sensitive screen—a mix of social and technological parameters that restricted the feeling of tactual engagement with the world. Hailing a tactile/haptic subject frustrated by constant cultural prohibitions on touching, Nintendo attempted to make the device's usage synonymous with the reawakening of a neglected sense modality. This notion of touch as forgotten—this nostalgia not just for lost sensations, but for the lost sense of touch itself—locates the senses as crucial sites of not just personal but also cultural and economic transformation. As Nadia Seremetakis describes, the nostalgia for senses lost indicates the haunting of the sensory present by the feeling of absence, of not quite being able to recall the thing that has slipped away.[15] The boundedness of the senses to the social temporality of modernity—the link between modes of perception and changing practices of consumption and production—prompts us to treat sensory memory as a vital site for registering cultural transformations. In Nintendo's strategic positioning, the touchscreen became a site of disruptive remembering, as it hailed an ideal user who felt consistently repressed by the prohibitions on touching in polite and civilized society. Modern life entailed a constant denial of the impulse to touch; every object of desire was off-limits to the hands, only accessible through an unsatisfactory—and decidedly nontactile—visual mode of apprehension.[16] The DS functioned as a way of liberating a haptic subjectivity so thoroughly repressed that its existence had been forgotten and denied:

> Touching is . . . thrilling, exciting, fun, weird, interesting. Sometimes a bit taboo. It's how we connect—with each other, the stuff around us and now, our games. Here, we celebrate the most under appreciated of your five senses. We make contact. We get in touch with touching. Because with the Nintendo DS, touching is good.[17]

Nintendo's narrative positioning of touch celebrated the touch-screen for its capacity to activate not merely a particularly limited form of technologized touching, but to mobilize through the interface the sense of touch as a more holistic (forgotten) mode of being in the world. The DS promised to allow tactile contact—tactile knowing—not just of game worlds but also of touch itself. Technologies of perception reform the senses through both our material interactions with them and through the discursive framing of these new material interactions—what Seremetakis describes as the "tales, myths, and memories" of the senses.[18] In the DS ads, the fading memory of touch no longer lingered at the fringes of consciousness, but became instead central to the commercialization of a new media interface, as consumption was framed as the pathway to remembering and restoring touch.

However, the rediscovered touch was one that had drastically shifted form while concealing the markers of its transformation. Print ads for individual DS games repeatedly depicted a single finger extended across the page, immersed in the game world. The finger succeeded in penetrating and entering the image, calling attention to the new relationship between the player's body and the game text. In the ad for *Advance Wars: Dual Strike* (Figure 5.2), for example, the finger morphed into a B-52 bomber, stretched across the sky, with a string of bombs falling from the finger-plane. In projecting militarized imagery onto the hand and finger, the Advance Wars ad reinforces Nintendo's model of touch as domination while simultaneously recalling a particular military history of armoring and gendering the sensorium. Analyzing technologies of militarized perception, Paul Virilio situates the cockpit as a strategy for managing the assault of the environment on the pilot's senses.[19] Whereas military pilots in the first generation of airborne combat were almost fully exposed to the wind and the noises of the aircraft and battlefield, the enclosure of the pilot within the sealed space of the cockpit shielded pilots' senses from the assault of these forces. Developed by the end of World War II, the cockpit was a technique for controlling the immediate perceptual field of the bomber. However, these gains were not won without costs, as Virilio details:

> the pressurized cockpits of US Superfortress bombers had become artificial synthesizers that shut out the world of the senses to a quite extraordinary degree. However, the effects of technological isolation were so severe and long-lasting that Strategic Air Command decided to lighten the dangerous

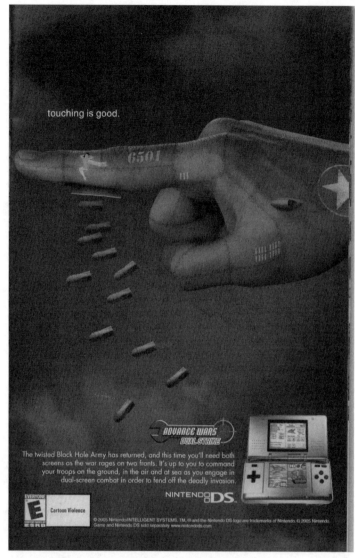

Figure 5.2. Print ad for *Advance Wars: Dual Strike*. Copyright Nintendo Entertainment, 2004.

passage of its armadas over Europe by painting brightly-
colored cartoon heroes or giant pin-ups with evocative
names on top of the camouflage.[20]

Much like the pressurized cockpit, the DS interface provided a
form of sensory armor that normalized, regulated, and filtered out
feeling to allow detached, careful, and precise manipulation. To
manage the harsh and impossible sensory stresses of aerial combat,
the cockpit shielded the body in an artificial environment. Any
information allowed to penetrate this technological bubble was
thoroughly controlled and instrumentalized, aimed at domination
rather than mutual penetration. Even the female figure painted
onto the nose of the plane served a strategic function, humanizing
the harsh and foreign environment that loomed dangerously outside the windows of this controlled space. The DS promised a similar
extension without co-penetration—a militarized tactility that
shielded the touching subject from feeling the touched object. The
electrification of the senses here did not involve a total but rather
a strategically incomplete sensory awareness designed to allow
manipulation without fear of an uncontrolled and unregulated
tactility contaminating the sanitized tactile field. Touching is good
only insofar as it is a touch that eliminates the possibility of reciprocal exchange and provides a technological framework for manipulation without fear of sensation. The strategic limits of the interface
promised to shield the user from haptic implication in the world
manipulated and mastered by touch: what the newly empowered,
anesthetized finger could touch, it could not feel.

Apple's "Touching Is Believing" print ads, circulated before the
2007 release of the iPhone and iPod Touch, similarly foregrounded
the tactile encounter between finger and screen in an explicit challenge to the longstanding cultural linkage between vision and
faith. The now-iconic ad showed a disembodied finger reaching
out to make divine and transformative contact with Apple's new
communication machine, as the tagline "Touching Is Believing"
playfully reversed the denigration of tactual knowledge. Immediately on the release of the ads, bloggers seized upon its religious
overtones, playfully suggesting the term "Jesus phone" to designate

the forthcoming device. Building on the ad's visual resemblance to Michelangelo's "The Creation of Adam" and Caravaggio's "The Doubting of Saint Thomas," and on the cultish devotion Apple fans showed toward the both the company's products and the brand itself, some commentators criticized its slavish technoutopianism, while others praised its complexity as a metaphor.[21] While I do not want to downplay entirely these religious and mystical overtones in Apple's positioning of the iPhone as a divine artifact revealed, I emphasize instead the way Apple beckoned customers to touch and interact with the new device. The metaphorical richness of the ads served as a distraction from their more mundane and pragmatic invocation to touch a screen that had been made sensitive to dexterous manipulation. The afterimages of the finger's movement suggest a digit in motion across the screen, not just making contact at a single point, but also actively and nimbly gliding about the screen's surface. As with Touching Is Good, it simultaneously attempted a retraining of interfacing habits and a revaluation in touch's position in the cultural hierarchy of the senses. Although the iPhone was Apple's first entry into the world of telephony, Touching Is Believing did not embrace the same model of telephonic touching offered by AT&T in Reach Out and Touch Someone. The device was not presented as a transparent medium through which the finger could pass to touch a remote communicative subject but instead was rendered as an object of tactile knowledge itself.[22] The screen served, rather, as a site of disruptive and "revolutionary" contact—a promise to awaken and recall an alternative epistemic orientation, where manual and digital rather than visual inspection would provide the most reliable pathway to knowledge and belief. Further, Touching Is Believing did not frame the touchscreen as a surface the finger penetrated effortlessly to come into contact with the image on the other side of the flat glass (the screenic image is barely visible in the ad). Instead the screen was framed as a space where cultural hierarchies of the senses would undergo an embodied, materially grounded renegotiation as a consequence of the meeting between finger and device (Figure 5.3).

Read in conjunction with each other, these campaigns made a naked declaration about both the screen and its material housing:

Figure 5.3. Print ad for the first-generation iPhone. Copyright Apple, 2007.

not that it can be touched, but that it *should* be touched. Touching the screen would allow for the pleasurable and thrilling manipulation of digital images, while also reactivating a dormant epistemic modality. In accustoming users to touching previously noninteractive screens, the ads served a normative function, making it not only acceptable but also good to touch the screen. As with the Touching Is Good ad (see Figure 5.1), the fingerprint on the screen no longer indicates an unwanted contamination of the image by the finger. Rather it is now the residue of a vital and productive contact with the machine. In this cultural retraining, the user was encouraged to understand the bodily habit of touching screens as the preferred mode of data manipulation—a more natural, holistic, effective, and instinctual mode of interacting with virtual objects that mobilizes a fantasy of precise control not just of digital images but also over the whole world of things they indexed and represented. Good touch is active touch: it is touching with the finger without the fear of being touched back, manipulation without countermanipulation, touching with feeling. Good touch is the ability to master and dominate, shielded by the interface. Good touch is control.

In the face of these empowering and desubjectifying apparatuses, the old regimes of interfacing became inadequate, stale, confining mechanisms that governed their users. The tacit critique of visuality collectively advanced in these ads sought to overturn our commonsense association between image-making technologies and mediatic modernity, with the embrace of touchscreen computing enabling a more perfect coupling of human and machine. But from a material standpoint, the new screenic touch was marked by a crucial absence: the screenic abstractions the user touched had no tangible reality—the flat glass screen provided no data for touch, encountered simply as a homogeneous and dedifferentiated tactual space. The task of these ads, then, involved making the nontactile, touching-without-feeling of touchscreen interaction constitutive of touch. They sought to make this new touch an adequate proxy for the model it displaced.

Are Touchscreens Haptics?

In thinking through the genealogy of haptic interfaces, the touch-screen presents a challenge. Although positioning touchscreens as a type of haptic interface seems to be almost intuitive, the technical development of the touchscreen occurred primarily outside the formal discipline of computer haptics. At first glance, this may seem a minor point, perhaps even one that runs the risk of tyrannically mapping the haptician's classification scheme onto media archaeological analysis. But in interrogating the touchscreen's place in a genealogy of haptic interfacing—asking, in effect, "does the touch-screen belong here?"—I am establishing a productive tension between the touchscreen interface and its haptic counterpart. This haptocentric strategy resists the primacy of the visual frequently operating in lineages of the touchscreen[23] and computer media more generally,[24] instead taking touch to have a material and discursive history independent from (though in dialogue with) the visual.

To start, I consider the case for locating touchscreens in the tradition of haptic interfacing. At the material level, the touchscreen invokes the fingers in processes of interacting with, manipulating, and navigating computational space, imparting a new set of embodied gestures to users. Such gestures are simultaneously relationships with computational data itself and stand-ins for social relationships; the notorious act of "swiping left" on the dating app Tinder, for example, is a variant on what Verhoeff and Cooley describe as a "navigational gesture"[25] that allows the user to move through a database and also as a swift, socially situated declaration of individual aesthetic preferences. The addictive popularity of touchscreen videogames, such as *Angry Birds, Candy Crush, Cut the Rope,* and *Fruit Ninja,* derives in part from the haptic pleasures that come from wildly tapping and franticly swiping at the screen. The movements of the hands and fingers are acclimated to particular demands of sensitive screens, as a new set of fundamentally haptic bodily habits allows users to become legible to computing machines. And the intuitive ease with which young children manipulate touchscreen images—both a source of jubilation and of moral panic[26]—seems to confirm advertisers' claims that touch interfacing circumnavigates the need for cultural training, tapping into an innate, naturalistic storehouse of gestures.

Further, the embedding of touchscreen interfaces in handheld computing devices locates questions of the fit between device, hand, fingers, and touched screen, as Heidi Rae Cooley suggests, centrally in both the product design process and the user's experience with the device.[27] For Cooley, the active haptic movement of the device in the user's hand is the outcome of a careful, biomechanical industrial design process that accounts for the physiological particularities of human hand movement, including the way that the fingers and thumb articulate as they grasp and manipulate the device, and the seating of the device in the palm as the user angles it toward their gaze. Fit does not designate a characteristic of the device; rather, it specifies a relationship between bodies and devices underpinned by a technics of design. Decisions about which materials to use in crafting the touchscreen's housing, the device's angles and contours, and the placement of the hard and soft keys around the touchscreen each become crucial components of brand identity (what I have described elsewhere as *ergonomic branding*),[28] product differentiation, and the embodied relationship between user and device. The design of accessories for handheld touchscreen computers, too, is substantially informed by tactile considerations—cases and "skins" for smartphones and tablets are intended to protect the devices while also modifying the way they fit and feel in their users' hands. Plastic and glass screen protectors are rated based on their ability to allow the fingers to move frictionlessly across the screen, as if erasing the protective layer that mediates between skin and screen. Touchscreen computing, in short, involves a whole host of quintessentially haptic material interactions and relationships between user bodies and screenic devices.

The case for excluding touchscreens from a genealogy of haptic interfaces hinges on considerations of their discursive framing, technical history, and experiential materiality. While touchscreen ads constantly invoked touch as an experiential and epistemic category, explicit uses of the term "haptics" were conspicuously absent from these advertisements. As the next case study I take up illustrates, the marketers tasked with advertising interfaces that employed haptics technology itself were not shy about announcing their devices' positioning in a formally constituted genealogy of haptics. These marketers frequently provided explicit definitions of the term that capitalized on its neologistic and futurological connotations while simultaneously gesturing back to the word's root in ancient Greek. By excluding references to the more specialized term, ads for touchscreens seem to tacitly acknowledge that touchscreens are not haptic screens.

Concerning their technical lineage, the methods used to make screens sensitive to human touch developed explicitly outside of the "new discipline of computer haptics," troubling what might seem at first glance like a commonsense kinship between the two categories of interfacing. A full-scale rehearsal of the various mechanisms screens employ to register touch is beyond the scope of my immediate interests, but generally speaking, the engineering of touchscreens involved techniques for registering user input as a one-way flow of information to machine, without a mechanism for feeding data back through the tactile channel.[29] Touchscreens operate by mapping the point of contact between user and screen onto x-y coordinates, allowing the screen to become sensitive to the movements of the user's finger across the screen's surface. The specific mechanisms screens deploy to register the user's touch directly impact the types of touch required for the user to be legible to the screen. For example, resistive (or pressure-sensitive) touchscreens map the user's touch through the deformation of sensitive layers of film concealed beneath the surface of the screen, requiring a hard and deliberate press of the finger against the top layer of the screen. Such screens are equally as sensitive to pressure applied by a finger as they are to pressure from a mediating agent, whether it be a stylus, a wooden stick, a stone, or whatever other ad hoc object the user prefers. In contrast, the far more sensitive electrocapacitive screens prevalent in most mobile touchscreens use a charged electrical field across the surface of the screen to map the user's contacts. Capacitive screens depend on the electrical charge stored in the human body—discharged on contact with the device—to strategically interrupt the electrical field of the screen. Through the careful plotting of these interruptions on a dense grid of sensors, and their processing by proprietary algorithms that differentiate between gestures, the user's touch becomes intelligible to the device. As such, the input mechanism matters immensely: a nonconductive mediating agent (a wooden stick or fabric glove) will not trigger an interruption of the screen's electrical field (prompting the emergence of an industry devoted to the manufacture of capacitive "touchscreen" gloves). Unlike the shocking contacts between humans and machines so crucial to the cultivation of an electrical epistemology in the eighteenth and nineteenth centuries, the transfer of electricity between body and touchscreen produces, by design, no tangible effects; it is a meeting of organic and inorganic electrical fields registered only by the machine. My point here is that computer haptics was driven by the animating mission of making computational worlds perceptible to the user through machine-generated sensations, whereas those interested in designing touch-sensitive screens proceed with a different—though related—aim, one part of an alternative tradition in interface design that sought to invoke touch as a unidirectional rather than bidirectional information channel.[30]

The question of materiality follows directly from this alternative tradition. By replacing data-rich buttons, keys, scroll wheels, and knobs with the flat, homogeneous tactile space of glass, touchscreens seemed to diminish rather than enhance the tactility of human–computer interfaces, effectively shifting the burden of registering inputs from the dense clusters of nerve receptors in the pads of the fingers onto the eyes. According to hapticians, the graphical user interface had already initiated this dematerializing process by transforming conventional physical workspaces into spaces of pure visuality (primarily through the desktop metaphor). But before the touchscreen, there were still the data-rich tactile cues provided by keyboard–mouse interface, in spite of their purported inadequacy when compared to a robust haptic interface.[31] Touchscreens, then, seemed to be a step backward both from earlier efforts at offloading the labor of processing machine-generated data from the eyes and ears to the skin and from the predigital media interfaces that relayed vital information about their constitution and operation through their tangible materiality. The pages and binding of a book, the dial on a telephone, and the buttons and knobs on a radio are encountered through complex processes of touching, involving a dense network of receptors in the skin, joints, and muscles that work together to configure the experience of media both as functional and aesthetic objects. As Zoe Sophia argues in her essay on container technologies, each also has generally been treated as superfluous and taken for granted, inconsequential to the disembodied meaning of the texts media containers facilitate access to.[32] Touchscreen interfaces participate in a great tactile homogenization of the materiality of media containers, resulting in the labor of manipulation shifting from fingers, hands, wrists, and arms onto the eyes and ears; the fingers no longer register the press of a button, the turning of a page, or the twisting of a knob. The touchscreen, in its pure form, uses touch as a unidirectional channel, relying on the eyes and ears to register feedback from the computational system.

Those in computer haptics offered a practical response to this problem of the touchscreen's tactile dedifferentiation. Motivated by Ramstein's aforementioned view that "from a tactile standpoint, the touchscreen is literally dead,"[33] designers of the haptic feedback mechanisms employed in touchscreens began attempting to write something akin to the materiality of buttons, keys, and knobs back into the flat glass screen's lifeless space. Using small, precisely controlled motors, designers produced vibrational cues that simulated, however imperfectly, the sensations produced by hitting buttons and pressing keys. Initially little more than jolts that moved through the screen to the fingers, advances in the actuators used to produce vibrations allowed for the production of increasingly complex sensations (see Figure 5.4). Similar haptic feedback had already been used in mobile communication devices as a means of alerting users to incoming calls and messages, so using these same mechanisms to produce vibrational feedback to simulate the press of a button proved to be a fairly seamless transition.[34] Owing to the rapid incorporation of haptic feedback mechanisms into touchscreens, it

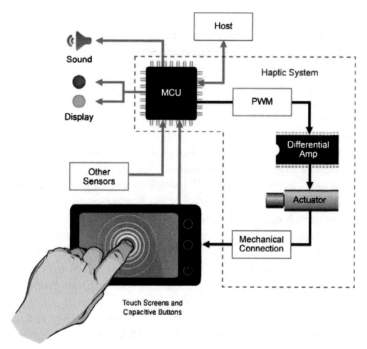

Figure 5.4. Building blocks of a haptic system. Image from Immersion Corporation, "Enhancing Your Device Design through Tactile Feedback," white paper (San Jose, Calif.: Immersion Corporation, 2011), 3.

became increasingly impossible to find a touchscreen that did not use some form of touch feedback, no matter how limited, as a stand-in for the materiality of buttons and keys.

This move to write touch into the screen using machine-generated sensations provides a practical answer to the question of the touchscreen's positioning in the genealogy of haptic interfaces. I take up the justifications for making screens that touch in the next section; for now, I suggest that the widespread adoption of haptic feedback in touchscreens inseparably fused together the two technologies. In this dialectic relationship, the discipline of computer haptics became essential to the adoption of touchscreens in mobile communication devices, resulting in a drive to create new sensations through haptic feedback that could compensate for the lost tactile complexity of buttons and keys. In turn, this fueled new investments in haptics research: the increased use of haptic feedback in touchscreens gave rise to a new industry devoted to manufacturing the complex electromechanical components that imbue the touchscreen's smooth glass with its apparent texture.[35] Software settings allow users to customize the intensity of haptic feedback cued by different actions. And while current-generation devices remain primarily limited to simulating the feel of button presses and using vibration patterns to alert users of incoming messages, the development of mechanisms that will allow the screen to emulate the texture of physical objects (designers frequently cite the skin of an orange as a texture that high-definition touchscreen haptic feedback will soon simulate) remains an ongoing project. Robyn Schwartz, associate director of Research Retail Analytics at IBM, provided a deterministic variant on this narrative in 2012, predicting that "within the next five years your mobile device will let you touch what you're shopping for online. It will distinguish fabrics, textures, and weaves, so that you can feel a sweater, jacket, or upholstery—right through the screen."[36] The enduring importance of the touchscreen, in the genealogy of haptic interfacing, lies in its framing of the fingertip as a benumbed, clean surface on which new machine-generated tactile images can be painted. But touch's purportedly vital importance as an information-reception channel was not self-evident. Instead, haptics firms had to make the case that device manufacturers should invest substantial capital in licensing haptics software and hardware, while end users needed to be convinced that the inclusion of complex haptic feedback added to their experience with a device. The imbrication of touchscreens and haptic interfaces, in short, depended on the successful crafting of a desire for machine-generated touch sensations.

Immersion: We Are Haptics

The firms whose financial success depended on the proliferation and uptake of haptic feedback computing actively shaped the public conceptualization both of the relationship between touch and haptics and of haptic interfacing's use value in an economy now thoroughly transformed by the computerization of information circulation. Since haptic human–computer interfaces first became a site of research and commercial investment in the 1970s, no company has done more to both frame the public understanding of haptics and subtly push haptics technology into the hands of users than San Jose–based Immersion Corporation. Founded in 1993 by Stanford PhD Louis Rosenberg, Immersion had assumed a leading role in haptics research dating back to the early days of the virtual reality boom. From its efforts at incorporating haptic feedback into instruments used in robotic surgery to its work building force feedback videogame joysticks, Immersion's investments in haptics during the 1990s ran the gamut of computing applications.[37] With a name seemingly pulled from an early William Gibson novel, Immersion's long-term branding plan aimed at making the company synonymous with haptics, and in turn, with making haptics essential to a move into a future of human–computer interaction, where more complex touch feedback systems would erase the gap between users and the computer-generated environments they inhabit. A recent update to its corporate Web site made its aspirations explicit, featuring the tagline "Immersion. We are haptics."[38] Immersion's legal practices have also served this aim of cementing its association with haptics, as the company has pursued an aggressive patent acquisition and intellectual property protection strategy intended to secure its centrality to a host of emerging haptics applications.[39] The company's patents—which number in the thousands—are so vast in scope that they lay claim to the entire skinspace of the computer user (as shown in Figure 5.5, which is taken from one of Immersion's most valuable patents).[40] Though their practice of filing intellectual property violation lawsuits against prospective licensing partners has been met with some derision[41] and raises crucial questions about the relationship between intellectual property and the mediated sensorium, it has allowed the company to extend its tentacles widely throughout the hardware and software infrastructures of digital communication, particularly in the areas of mobile computing and videogaming. The company boasts that its TouchSense 3000 vibration technology for mobile phones has been

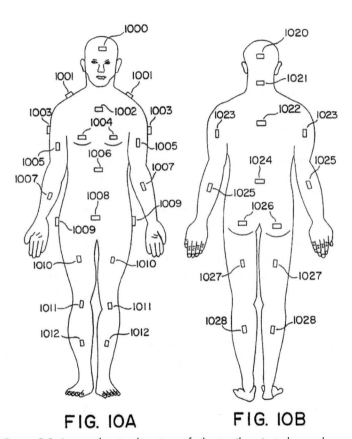

FIG. IOA   **FIG. IOB**

Figure 5.5. A map showing locations of vibrotactile units to be used in the transmission of haptic feedback to computer users. From Mark Tremblay and Mark Yim, "US Patent 6275213 B1 —Tactile Feedback Man–Machine Interface Device," May 1, 2000.

deployed in more than 300 million devices,[42] and worldwide sales of game consoles bundled with rumble hardware licensed from Immersion easily exceeds 500 million units.[43] Its ongoing partnerships with Samsung, Huawei, Kyocera, LG, and Motorola ensure that Immersion's software solutions will continue to modulate the vibrations emitted by mobile and wearable computing devices worldwide for some time to come.

I raise these details to show the sociocultural, economic, bureaucratic, and material levels at which Immersion labored to shape and articulate haptics as technoscientific touch. This process of remaking touch as haptics unfolded in the iterative design and testing of haptics hardware and software; in the drafting, acquisition, protection, and enforcement of haptics patents; in the marketing and licensing of Immersion's multifaceted haptics applications; in the adoption of best practices for coding haptic effects and haptic languages; in the deployment of new motors, actuators, and the new haptic algorithms that controlled their operation; and finally, in the often unconscious training of end users to read the machine-generated sensations that emanated constantly from their devices. Many of the developments in this process occurred with little fanfare. For example, the handset manufacturers' adoption of Immersion's VibeTonz system of customizable, assignable vibration alerts in the early part of the 2000s brought a new complexity to the tactile signals used in mobile phones, but the feature failed to generate widespread attention. Other developments were far more conspicuous, such as the protracted legal battle between Immersion and Sony that resulted in Sony's third-generation PlayStation console being released initially without its signature rumble mechanism. Regardless of the attention any one episode attracted, taken together, they indicate the extent to which Immersion succeeded at enframing haptics within a new network of standards, protocols, and languages distributed throughout the expanding ecosystem of digital computing.

Immersion's machinations were underpinned by a multistage discursive project, where it attempted to convince expanding sectors of investors, consumers, and potential corporate partners of touch's importance to computing. In press releases, interviews, product demonstrations, conference presentations, and documents distributed via its corporate Web site, the company's marketers, executives, and engineers advanced the case for the urgency of haptics. Building this narrative involved repeatedly rehearsing a series of related arguments, with Immersion contending first, that touch plays a vital but unacknowledged and neglected role in human experience and second, that touch's neglect is primarily an effect of contemporary computing. As the computer burrows deeper into the fabric of social and economic activity, Immersion suggested, the crisis brought on by the neglect of touch worsens and intensifies. The crisis can be alleviated—touch can be successfully deneglected and rediscovered—through the incorporation of haptic feedback into extant and emerging computer interfaces. Finally, Immersion suggested that, with its stable of patents and expert developers, the company was uniquely suited to this vital task of writing touch into the landscape of contemporary computing.

Immersion's careful crafting and propagation of these arguments in its white papers illustrates the intentionality with which it portrayed haptics, while also showing how the project of haptic interfacing was continually rearticulated and complicated in response to shifting trends in human–computer interaction. Although strands of these arguments in favor of touch feedback echoed and extended those initially offered by engineers during the epoch of haptic interface a decade earlier, by addressing their appeal to investors, designers, consumers, potential corporate partners, and end users, Immersion pushed the theme of touch's immediate use value to the forefront of its narrative. These papers functioned as sites fundamental to the articulation and formation of a new subjectivity defined by touch's rediscovery via haptic interfacing, participating not in a neutral recall of basic accepted facts about the functioning of the human haptic system and its technological extensions but rather in a normative social process of imprinting structure onto sensations. The documents naturalized and familiarized the model of touch discovered by passing the tactile system through an unnatural, highly technical series of scientific filters.

Immersion released the first and most concise in this series of documents—bluntly titled "What Is Haptics?"—in 2003, four years after becoming a publicly traded company and fresh off entering into a series of high-profile licensing agreements with corporate partners that included BMW, Microsoft, Sony, and Apple. The paper intertwined a plain-language explanation of touch's physiological structure with a framing of contemporary computing interfaces as missing the analogous structures that would allow for the feeling of onscreen events. Immersion deployed a functionalist model of touch that stressed its role as an always-on system that constantly fed back information about the physical world to the mind. Immersion described the complex network of mechanoreceptors distributed throughout the body as transmitters that sent "rich information" to a primary sensory cortex that served as a processor for this dizzyingly vast quantity of data. When sitting down in front of a "computer, cell phone, PDA, or any digital interface," the substantial data normally received by the primary sensory cortex is simply not present, effectively numbing touch, and making routine tasks often insurmountably difficult. For Immersion, forgetting the "wonderful, sophisticated sense of touch" was an effect of existing media technology. Users' brains had been lulled into a deep sleep by the ocularcentrism of computer displays, and Immersion promised to "rouse [the brain] from hibernation" by "adding back the tactile and kinesthetic cues" normally present in real-world activities.[44] The resultant configuration would produce a more satisfactory, successful, productive, and engaging pairing between human and machine by reactivating a dormant tactile mode of experience. According to this argument, the problem of touch's neglect originated in the raw materiality of informatics: lacking haptic feedback, the graphical user interface simply failed to act on the complex information receptors that constituted touch. A similarly material process, then, provided the key to deneglecting touch, as it could be awakened through the application of machine-produced sensations that would provide stimulation—or information, in Immersion's framing—to the slumbering tactile nerves.

To buttress these claims, Immersion soon added a "Haptics Glossary" to its site, providing brief definitions of terms that related both to haptic human–computer interfaces (such as "haptic effects," "rumble feedback," and "actuator") and to the physiology of touch (including terms like "mechanoreceptors," "Meissner's Corpuscles," and "Merkel's discs"). The glossary indicated the extent to which computer haptics had succeeded in fusing the scientific study of touch with interface design, while reinforcing a conceptualization of the user's tactile system as a network of differentiated and specialized nerves that functioned as receivers of information. Further, the glossary offered an updated definition of haptics that annihilated the distinction between touch and the technologies designed to act on it. Whereas in "What Is Haptics?" Immersion defined the term variously as "a Greek word meaning the science of touch" and "the science and physiology of the sense of touch," the "Haptics Glossary" subtly added a new layer, describing haptics as "the physiology of human touch *and the technologies used to engage it more fully.*"[45] In this sleight of hand—corporate propaganda masquerading as public pedagogy—Immersion erased the difference between the scientific study of touch and its stimulation through technology, while also painting human touch as a sense modality purportedly marked by a fundamental dearth of engagement that necessitates technological intervention. As I suggested in Interface 2, haptics—as the doctrine or science of touch—had been technological since its inception: from Weber's simple compass experiments to the mapping of hot and cold spots using targeted bursts of electricity, the science of touch depended on artificially generated sensations to quantify and subdivide the tactile processes. But Immersion, foreshadowed by the computer haptics movement of the 1980s, wove technology into the category of haptics itself, framing touch not as something that existed outside of and above the social, but embracing instead a touch that was necessarily incomplete and disengaged without the productive intervention provided by (Immersion's) technology.

With the growing reliance on touchscreens, the crisis facing touch seemed to be worsening, in spite of Immersion's vast efforts to forestall it. For all of their purported tactile shortcomings, the range of interfaces being displaced by touchscreens—keyboards, computer mice, clickwheels, keypads, buttons, and dials—at least provided touch with some valuable information. Responding to this dematerialization, Immersion published a series of six white papers that articulated the value that haptic feedback added to a variety of computing applications, including handheld mobile interfaces, in-dash automobile instrument panels, videogames, point-of-sale systems, and surgical simulators. Extending the narrative it seeded in "What Is Haptics?" Immersion offered two competing visions of computer users: a negative portrayal, in which users were deeply troubled by their constant interactions with haptics-less computing, and a positive depiction of users whose lives could be drastically enhanced through the ubiquitous deployment of something the company termed "haptic touch" in computer interfaces.

The negative vision was no generalized and baseless fear-mongering. Immersion supported its claims about the perils of computing without feeling by summarizing a host of user performance and satisfaction studies that quantified the differences between interfacing with and without haptic cues.[46] Complete with graphs that visualized the information presented (see Figure 5.6), Immersion's 2010 white paper "The Value of Haptics" contained references to more than thirty publications on touch computing, with the sheer volume of research devoted to the subject indexing the severity of the problem. An extension of the Taylorist tradition of monitoring, quantifying, and routinizing labor, the studies cited in the white paper charted users' behavior, output, task completion rates, and emotional well-being. The wide-ranging dangers of haptics-less computing detailed in the document included increased rates of errors when entering and manipulating data, a lack of satisfaction with the human–computer interaction experience, inefficient data entry, an inability to receive messages privately via the tactile channel, distracted focus due to competing messages crowding the audiovisual channels (particularly important when attempting to manipulate computer interfaces while engaged in attention-intensive tasks like driving), an inattention to incoming messages (listening for the ringing cellphone in a noisy environment), eye strain resulting from constantly reading on small screens, and a deficiency of emotional attachment both to other networked communicative subjects and to the device itself. Immersion painted a bleak picture of computer users beset by a range of

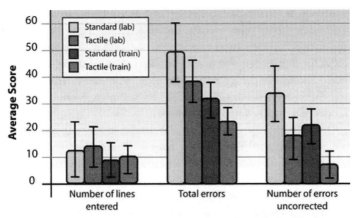

Figure 5.6. Improvements in user experience with tactile feedback. The graph was provided as evidence that haptic feedback increases typing speed, reduces errors, and contributes to higher error correction. Adapted by Immersion Corporation from Stephen Brewster, Faraz Chohan, and Lorna Brown, "Tactile Feedback for Mobile Interactions," in CHI '07: Proceedings of the SIGCHI Conference on Human Factors in Computing Systems (New York: Association for Computing Machinery, 2007).

digital-induced woes, all brought about by the ocularcentrism of conventional interface schematics.

In other words, Immersion appealed to consumers as information subjects with vested interests in maintaining healthy, productive, and personal relationships to the various digital communication devices they labored with. This connected self needed to be cared for as a function of neoliberal economics: not identifying as a subject of this new machinic haptics would leave the information-manipulating subject disadvantaged in a competitive marketplace, where success depended on the ability to process, circulate, and respond to data quickly and efficiently. The sensorium—and especially the tactile channel—had to be always on, always open to receiving the vital messages that flowed through networked information machines. The model of haptic subjectification operating in Immersion's narrative involved asking individual consumers and those designers working on mobile user experience to identify the existence of a haptic subjectivity that was a subcomponent of the informatic subject more broadly. Immersion framed the story so that the mere acknowledgment of this haptic subjectivity brought with it the automatic recognition that it existed in a state of crisis—the haptic component of the self required rehabilitation and careful attending to through the layering of haptics technology onto existing modes of human–computer interaction.

In the positive vision, Immersion described a society poised for a dramatic epochal shift, on the cusp of moving into an era that would render obsolete this notion of the self as an information circulator. The coming age would be marked not by the circulation of information, but rather by the creation of experience. Crucial to this movement was the awakening of a haptic subjectivity, accomplished by incorporating Immersion's product suite into computing interfaces, as touch's rediscovery via haptic interface would enable the unleashing of a new symbiosis between humans and computational machines. Ontologizing touch as "the most direct means of communication" and "the only sense capable of simultaneous input and output," Immersion repeated claims mobilized around the touchscreen by suggesting that haptic feedback would allow for "more natural" interactions with data. The impending change brought by haptics would be more than merely quantitative—more than the addition of simple tactile data through "the 'dumb buzz' of today's cell phones"[47]—and instead would entail a wholesale upheaval in the existential mode of digital life. Weaving this "haptic touch" into computing via Immersion's range of digital solutions would provide "the missing piece, the sensory element that will transform information into *experience*."[48] The narrative embraced a crude, deterministic model of media evolution, where the horizon that haptics promised to eclipse was one seemingly hard-coded into audiovisual interfacing. For all the benefits the shift brought, the transformation of mobile devices from "mere phones into the ultimate digital companions for anytime, anywhere information communications, productivity, and entertainment" had exposed the limits both of "Information Age technology" and of "exhausted consumers'" overtaxed information-processing capacities.[49] The hapticization of computing would correct this imbalance, and by doing so, provide a productive new centrality for embodied sensory experience:

> Engaging that action/reaction potential of haptic touch, technology in the "Age of Experience" will bring to life completely new market and economic opportunities in healthcare, telecommunications, entertainment, safety, education and commerce, to name a few. Just as movable type democratized literacy and fueled massive social change in the 15th century through growth of written information, haptic experience will pave the way to something new in the 21st Century: new ways of learning, exploring, sharing, and understanding the world through direct sensory experience—transmitted, simulated, or recorded. We will move beyond the mere digital exchange of information "for the brain" to a more holistic, all-sensory experience "with the body."[50]

The era Immersion promised to usher in "by layering the primal yet vital sense of touch into our digital lives" embraced and furthered the logic of analog medialization, claiming that haptics was not merely additive, but transformative and positively disruptive.[51] Technology, media historical time, and the human senses here were imbricated in a dialectical process that locates human–machine haptics as the endpoint in an interplay of industrial, economic, computational, and social forces—what we can understand, following Bernard Stiegler, as a perpetual genesis of "technical individuals" that paradoxically makes constant claims to having arrived at finality.[52] It is not enough that the addition of haptics transforms media interfacing. It also must transform (to again recall McLuhan) the humans that media interface with and the social and economic practices that arise from the agitations of their newly reconstituted sensing selves.

It is tempting, and perhaps not wholly inaccurate, to reduce all of this to just routine commodity fetishism, part of what Jackson Lears identifies as a "therapeutic ethos"[53] used to foster demand for new products for over a century. That is, Immersion's trumped-up claims about the perils of haptics-less information manipulation and the promises of haptics to instantiate a rupture between the present and a (utopic) future seem to be hyperbolic and inflationary rhetoric typical of advertising copy. The focus on experience as the driver of consumption was very much in the air by the time Immersion crafted these documents.[54] Claims that computing technology could be used to enhance and augment the human intellect also had been central to the visions outlined by early pioneers in the field, including Alan Kay, Douglas Engelbart, Ivan Sutherland, and Frederick Brooks. Nor was marketers' equating of forward progress in computers with utopian social transformation unique to Immersion. From Apple's famous "1984" commercial for its Macintosh personal computer to the "digital utopianism" of the 1990s, the association between new computing machines and social progress had become automatic.[55] I suggest that by grounding the technoutopian potentiality of computing in the conjuring and cultivation of a new haptic subjectivity, Immersion's protracted campaign simultaneously furthers and deviates from the logics mobilized by these prior narratives. By positioning a single sense modality at the center of computing's transformative possibilities, Immersion both prompts a revaluation of touch and calls for the embrace of a reconstructed tactility made adequate to the challenges computing places on the senses.

With the continued spread of touchscreens and progress in both actuator technology and haptic control algorithms, Immersion's efforts remain ongoing, producing new human–machine languages that depend on the cultivation of trained haptic subjects to articulate their value and attend carefully to their vibrations (see Figures 5.7 and 5.8).[56] Immersion's commodification strategies position haptics as a way to add value both to mobile computing devices themselves and to the products advertised on mobile touchscreens. In this latter figuration, touch becomes a way to navigate a pathway through the "advertising clutter" strewn about the visual and audio channels.[57] Haptic trailers for movies and television shows, first used by Immersion in a teaser for the Showtime series *Homeland* in 2014, employ its TouchSense Effects to differentiate ads themselves as products. This was not the simple transcoding of sounds-as-vibrations long employed in rumbling movie chairs. Instead it entailed the writing of haptic effects designed specifically to provide a tactual narrative braided into the audiovisual one

	ERM	LRA	Piezo Module	EAP
Actuator Types and Characteristics				
Form Factor	Bar or hockey puck	Hockey puck	Matchstick	Flat panel
Approximate Size	11 x 4.5 dia. mm	10 x 3.6 mm	3.5 x 3.5 x 42 mm	45 x 38 x 0.8 mm
Power Requirements	130-160 mA RMS @3V	65-70 mA RMS @3V	300 mA RMS @ 3V	Ask vendor
Frequency Range	90-200 Hz (non-uniform strength)	150-200 Hz, single frequency (e.g. 175 Hz)	150 to 300 Hz usable	90 to 125 Hz (resonant peak), 50-200 Hz usable
Mechanical Time Constant	50 ms	30 ms	<5 ms	<5 ms
Durability	Variable	Very durable	Very durable	Excellent
Fidelity of Sensations	Low	Medium	High	High

Figure 5.7. Actuator types and characteristics. From Immersion Corporation, "Haptics in Touchscreen Handheld Devices," white paper (San Jose, Calif.: Immersion Corporation, 2012), 6.

Haptic System Overview

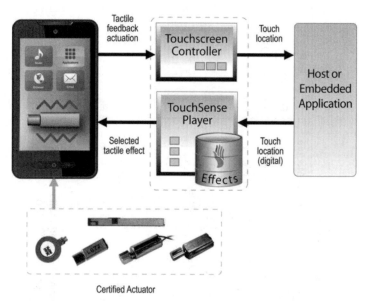

Certified Actuator

Figure 5.8. TouchSense 5000 haptics system overview. From Immersion Corporation, "Haptics in Touchscreen Handheld Devices," white paper (San Jose, Calif.: Immersion Corporation, 2012).

provided by conventional video commercials. The coding process resembles the one employed in the writing of rumble for console controllers, which Immersion had developed expertise in progressively from the late 1990s on. In this case, however, Immersion framed haptic effects not as the means of adding depth to gamic or cinematic narratives but instead as a way to add value to the advertisements themselves. Soon after, Immersion partnered with Chinese entertainment company LeTV to bring haptic effects to a range of movies distributed through LeTV's mobile online video platform, and with Lexus and Peugeot to develop TouchSense-enhanced ads for their new automobile models. The addition of a "haptic track"—a specially coded file that "triggers the actuator on the mobile phone to generate tactile effects synchronized to the video content"[58]—distinguishes these new ads from their conventional counterparts. Further, the reception of haptic effects depends on the user engaging in a new bodily configuration to the mobile device: learning to hold it tightly, to press the sensitive pads of the fingertips against the screen, and to make intelligible the dynamic relationship between carefully assembled images, sounds, and vibrations emanating from the device.

My point through all of this is specifically *not* to suggest that Immersion's attempted subjectification program has been executed successfully. Instead I have highlighted the multilayered and protracted operation of Immersion's technoscientific haptics, as it diffused throughout a global network of actors that includes device manufacturers, software designers, electrical engineers, legal teams, content providers, marketers, and various groups of socially situated end users, each of whom was asked to respond to a purported crisis of perception brought on by the accelerating spread of visualist computer interfaces (including, in Immersion's framing, touchscreens). While haptics appeared suspended in its state of perpetual immanence, beneath the surface—concealed from view and yet tangible in the ubiquitous vibrations of mobile communication devices—it slowly mutated, gaining formal and concrete expression in these institutional, economic, technical, and commercial spaces. Haptics as the scientific study of touch slowly became haptics as the technological stimulation of touch, with one corporate firm serving as the prime mover in this strategic rearticulation.

The Teledildonic Imagination

As I have suggested throughout *Archaeologies of Touch,* the story of haptics has generally been one of fragmenting, atomizing, and subdividing touch into increasingly specialized component parts while also separating out those sensations, such as pain and pleasure, that interfered with touch's capacity to functionally discriminate between stimuli. However, the development of computer-controlled machines to facilitate remote, computer-mediated sexual contact entailed the explicit mobilization of the same technoscientific haptics used in other telemanipulation and virtual reality systems, serving to write sexual pleasure back into the discourse of haptics. Commonly designated by the term "teledildonics"—a neologism attributed to computer visionary Ted Nelson[59]—computer-mediated sexual encounters have long been the stuff of cybercultural legend. Like haptic interfaces more generally, their story is given structure by the narrative threads of perpetual immanence, the master device, and the logic of analog medialization. Their popularization as a cultural imaginary, too, is intertwined with the overarching visibility Howard Rheingold gave to touch interfacing with the publication of *Virtual Reality*. Rheingold's engagement with the subject began in a short post to the Whole Earth 'Lectronic Link (later reprinted in a 1990 *Mondo 2000* article)[60] that he revisited and expanded in a chapter included in *Virtual Reality* titled "Teledildonics and Beyond." Citing the technical difficulties involved in producing an interactive, one-to-one computer-generated copy of a remote body, Rheingold cautioned his audience against expecting the rapid proliferation of teledildonics. Those who read his work, however, frequently came away with the impression that such fully embodied, real-time sexual encounters were imminent to 1990s-era computing. Rheingold himself rather conservatively estimated that "portable telediddlers" were thirty years from achieving ubiquitous adoption. And while "teledildonics," by suggesting heteronormative and androcentric notions of female sexual pleasure, presented a terminological problem that he was keenly aware of, Rheingold preferred to keep the name in spite of its "archaic" connotations, instead of suggesting the "more sober" designation of "interactive tactile telepresence."[61]

We are nearly at that magical thirty-year point, and portable telediddlers have yet to achieve the ubiquity Rheingold predicted for them. But in the brief treatment of teledildonics I provide here, my interest lies not so much in cataloging and chronicling the various failed instantiations of the technology,[62] or with speculating on the reasons that the technological real has not yet comported itself to Rheingold's fuzzy timeline. Nor do I want to intervene in the long-running debates about the merits and demerits of telepresent and machine sex.[63] And while it is crucial to recognize from the outset that much of the language used to frame teledildonics is shot through with heteronormative and androcentric notions of female sexual pleasure, teasing out these themes will not be the focal point of my analysis here. My concerns are instead circumscribed to the specific relationship between haptics as a techno-cultural means of commodifying touch and teledildonics as a specific embodiment of touch's reordering, which I tease out through a case study of the RealTouch remote sex system. This detailed glimpse into the recent past of teledildonics is intended to show how these touch machines have spread in practice, situating them as technologies—like haptics more generally—that belong to the past and present as much as they do to some just-on-the-horizon future. In addition, my examination of the RealTouch system revisits the idea raised in Interface 4 that haptic interfaces embody strategic segmentations and reassemblages of touch, with the interface providing the science of haptics a material expression. Finally, the RealTouch's discursive positioning as a haptics technology, accomplished through specific references on its Web site to haptics as the science of touch, served the strategic aim of adding value and legitimacy to the company's cybersex machine. Speaking to these themes—uncovering the intertwined material and discursive construction of the RealTouch—requires me to indulge in some graphic descriptions of the device's operation. My goal in providing these details is specifically not to shock, but rather to illustrate, by reference to the particulars of the device's technical and social construction, the way that haptics gained expression through teledildonics. Attending to these details allows me to show how the power of haptics—a formally constituted technical discourse of touch capable of segmenting, ordering, and calculating the body— took hold over sexuality in the process of making a gendered model of sexual pleasure transmissible through computer networks.

Manufactured and sold from 2008 until 2014, the RealTouch system represented a significant leap forward in both the complexity and feasibility of teledildonics, serving as a way to commercialize both the device itself and the content that would be delivered through it. The machine consisted of a USB-connected, computer-operated penile sheath that acted on the male's genitals through a

combination of three different mechanisms: 1) a "precisely controlled" dual-belt drive system that allowed the device to "simulate
a wide variety of sexual positions," 2) a heating element that
warmed it "to actual human body temperature," and 3) a reservoir
of strategically released lubricant used to generate "natural levels
of wetness" (see Figure 5.10). As the wearer streamed a pornographic film specially encoded with a layer of data to control the
RealTouch, the device's dual belts—"comprised of a soft, supple
material called VersaFlex engineered to feel just like actual
skin"—spun and tensed to simulate the onscreen sexual act. Using
four motors, it varied the speed, intensity, and tightness of the belts,
allowing it to emulate the feel of "the mouth, vagina, or anus of a
real human." The RealTouch expressed each of these orifices as a
set of numerical values that controlled the device's mechanisms.
By synchronizing these values with data that the computer rendered through the monitor and speakers, the RealTouch established
semiotic linkages between image, sound, and touch. Or, as Real-
Touch expressed it in a promotional image on its Web site that fully
embraced the logic of analog medialization: "You can *watch*. You
can *hear*. Now you can *FEEL!*" (see Figure 5.9).[64] RealTouch's software differentiated the stream of tactual data itself, with separate
code for governing the oscillations of the belts, controlling the
"well-timed" release of lubricant, and regulating the heating element's emanations. The RealTouch thus worked not only to inscribe
the differentiation of the senses at the material level, but also to

Figure 5.9. "Feel the Action": Promotional image explaining the value
the RealTouch adds to existing modes of viewing pornography.
From "RealTouch: How It Works." Courtesy of RealTouch, powered by
AEBN.COM.

Interior

Made from a highly-specialized, custom-made material, the RealTouch interior is soft and supple. Its realistic texture makes it the closest thing to actual skin for an unmatched level of excitement and comfort. Using a process called relative motion, RealTouch gently strokes your entire length at variable speeds, effectively stimulating touch receptors in the skin. You'll enjoy a heightened level of sensitivity while getting the results you desire.

Heat

Inside are two heating elements that gradually warm the RealTouch to actual body temperature during use. RealTouch distributes the warmth evenly throughout, penetrating your body deeply to sooth and relax, while reproducing the sensation of real skin-to-skin contact.

Lubrication

The lube reservoir holds a generous quantity of recommended RealTouch lubricant, releasing it in precisely the right amount, at precisely the right time, every time you use RealTouch. Lubrication adds not only a dimension of realism to the RealTouch experience, but it intensifies and enhances the sensation of pleasure during use, helping to prolong your performance for maximum satisfaction.

Tightness

RealTouch has a specially tailored orifice which gently flexes and squeezes to create a comfortable seal around you during use. Able to accommodate men of almost any size, the orifice and its dynamic feedback provide a snug fit to feel like actual penetration.

Figure 5.10. Diagrams of sensation-producing mechanisms in the RealTouch. From "RealTouch: How It Works." Courtesy of RealTouch, powered by AEBN.COM.

embody the nineteenth-century division of touch into mechanically isolatable subcomponents. Moreover, the device's material configuration brought with it a male gendering of the feeling subject in this new mediatic order of sensations, as the "you who can feel" must necessarily be anatomically compatible with the machine's hard-coded specifications.

This complex mechanism for producing the cybersexual real functioned simultaneously as a technical system and as a business model. The Adult Entertainment Broadcast Network (AEBN), a Charlotte-based video-on-demand company whose stable of pornographic videos featured in excess of 100,000 titles, developed the device as a way of adding value to its streaming service. AEBN set an army of coders to work adding tactual effects to selected videos in its library, quickly amassing a storehouse of RealTouch-enabled video. Customers who purchased the device (which initially retailed for $200 and included a four-ounce lubricant packet) were locked into the RealTouch Network's walled garden of content. Through this hapticification strategy, the RealTouch also worked to provide a defense against the growing threat that piracy posed to the adult video industry; as AEBN CEO Scott Coffman explained in a press release, the haptics-encoded videos, by "delivering the sensory dimension of touch," would prove impossible to pirate. "You can pirate the movie," Coffman claimed, "but you can't pirate the experience. It would be like trying to steal a roller coaster."[65] The massive undertaking of adding haptics to the videos required the coders to move scene-by-scene through the library, imagining what the scene would feel like on their own genitals, and then assigning that complex of sensations a string of numerical values to control the operation of the motors, the release of the lubricant, and the temperature of the device's interior. Like the RealTouch itself, the immense labor involved in producing the teledildonic real—in haptifying AEBN's library—was also gendered male, on the assumption that male coders would be the ones best equipped to design sensations for the machine. By 2012, the RealTouch Video-on-Demand library included more than fifteen hundred haptics-enabled scenes, organized by sexual orientation (the front page of the RealTouch site asked visitors to self-identify as gay or straight before displaying the video library), sex act, and model type—with nearly every scene available in the library employing a point-of-view shot. AEBN set streaming rates at \$.50 per minute, with videos playing only through RealTouch's proprietary software.

Later that year, the situation drastically changed when Real-Touch added a mechanism that allowed for real-time remote manipulation of the device. Dubbed the JoyStick, this dildo-shaped device used sensors to capture the movements of its operator, then translated the pressure and motion of those movements into code, which was transmitted digitally for rendering by the RealTouch. Employing seven sensors positioned strategically at different points on the JoyStick and featuring soft buttons that could modulate the temperature and lubrication of the remote RealTouch, the device permitted its operators to exercise a precise level of control over the male organ. The gendering of the labor involved in producing sensations for the RealTouch shifted instantly, as the company founded the RealTouch Interactive network, referred to by *Engadget*'s Daniel Cooper as "the world's first digital brothel."[66] An army of "Cam Girls"—women who perform live sex acts for the purpose of "live interactive masturbation"[67]—began contracting with Real-Touch to incorporate haptic feedback to their streams (according to Cooper's interview with RealTouch product manager EJ, the company attempted to bring male models into the fold, but "the gay crowd just wasn't interested in it," and the male models quickly exited the network).[68] Questions related to this new form of digital labor will be taken up below. For now, I highlight the process by which the RealTouch encoded, specified, and differentiated bodies at the material level as they passed selectively through the filter of the interface. The interface's affordances both shaped and were shaped by economic and labor considerations, as the device expressed the financialization strategies AEBN pursued with regard to RealTouch.

As much as it depended on the ingenious design of hardware that would realize these aims, the device's success also hinged on discursively positioning haptic feedback as a desirable and vital component of the pornographic experience. The RealTouch deployed haptics not only in the engineering of the device's hardware and the coding of its tactual effects, but also featured the term prominently in descriptively marketing the machine, with the neologistic and scientistic connotations that accompanied haptics figuring crucially in the device's positioning as the future of sex:

> Haptic Technology: Called the "Science of Touch", this advanced technology using relative motion allows RealTouch to synchronize its movements in realtime with step-by-step instructional videos, adult videos and many other features to come.[69]

In this narrative framing, haptics—as the science of touch—provided the means of accessing a faithfully synthesized real. The precision with which the device simulated the real was attained through a careful, technicist engineering process valorized repeatedly in the RealTouch's promotional materials. Images of the device broken down into its component parts (see Figure 5.10) illustrated the complexity of haptic stimulation it promised to render. This act of laying the machine's insides bare—exposing the motors, belts, and cartridges responsible for its operation—sought to construct a haptic subjectivity similar to the one mobilized by Immersion's documents. The new subjectivity worked by clearing away the old one, by branding the displaced subject obsolete, unfree, and miserable. The old subject, in short, was one defined by its lack of touch; as RealTouch explained on the front page of its Web site: "you'll never watch adult videos the same way again. Go beyond the limits of sight and sound and feel what you've been missing!" By rendering through haptics, the device offered to erase the distance between the desiring male body and the bodies of the females and males captured in AEBN's vast archive (or connected to the RealTouch Interactive network), bringing the wearer into contact with "some of the hottest models around at their sexual peak."[70] The haptic subjectivity conjured here celebrated the technicity of the interface as the means to its effacing; embracing the science of touch assures the new subject that it will experience the pleasurable erasing of the interface.

The layering of a gendered, calculable haptics onto extant media texts, then, facilitates the subject's passage into a new era of interfacing, defined by its capacity to bring about a specific type of physical contact between networked subjects. In doing so, it encodes ideologies about the desired functioning of the networked bodies in this futurological ordering. As EJ explained pointedly in the 2014 HBO documentary *Sex/Now*:

> We're going to take sex over the Internet into the future. Sex over the Internet started with still images, then you could download a video clip. So what is the next thing? The next thing is being able to actually have sex over the Internet. . . . We've always said that on that day when a girl in Romania can reach out and touch your penis, that's the beginning of something completely new.[71]

The move into the technomediatic future—the progressive, analog medialization of the senses—is defined here by the capacity of haptics technology to collapse and erase the gap between bodies positioned by geography, nationality, gender, class, and age (given the rampant use of child labor in the sex industry, EJ's use of the term "girl" here is even more problematic than it would be in another context). As with other haptics technologies, the interface is a strategy of replication, a carefully considered process of instrumentally selecting which parts of the body and which sensations ought to be—and which ought not to be—translated through the interface. The material and discursive positioning of the haptics combines here to reduce the female to an instrument for male pleasure: For EJ, the future will have arrived once the female can reach out to manipulate the male receiver's organ for *his* sexual pleasure. The RealTouch Interactive's unidirectional flow of haptic data, from female manipulator to male manipulated, might seem almost empowering on one level—allowing the female to reach out and control the male without fear of a reciprocal touching back—were it not bound by its unfolding in the commercial structure of AEBN's "digital brothel." Although EJ suggested that the device may find a use outside of AEBN's sexual marketplace (such as enabling distanced sexual contact between romantically involved couples), the material configuration of the machine betrayed its primary function as a device for the commercialized production of androcentric pleasure.

For the laborers in AEBN's network, haptics provided a different promise and attempted to mobilize an alternative haptic subjectivity. Neoliberal capitalism encourages the worker to identify as an entrepreneurial self—to manage and care for the self as an instrument for the production of value.[72] Sex work in general, and cybersex in particular, unfolds in this broader context, and one of the earliest utopian narratives articulated around teledildonics concerned its potential to improve labor conditions for sex workers (it is worth noting here that when Rheingold initially wrote on the subject in 1990, his home city of San Francisco was a hotbed for sex worker unionization). The layering of haptics onto extant technologies of cybersex labor becomes a means of differentiating sex workers in a competitive marketplace. Joining the Real-Touch Interactive network opens up the possibility of forming new affective bonds with customers—a cybersexual experience made more *real* and affectively engaging through the materiality of remote manipulation. The crucial question here concerns the potential of new digital technologies to transform labor practices—the interface's capacity to liberate and empower those who fuse themselves to it. But technologies function as convenient stand-ins for and distractions from the broader economic and legal structures enframing them. EJ mobilized precisely this vocabulary when claiming that RealTouch "tried to empower models"[73] by allowing them to fix their own rules and pricing structures (according to Cooper, the rate for oral sex via the RealTouch ranged from $30 to $60). In spite of the company's assurances, these workers remained bound by RealTouch Interactive's networking infrastructure; whatever "empowerment" the device brought rubbed up against the legal and commercial framework that constrained the model's freedom to use the device on her own terms, outside the labor network established and maintained by RealTouch.[74] Like other new technologies, the RealTouch promised to desubjectify—to free and empower by clearing away the old subject—while producing new mechanisms of subjectification. Haptics orders touch simultaneously as a technical problem to be confronted through design and as a cultural problem to be confronted through strategies of marketing and financialization.

In addition to allowing its male user to be manipulated by pornographic videos and remote models, AEBN also marketed the RealTouch as a device that could improve sexual performance and prowess. Allowing the "Male Enhancement Application" to take control over the machine would help the user increase their size, stamina, and performance "naturally, without the use of pills, drinks or pumps." This was not a new function for touch machines; since the mid-1800s, men had connected strange electrical machines to their genitals in the hopes of regaining lost sexual energy, and the RealTouch continued in that tradition.[75] But the claim that the RealTouch provided a "natural" method of enhancement was curiously at odds both with those prior technologies and with the self-aware fetishization of the RealTouch as a technological wonder. Here, the ability of haptics to slip effortlessly between connotations—to designate both a touch that submitted productively and pleasurably to technology and a touch that was fundamentally human and pretechnological—proved essential to this framing of the device as a naturalistic machine for restoring potency to the depleted male organ.

The RealTouch's potential to revolutionize cybersex will remain unrealized for the moment, not due to any categorical cultural rejection of pleasurable manipulation by computer-controlled machines, but instead as a result of a complex confluence of circumstances surrounding patents that AEBN had licensed following a 2010 lawsuit. According to Cooper's account, RealTouch ran short on capital—and consequently, had to stop manufacturing the device (and the ever-important replacement parts for existing devices)—just before cultural fascination with the device caught fire (the HBO documentary that the RealTouch featured prominently in would unfortunately only air in 2014, once the machine was no longer available for sale). Teledildonics patent infringement lawsuits are a tangled mess, and the specific chain of events that caused AEBN to abandon the RealTouch merits further investigation.[76] But the existence of the battle itself provides an enduring illustration of the history haptic interfacing has already acquired, in spite of its repeated positioning as a technology that belongs to an unrealized future. It shows haptics to be enmeshed in and inseparable from a comingling of legal infrastructures, business interests, labor practices, manufacturing networks, software suites, and hardware platforms. And further, it shows these various protocols underpinning haptics to be sites of contestation between competing, heterogeneously situated actors who each seek to deploy and mobilize haptics as an instrument for their own ends.

The axiomatic and reductive claim that "sex drives technology"—echoed by EJ in *Sex/Now*—seems not to quite hold up in the case of the RealTouch. Instead, the substantial institutional investments made in haptics from the late twentieth century on staged the development of the RealTouch. Ramon Alarcon, the device's inventor, had previously worked briefly as an intern at NASA's Ames Research Center and spent nearly six years as the director of Immersion Corporation's gaming and entertainment business unit, where he brought over twenty gaming-related peripherals to market, and oversaw the establishment of licensing agreements with Immersion's many corporate partners. Other members of the team who worked on the project over the years received graduate degrees in mechanical engineering from Stanford, which, like Ames and Immersion, has also been a hotbed for research into haptics.[77] While the RealTouch represented a significant technical step forward for teledildonics, it built on a foundation of training institutions laid by government grants and private investments in haptics.

Before moving on to the chapter's concluding section, let me make one final comment on the relationship between electromechanical technologies of sexual pleasure and the history of touch machines traced throughout this book. During the twentieth century, as researchers worked to perfect vibratory communication systems, a parallel development in sexual therapeutics employed unbalanced, weighted motors—similar in function and material configuration to those used in the Vibratese apparatus— to produce vibrations that would induce "hysterical paroxysm" in females. Marketed and sold as what historian Rachel Maines famously referred to as "socially camouflaged technology," physicians initially developed these machines in the nineteenth century as labor-saving devices to automate the induction of female orgasm, participating in what Maines described as "medicalizing the production of the female orgasm."[78] Maines framed therapeutic massage as an industry that electromechanical vibrators helped initially mechanize and deskill, before they eventually brought about the physician's removal from the process altogether as "treatments" migrated into the home. The design of these machines, sold openly on the shelves of drugstores and marketed in advertisements found in popular, female-targeted magazines, remained relatively stable for roughly a century. But with the miniaturization of computers and the increasing availability of cheap microchips, what one adult publication referred to anachronistically as "cyberdildos"[79] (or more accurately: "digitally enabled designer sex toys")[80] began to increase in complexity and function, as product designers added programmable, customizable patterns of vibration and motion. The patterns these machines dynamically rendered for their users started to resemble those that Geldard and his colleagues had used to write language onto the body in the 1950s, and sex toys' parallels with the vibrations produced by cellphones and videogame controllers quickly morphed into a source of constant cultural amusement and fascination.[81] While the early years of teledildonics were marked by the continued dominance of what Maines called the androcentric model of female pleasure, these more recent developments have self-consciously attempted to move to female-centered user design and product marketing, what Bardzell and Bardzell describe as an "embodied design process" that focuses on the designer's own experience with the product.[82] As with the RealTouch, many of these product designers were professionalized outside the sex toy industry, with backgrounds at companies like Apple, Nike, Motorola, and Ericsson. In closing, my point is to suggest that the general disaggregation of sexual pleasure from haptics is currently being actively undone by a combination of shifting design practices, changing attitudes toward female sexual pleasure, and the exertions of those makers and modders attempting to overturn the old patriarchal and heteronormative assumptions encoded in the engineering and marketing of sex toys. Throughout all of these shifts, the mainstream haptics community continues to largely disavow the close links between haptics and the industry that manufactures these computer-controlled touch machines.

Neglected No Longer

Following the epoch of haptic interface, touch seemed poised to assume a vital new role in the sensory configuration of computing, as it promised not only to usher in a new era of human–computer interaction but also to bring about a transformation in the cultural sensorium, where a touch salvaged by technological prostheses would assume its rightful place alongside vision and hearing. Instead, this future always appeared perpetually on the horizon, with no single instantiation of touch technology able to produce the foretold upheaval in the epistemic ordering of the senses. The examples presented in this chapter show that, however much haptics may have fallen short of the impossible promises mobilized around it, attempts to bring touch to computing have nevertheless fundamentally and profoundly altered the category of touch itself, enclosing it in new languages, legal apparatuses, algorithms, financialization schemes, and cultural techniques. These are not unheralded alterations but instead cultural, material, and institutional expressions of the scientific and technical positivities accumulated around touch since its initial structured electrification in the eighteenth century. Under overdetermined and carefully calculated conditions, touch has been consistently attended to by the designers of computational media, with the outcomes of their exertions felt in the ubiquitous vibrations constantly emanating from everyday objects.

Although touch may at times seem to be, as John Durham Peters claims, "stubbornly wed to the proximate"—a sense that "has no remote capacity" and "resists being made into a medium of recording and transmission"[83]—the examples provided here show that it is also something extended, captured, stored, and transmitted (even if in diminished form) by computational systems. It may be tempting to think of the repeated commercial failures of haptics technologies as evidence in support of the thesis Peters offers, as empirical validation of an intuitive and axiomatic claim about touch's psychobiological capacity to resist mediation. In this figuration, the marketplace serves as a laboratory for conducting an ongoing series of experiments designed to test the hypothesis of touch's capacity to acquire its own storage and transmission media, with the results thus far confirming the longstanding belief that touch resists virtualization. However, such a pronouncement would be premature and constraining—the projection of commercial standards for determining success and failure onto ontological models of the senses. It downplays the complex interplay across culture, technology, and physiology in shaping what counts as an acceptable proxy for the direct or "immediate" sensing, assuming that such standards are absolute and fixed rather than relative and elastic.

Moreover, such a move, in foreclosing touch's potential to act on and be acted on remotely, risks diverting our attention from the afterlives of these various "failed" technologies (which have not been in the ground long enough yet to be pronounced "dead media" or even "zombie media").[84] In a curious but inevitable act of technological repurposing, Novint's Falcon haptic interface (see Figure 5.11) currently enjoys a new life as a teledildonics device. Originally marketed in 2007 as a videogame controller that could add realism to virtual worlds via complex, customizable haptic effects, the Falcon's commercial failure left it an orphaned machine.[85] By replacing the Falcon's standard ball grip with a variety of sex toys, modders playfully transformed the Falcon from a gaming peripheral into a cybersex instrument. Initially little more than a curiosity, the startup company FriXion developed an application program interface intended to facilitate the remote manipulation of sex toy–equipped Falcons in real time. Billed as "a new kind of social network platform," FriXion uses "real-time bidirectional feedback telemetry to achieve convincing and organic intimacy."[86] Through this strategic reappropriation, a technology developed initially for scientific visualization and later downscaled for use in videogaming now becomes a means of allowing remote subjects "to touch and be touched at any distance." The Falcon's easy availability[87] facilitates this creative bricolage or hybridization

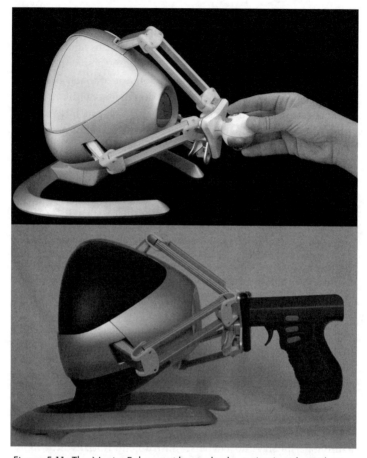

Figure 5.11. The Novint Falcon, with standard grip (top) and pistol grip (bottom). Courtesy of Novint Technologies.

of interfaces, allowing bootstrapping tinkerers to raise the fundamental question of "what happens if we splice a Fleshlight or a dildo onto the Falcon?" What new aesthetic and social possibilities arise from this creative remixing of hardware?

I suggest therefore that it is not just the abstract technoutopian dreams of mediatic haptics that remain as traces of these previous exertions, but also the material outcomes of designers' labors (the various pieces of hardware and their affiliated subcomponents, the APIs, software developer kits, and best practices manuals) and the cultural networks that mobilized around these efforts (user communities, informal professional networks, and research teams). In other words, what remains is the haptic subject as an ongoing historical project, equally discursive, material, institutional, and infrastructural—a promise expressed in both ideas and things to make whole information subjects fragmented by prior interfacing schematics. It is the pronouncement that touch can be both expressed through and transformed by its liberating embrace of technoscience. That old nonhaptic subject—the subject that assumes touch cannot be mediated—can be washed away and discarded, replaced by a haptic subject that believes vitality, immediacy, and materiality can be restored to media worlds by suturing new sensation-producing machines onto the body. This haptic subject embraces the promise of therapeutic rejuvenation through the infusion of machinic energies—it becomes haptic by learning to be sensitive to subtle differentiations in the vibrations emanating from digital devices, by understanding itself as a receiver and decoder of tactual messages, and by desiring to be made into a subject of haptic interfaces. The hope designated by this new subjectivity—that haptic interfaces contain the utopian promise to alleviate and undo the alienation of the body initiated by scopic technologies—indicates a desire to regain a natural and holistic mode of being absent from contemporary interface schematics. As Caroline Jones argues, "the hierarchies placing sight at the top of our sensory aristocracy are anxious narratives,"[88] always under the threat of being undone by new and disruptive technologies. Haptic interfaces express a series of self-conscious attempts to dethrone vision, to use communication technology as a means of overturning the established mediatic hierarchy of the senses.

This haptic subject, as I have maintained throughout *Archaeologies of Touch,* owes its apparent novelty to a passive strategic forgetting of touch. The refrain that touch has been a "neglected" sense—by aesthetics, by psychology, by media, by philosophy, by science—obscures the long history of concerted attempts at attending to touch. Haptic interfaces can only be treated as a pathway to desubjectification by embracing a historical account of the senses that locates touch as external to mechanisms of scientific management and technological control. Unlike vision and hearing, long subject to manipulation and deception by mediatic and representational technologies, the touch mobilized by haptic interfacing appears to be unspoiled and innocent; as the "hardest sense to fake" and one of "our only guarantees of sincerity,"[89] its appeal is grounded in the myth of its resistance to technoscientific capture. But this narrative elides the vast accrual of technical and scientific knowledge about touch that underpins the operation of the liberating new touch machines. It distracts from all the ways in which haptic subjects were called on to productively register the difference between machine-generated tactual sensations. Apple's 3D Touch technology, which debuted in the iPhone 6s, employs a "next-generation" haptic feedback mechanism capable of producing vibration events that last only 10 milliseconds. But passage into this next generation is imperceptible to vision—the Taptic Engine that produces this finely tuned feedback is buried deep inside the sealed-off guts of the device.[90] The tagline Apple uses in its ads for the new product calls on users to register this difference by touch: "not much has changed," Apple tells us in a playful allusion to the iPhone 6s's visual similarity to its predecessor, but "it just feels different." This tactile transformation is so totalizing, it has so thoroughly worked its way through the ecology of the device's operating system, that "pretty much everything you do feels different."[91] And as this drastic experiential upheaval is attributable to a technological change inaccessible to the eye, Apple mobilizes a mass haptic subject to register the transformation of machinic stimuli. Isolating the just-noticed difference between tactual sensations—the driving method in Weber's initial experiments on touch in the 1820s, and later, the basis for psychophysics (and further still, according to Kittler, the underlying technical principle enabling the electrical coding of data for the senses)—here becomes a vital task consumers are asked to perform in order to recognize the passage into a new generation of haptocentric interfacing with the iPhone. Haptics derives its appeal from the denial of this continuity, from its positioning as simultane-ously ancient, ahistorical, and futurological. Its paradoxical promise to desubjectify—to enable the emergence of a new and free subject—is grounded, on the one hand, in its strategic refusal to acknowl-edge the history of subjectification undergirding its emergence, and on the other hand, in the fetishistic celebration of technology's capacity to undo its own deleterious effects. The new visibility touch acquires through its expression in computer interfacing, then, prompts us to reevaluate claims of its purported neglect, bringing into view a touch that is neither ignored nor cast aside. Instead it is constantly attended to under specific conditions, constantly rearticulated in response to the new sets of instrumental demands capital places on it, and constantly observable through the repeated attempts to pro-ductively mobilize the haptic subject.

Coda

Haptics and the Reordering of the Mediated Sensorium

As important as haptics potentially is for VR, it's embryonic right now. There's simply no existing technology or research that has the potential to produce haptic experiences on a par with the real world. So any solution will have to come from breakthrough research.
—Michael Abrash, chief scientist, Oculus VR

To conclude *Archaeologies of Touch* is to necessarily impose an arbitrary endpoint on touch's ongoing technogenesis at the precise moment when the field of computer haptics is enjoying a surge of both financial investment and popular press attention that exceeds even the high-water mark it had previously established during the virtual reality boom of the late 1990s. And as with the late 1990s, of these burgeoning research and development clusters—which include touch feedback–enabled prosthetics, vibration alerts for wearables, cybersex interfaces, and haptic effects systems for touchscreens—none holds greater potential for realizing the core promise of haptic technology to fully embody users in computer-generated environments than virtual reality. Accordingly, I revisit Michael Abrash's 2015 address at the second annual Oculus Connect developer's conference, initially taken up in the Introduction. With the much-anticipated release of the Oculus Rift still on the horizon, Oculus Connect was intended both to hype the new virtual reality platform that had been acquired by Facebook eighteen months earlier for $2 billion, and to showcase the new Oculus Touch motion controller recently developed for the system. As combination developer gatherings/press events go, Connect was nothing exceptional: the company's leaders wowed the audience with impressive gaming demos, and they crafted a soaring techno-utopian narrative around the Rift intended to make the audience of software developers feel as if they were contributing not just to the success of a commercial product but also furthering the over-all forward progress of humankind in the process. Carefully prepared demonstrations showed the new machine operating smoothly in the optimal conditions of the test lab, with the implied promise that this performance could be replicated outside the lab's confines.

In an event dominated by discussions of the Oculus Touch, Abrash executed a strange maneuver: in spite of Oculus's decision to feature the neglected sense prominently in its branding, the Oculus Touch would be nothing haptic. The user's movements, captured deftly by the two handheld controllers that made up the Touch system, would only feed *into* the computer, with minimal vibrotactile cues fed back to the user through the controllers.[1] Even with robust, high-fidelity haptics conspicuously absent from the forthcoming first-generation Oculus Rift, Abrash conceded that touch feedback was crucial to creating compelling experiences in virtual reality. Twenty-five years after Rheingold had rivetingly extolled the wondrous, quasi-psychedelic virtues of haptics in *Virtual Reality,* and fifty years after Sutherland laid out his blueprint for a computer-controlled environment so realistic that it could bind and kill anyone who dared enter this ultimate display,[2] Abrash simultaneously positioned haptic feedback as both vital to and missing from present-generation virtual reality systems. It was a strange admission that seemed to suck some of the air out of the overinflated virtual reality balloon, unexpected at an event intended to generate enthusiasm for an emerging platform.

But Abrash, more of an engineer than a marketer and more interested in outlining a vexing design challenge than in overselling his employer's quixotic solution, succeeded in identifying a problem that Sutherland had papered over: the sense of touch, as purportedly laid bare by the science of haptics that Abrash referenced extensively in his keynote, is just maddeningly and frustratingly *complicated*. It involves a dense network of receptors distributed throughout the skin, muscles, and joints that are difficult to accurately target and stimulate. Abrash circled around to the futility of this enterprise, noting that, particularly where haptics is concerned, "a lot of the perceptual system is still a mystery." The endpoint was as clear for Abrash in 2015 as it had been for Sutherland in 1965: "what we really want in the long run," with the incorporation of touch into virtual reality, is to make the hands "act as the dexterous virtual manipulators that they are in the real world." Accomplishing this end, Abrash pointed out, would require advances in three related areas: knowledge of the perceptual psychology involved in touch, "breakthrough research" in engineering "new haptic technology," and finally, "a whole new interaction language around that haptic technology." With this last move, Abrash joined Diderot, Gault, Geldard, Margaret Minsky, and a whole chorus of voices from the past three hundred years in calling for the establishment of a new, standardized, technologically enacted, physiologically grounded language of touch, one that would upend touch's place in the epistemic ordering of the senses and allow it to achieve a new utility in the economy of human communication. The first haptic interface for virtual reality that "really worked" would be, Abrash declared, "world-changing magic."[3]

By signing onto this crusade, Abrash demonstrated the power of the haptic subject both to order touch and to specify its limits. The accumulated knowledge about the haptic system, generated from repeatedly stimulating the haptic systems of experimental subjects in the structured conditions of the lab, simultaneously revealed the most effective means for deceiving touch through artificial stimulation and the immense difficulties of assembling these machinic stimuli to form a cohesive and totalizing haptic simulation. The haptic subject's uniquely disruptive power can only be activated by overcoming the design challenge presented by touch's hard-coded psychophysiology. The haptic subject exists, then, as both a driver and marker of technological change, simultaneously informing the imaginary and materiality of touch machines. This technological subject—the product of a long history of touch that reaches back into the eighteenth century—establishes both the boundary conditions of haptics and the means to push beyond them. The haptic subject seeks to be transhistorical and apolitical but is marked at every stage of its emergence by the political and ideological goal of bringing touch under the dominion of scientific, technical, and economic management regimes.

The Future of Networked Haptics

But what if the Oculus team (or some other enterprising collective of engineers) succeeds in their attempts at breakthrough research, and unleashes the long-promised, world-changing magic of haptics? Given all of the reasons Abrash outlined and more, the odds are against them: for a high-fidelity haptics system to enjoy widespread adoption akin to radio or television, it will have to function properly in its stated goal of delivering convincing touch sensations, no easy task in and of itself. But in order to justify the development of a shared cross-platform programming language, it will also have to thrive in the competitive marketplace of consumer electronics. Haptics will need to find a killer app—that one activity that robust, high-fidelity haptics would prove indispensable to. Thus far, the expectation has been that gaming would provide such an application, but efforts to incorporate haptics that are more robust than rumble have been met with commercial failure (see Novint's Falcon and TN Games' 3RD Space Vest, both released nearly a decade ago and now all but forgotten). Facebook, in purchasing Oculus, gambled that future social networks would emulate and replicate face-to-face interaction, with the possibility of touch interaction enhancing the affective bonds between users. However, at present there have been no indications of widespread dissatisfaction with the existing two-dimensional mode of interaction—and nothing that suggests an imminent demand for haptic immersion in online social networks.

But assuming that researchers overcome these myriad obstacles, and assuming that marketers are able to produce a desire for full-body haptic feedback systems, it is possible to forecast a scenario where such systems rapidly achieve the ubiquity that television has enjoyed from the late 1950s onward. What then? Perhaps we will find that touch has presented one of the last defenses of the real against the onslaught of the virtual. If simulation technologies become so precise that they can emulate the feeling of a hug from a distant loved one (as CuteCurcuit's Hug Shirt attempted a decade ago), or so accurate that they could reconstruct the exact sensation of an ocean breeze blowing against the face, the need to travel for social, recreational, or business purposes—to move physical bodies across space—could diminish or melt away altogether. Here, we can recall Atkinson et al.'s suggestion in their 1977 article on force feedback computing, where they asked "why transport people, with the resultant waste of energy and human time; why not transport signals instead?"[4] Perhaps the *inability* to faithfully and convincingly simulate touch has allowed the real to maintain its value even in a hyper-mediated and simulated culture. The senses that media fail to extend, then, are as important as those that they extend, as those excluded senses help to establish the cultural border between the real and the unreal, between the authentic and the inauthentic. At present, even as the real is constantly mediated—whether by language, by writing, by photography, or by computational media—physically "being there still matters."[5] A fully realized haptics system would threaten to erode the remaining bits of the real—to push beyond the limits of extant simulation technologies—by rendering that which previously resisted mediation as machine-legible and transmissible data. Any accident of invention that allows for a near-perfect, one-to-one extension of the haptic system in distant and computer-generated environments holds the potential to initiate sweeping social and cultural consequences.

Along with its newfound powers, the body in this scenario would be subject to a host of vulnerabilities. Intimate data about how users touch, and how they like to be touched, would be stored on remote servers, providing a searchable (and hackable) archive of their activity.[6] Given that touch is the most highly regulated of our senses, existing laws would need to be rewritten in order to account for the possibilities of unwanted and nonconsensual digitized touch. Such laws would also govern the range of sensations haptic interfaces are allowed to reproduce, in order to keep them from fulfilling Sutherland's vision of an ultimate display capable of injuring or killing anyone who dared enter it. And just as novel visual technologies like photography and cinema introduced new perceptual and cultural disorders, machine-rendered touch would bring with it a host of unforeseen problems and anxieties (simulation sickness, a common experience in virtual reality, and phantom vibration syndrome, which has plagued users of mobile devices since the days of vibrating pagers, are early instances of such disorders).

These machines would also bring new forms of touch, as haptic simulation systems enable the remixing and remapping of touch sensations. Here we can again recall Rheingold, who suspected that it would eventually be possible "to map your genital effectors to your manual sensors and have direct genital contact by shaking hands,"[7] resulting in the transformation of social touch. Less prurient uses of such remixing techniques, foregrounded by Chris Salter's recent Haptic Field project,[8] could redraw and confuse the felt border between the self and the other, calling into question long-held assumptions about the operation of the human neurophysiological system. Thus far, the aesthetic possibilities of haptics have been mostly imagined within a realist framework that seeks to transparently reconstruct the haptic as a subcomponent of audiovisual simulation systems, as in Huxley's Feelies, Salvador Dali's tactile cinema, and touch-enabled 4-D film, but more experimental configurations that remap the organs of touch could go further to create novel tactile sensations, just as electrical machines and psychophysical instruments did in the previous centuries.

Such technologies could also have far-reaching economic consequences. Facilitating consumption from a distance by re-creating the tactile materiality of commodities—another outcome Noll forecasted in his writings on human–machine tactile communication—could further disrupt a commercial sphere already being upended by the migration of shopping to the Internet. Improvements in machine haptics—giving robots the ability to move, balance, and handle delicate materials as well as humans—could hasten the replacement of humans by machines in the workplace, as machines begin to succeed at tasks in which they presently fail. Mosher's hapless robot from Interface 4, frustrated by its inability to complete even the simplest of dexterous tasks, would gain new confidence and utility if it could learn to touch like the humans it was modeled after.

These scenarios illustrate the potentially totalizing impacts of haptics applications. Whatever forward path haptic technologies take—if some breakthrough research conjures the "world-changing magic" of haptics—they will demand a coherent, comprehensive, and sober response from media researchers. But in our response, we must resist the evangelizing and fetishizing rhetoric of industry propagandists, while also casting aside longstanding ontological claims about touch's inherent hostility to mediation and extension. We must acknowledge all of the ways in which touch has already been transformed by its interfacing with technoscience. At present, haptics applications have achieved widespread diffusion, opening up a new communicative channel between bodies and global flows of data, and providing proof of the concept that touch does not exist beyond the reach of media technologies. The approach I am advocating requires us to treat media touch as a site of continuing technogenesis that must be attended to in its particularities rather than its generalities. Immersion's nascent attempts at adding haptics to video advertisements on mobile devices—which began with its vibration-enabled trailer for Showtime's *Homeland* and continues with its new touch-enhanced commercials for cars from Lexus and Peugeot—suggest that the relocation of video from televisions to handheld screens has opened up new sensory pathways for the transmission of consumption messages. Haptic messaging systems for smartwatches—such as the Apple Watch's Taptic Engine—treat the skinspace of the wrist as an always-on communication channel that can be used for the transmission of coded messages, effectively establishing a material link between the wearer of the device and global flows of data.

Compared to the grand visions outlined above of a standard-ized, ubiquitous master device that would bring high-fidelity, full-body haptics to the masses (being realized in Sony's actuator-laden Synesthesia Suit, AxonVR's HaptX Skeleton, and the Teslasuit), these may seem like mundane, trivial, and anticlimactic outcomes. However, considered as part of the accelerating technoepistemo-logical momentum around haptics, these examples provide evidence for the claim that touch can no longer be considered a sense that has been neglected or left behind by mediation systems. Rather, by marking the immense human and financial capital devoted to the technical project of attending to touch, these examples show that our ideation of touch is both outmoded and inadequate to the changes touch has undergone as a consequence of its transformative interfacing with technoscience. They prompt us to investigate the specific conditions under which touch *has already been* admitted into cybernetic telecommunications networks—the new forms of affec-tive and informatic haptic intimacy that have been proposed, adopted, embraced, and in some cases even discarded at this stage in the history of human–machine interfacing.

A Haptic Media Studies

To meet the challenges posed by these ongoing transformations, we in the loosely constituted field of media studies can make touch an analog of seeing and hearing—and make touch media analogous to image and sound media—by applying to touch the same methods of investigation and theoretical frameworks that we have so productively employed to study the media of those so-called major senses. Through such a move, touch—in all its wonderful multiplicity—would become a site for dedicated empiri-cal investigation by those in media studies. Executing this shift in orientation will entail a reconsideration of what counts as the media historical archive, opening up the field to new points of contact between past and present, along with a new attention to the haptic-ity of so-called old media. This is not a radical suggestion. Media change always involves a reorganization of the archive—the discur-sive and material a priori—of media history. Like technologies more generally, both "new" and "old" media alike always challenge us, as Benjamin Peters points out, to make them intelligible, to craft histories around them, to (re)imagine the past in relationship to the present and to some mediatic future.[9] The historical record of media is always contingent on the present ontological understanding of what fits into that messy and ever-shifting category of *media*.

We can make touch analog, then, by pursuing something akin to visual culture studies and sound studies—a haptic media studies—that treats touch as an analog of seeing and hearing, from a methodological rather than ontological or technological standpoint. To borrow Gault and Geldard's phrasing, we should launch a sustained attack on touch media aimed at investigating the various attempts, successful and unsuccessful, at folding touch into the mediated sensorium. In its ideal form, this attack will involve a wide range of methods and approaches, including further historical investigation, feminist and queer accounts of touch and media, ethnographic study, formal analysis, disability studies, political economy, science and technology studies work on engineering practice, cross-cultural comparisons of media touch, and platform studies approaches to haptics hardware and software. This subfield will attend to the deployment of haptics technologies in a host of contexts, such as museum display, medicine, prosthetics, data visualization, videogames, virtual reality, wearables, mobile communication, cinema, assistive technologies, and advertising and marketing. And it should engage productively in cross-disciplinary conversations not just with the engineers and neuroscientists responsible for developing haptics applications, but also with those working in sensory studies, gender studies, literature, psychology, and philosophy, where scholars have produced excellent touch-centered scholarship over the past two decades. Finally, the story that I have told around haptics in *Archaeologies of Touch* identifies and weaves together a progressivist narrative about technologizing touch that circulates actively among hapticians. Future work in haptic media studies should investigate points of resistance to the hegemonic modeling of touch that underpins contemporary haptic interfaces.

Those of us investigating the relationship between media and touch find ourselves working in an area that lacks cohesion and stability; just as when Geldard issued his call to arms in 1956 for a dedicated program of research on touch communication, we are laboring on distant islands, without a readymade institutional home, or agreed upon set of baseline definitions.[10] We have no defined canon, no journals, no recurring conferences, professional associations, or scholarly interest groups. Investigations into touch media unfold in piecemeal fashion, scattered in isolated papers or panels at large disciplinary conferences and taking shape in discrete articles or chapters in unrelated journals and anthologies. Although much productive work has already been undertaken around touch by media scholars, such works lack a defined identity, a shared bibliography, and anything resembling an institutional home.[11] To borrow Sherrick's phrasing, haptic media studies, to the extent that it exists today, is parasitic on visual culture and sound studies. This is far from a catastrophic situation. As with research into tactile communication systems before Geldard's address, it has been possible to produce quality scholarship in spite of the field's scattered state. But, however workable, this circumstance does not lend itself to the generation of research that advances a cohesive agenda or even agrees on a baseline set of definitions and key works. What I am calling for here is a structured program of empirical and theoretical research around touch—oriented simultaneously toward its past, present, and future—that identifies tactility as a perceptual modality interior to technical and mediatic systems. "Haptic media studies" provides a provisional and somewhat pugilistic title, a strategy for orienting media studies to the existence of a long history—conjured by haptics as the doctrine of touch—of attempts to mold touch into an analog of seeing and hearing through the tactical application of machinic stimuli. This term "haptic media" expresses and embraces that transformational wish repeatedly mobilized around touch while explicitly declaring that touch is a sense that can have a life in media irreducible to its indirect activation by vision and audition.[12]

This struggle to transform touch into an analog of vision and hearing—and to awaken a haptic subject that can serve as a counter to the purported power that seeing and listening subjects gained through their mediatic expression—remains ongoing and active. New devotees and capital flows are continually recruited to the cause of engineering increasingly sophisticated mechanisms for communicating through and synthesizing touch. The history of haptic interfacing involves the constant interplay between a technoscientific knowledge production mechanism and a body that seems to be always exceeding and evading its limits. It involves repeatedly redrawing the border between the governable and the ungovernable—between the manageable and the unmanageable— as new machines, protocols, and apparatuses are set upon touch, each part of an ongoing experimental system designed to enframe touch, each driving toward the promised liberation and restoration of a human subject that seeks happiness through sensory extension and augmentation. It offers a break from the previous epistemic ordering of the senses. The exertions of hapticians are self-consciously directed toward empowering touch over and against the senses of seeing and hearing, with haptics technologies providing the means to upend an established mediatic ordering of the senses that treats the eyes and the ears as the primary agents of mediated perception. And while I stop short of embracing Robert Jütte's epochal claim that new technologies and practices of touch have ushered in a "haptic age," the scenarios outlined above illustrate the potential haptics holds to initiate a wholesale rebalancing of mediated sensorium. Such conditions would privilege a haptic epistemology that is grounded in a fundamentally machinic model of touch, where touch gains its new powers through its imbrication with computing machines. Acknowledging and specifying the haptic subject's active role in shaping the trajectory of media history presents a positive response to our current moment, providing a framework for confronting the future of mediated tactility.

Notes

Preface

1   This storyline took place in *Iron Man,* vol. 1, nos. 242–70 (New York: Marvel Comics, 1989–90).

2   David Michelinie, *Iron Man,* vol. 1, no. 244 (New York: Marvel Comics, 1989), 48.

3   David Michelinie, *Iron Man,* vol. 1, no. 248 (New York: Marvel Comics, 1989), 30.

4   Stark's recovery would later be complicated when one of Iron Man's enemies hacked the chip and hijacked control over his body, transforming the story into both an allegory about the dangers of technological dependence and expressing contemporary concerns over the coming networking of physical bodies. Anyone familiar with the bodies of comic book characters knows that they never stay in one shape for long. Alienated from his own nervomotor system, Stark donned a thought-controlled exoskeleton that allowed him to regain control over his body's movements.

5   Marshall McLuhan, *Understanding Media: The Extensions of Man* (Cambridge, Mass.: MIT Press, 1994), 130.

Introduction

1   Ivan Sutherland, "The Ultimate Display," in *Proceedings of the IFIP Congress,* 1965, 506.

2   Bruce Sterling, "Augmented Reality: 'The Ultimate Display' by Ivan
    Sutherland, 1965," *Wired,* September 2009, http://www.wired
    .com/2009/09/augmented-reality-the-ultimate-display-by-ivan
    -sutherland-1965/.

3   Sutherland, "The Ultimate Display," 508.

4   Michael Abrash, "Keynote," Oculus Connect 2 (September 25, 2015),
    https://www.youtube.com/watch?v=tYwKZDpsjgg.

5   This phrase, repeated often in both popular and scientific press
    discussions of haptics, implicates all haptic interface designers in a
    teleologically oriented project; even the smallest of steps contribute to
    the eventual discovery of this elusive end. For one example, see Helen
    Farley and Caroline Steele, "A Quest for the Holy Grail: Tactile
    Precision, Natural Movement and Haptic Feedback," in *Same Places,
    Different Spaces. Proceedings Ascilite Auckland 2009,* 285–95, http://
    www.ascilite.org/conferences/auckland09/procs/farley.pdf.

6   Kudo Tsunoda, general manager of Microsoft Game Studios,
    controversially dubbed rumble technology—included in game
    controllers since the late 1990s—"a rudimentary form of haptic
    feedback." In the 2010 interview with *Edge Magazine,* part of a
    publicity push that anticipated the company's release of its Kinect
    motion-capture controller, Tsunoda dismissed those who "hold onto
    rumble as the holy grail of haptic feedback." "Kudo Tsunoda: 'Rumble
    Is Rudimentary,'" *Edge Online,* July 8, 2010, https://web.archive.org
    /web/20100710091014/http://www.next-gen.biz/features
    /kudo-tsunoda-rumble-is-rudimentary.

7   Ben Anderson derives this term from Foucault's discussion of the body
    as an "object and target" in *Discipline and Punish* (New York: Vintage
    Books, 1995), using it as a way to account for the efforts made to bring
    affect under the control of technical and economic management
    systems. Ben Anderson, *Encountering Affect: Capacities, Apparatuses,
    Conditions* (Burlington, Vt.: Ashgate, 2014), 18–19.

8   Cathryn Vasseleu, "Touch, Digital Communication and the Ticklish,"
    Angelaki: Journal of the Theoretical Humanities 4, no. 2 (1999): 153.

9   Jonathan Crary, *Techniques of the Observer: On Vision and Modernity in
    the Nineteenth Century* (Cambridge, Mass.: MIT Press, 1990), 9.

10 Edward Bradford Titchener, "Haptics," in *Dictionary of Philosophy and
    Psychology, Volume 1,* ed. James Baldwin (New York: MacMillan,
    1901), 441.

11 G. Stanley Hall, "The New Psychology," *Harper's Monthly Magazine*, November 1901, 32.

12 Frank Geldard, "Adventures in Tactile Literacy," *American Psychologist* 12, no. 3 (1956): 115.

13 Confusion about what differentiates a tactile interface from a haptic one has continued to plague the field since *haptic* was adopted in the 1980s. Susan Lederman, who pushed roboticists and computer scientists to adopt the language of haptics, embraced the common perspective that tactile feedback refers to feedback that targets the receptors in the skin (such as thermal or vibrational feedback), usually obtained when the user's position is static. Kinesthetic feedback, by contrast, refers to feedback that results from the user's active movement of the interface through space. For Lederman, haptic interfaces are those that combine tactile and kinesthetic feedback. Writing in 2006, Lederman and Lynette Jones noted that machines that provide only vibration feedback or only kinesthetic feedback were being continually referred to in technical literature as haptic interfaces, and attempted to delineate more clearly among these various devices, while articulating the technical features and design challenges unique to each category. Lynette Jones and Susan Lederman, *Human Hand Function* (Oxford: Oxford University Press, 2006), 195–200. Throughout *Archaeologies of Touch,* I use the term haptic interface broadly, in a way that might be inconsistent with the type of specificity that Lederman pushed for. My goal in doing so is to link together a range of machines whose development has been informed by the technoscientific treatment of touch. I provide further analysis of this terminology in Interface 4.

14 Marvin Minsky, "Telepresence," *OMNI Magazine*, June 1980, 52. Emphasis original.

15 See "Haptics Research Labs," http://2008.hapticssymposium.org /haptics_labs.html.

16 Hans-Jörg Rheinberger, *Toward a History of Epistemic Things: Synthesizing Proteins in the Test Tube* (Stanford, Calif.: Stanford University Press, 1997), 20.

17 Ibid., 26.

18 In his 1980 essay "Telepresence," Marvin Minsky grounded the success of remote presence systems in the design of machines that could feed back complex data—especially including texture—from remote or virtual environments. His daughter Margaret's later works, building her Sandpaper force feedback system and mapping a forward path for

future haptics research, were positive steps toward realizing the goal he outlined.

19 Hiroo Iwata, "Epoch of Haptic Interface," in *Haptics: Basics, Approaches, and Applications,* ed. Martin Grunwald (Boston, Mass.: Berkhäuser Verlag, 2008), 355–62. When *Haptics-e: The Electronic Journal of Haptics Research* was established in 1999, in response to a documented spike in published articles devoted to haptics, the journal's governing board featured an international roster of luminaries with a range of disciplinary backgrounds, including psychologists (Klatzky and Lederman), roboticists (Susumu Tachi, Blake Hannaford, and Vincent Hayward), engineers (Karon MacLean), and computer scientists (Bill McNeeley).

20 Bernard Stiegler, *Technics and Time, 1: The Fault of Epimetheus* (Stanford, Calif.: Stanford University Press, 1998), 43.

21 Crary, *Techniques of the Observer,* 24.

22 Jonathan Sterne, *The Audible Past: The Cultural Origins of Sound Reproduction* (Durham, N.C.: Duke University Press), 2.

23 Mark Paterson has attempted to undo what he frames as the forgetting of touch by resuscitating philosophical and epistemic traditions that grounded notions of measurement and conceptions of space firmly in the body. Mark Paterson, *Senses of Touch: Haptics, Affects and Technologies* (New York: Berg Press, 2007), 59–77.

24 Crary, *Techniques of the Observer,* 81.

25 Robert Jütte, *A History of the Senses: From Antiquity to Cyberspace* (Malden, Mass.: Polity, 2005), 156.

26 Friedrich Kittler, *Optical Media* (Malden, Mass.: Polity, 2010), 172.

27 Friedrich Kittler, "Thinking Colours and/or Machines," *Theory, Culture & Society* 23 (2006): 42.

28 Jonathan Sterne, *MP3: The Meaning of a Format* (Durham, N.C.: Duke University Press, 2012), 19.

29 Gustav Theodor Fechner, *Elements of Psychophysics* (New York: Holt, 1966), xxviii.

30 Hui addresses touch extensively in pursuing the genesis of a "psychophysical aesthetics," confronting the lineage from Weber's initial experiments on touch to Fechner's formulation of psychophysics. She argues that a pragmatic desire to use psychophysics

to better understand aesthetics drove Fechner's turn to Weber's rigorous experimental method. Hui's concern with Weber's touch experiments, however, is rightly subordinated to her overarching thesis that the science of psychophysics was, from its origins, interwoven with concerns about the aesthetics of hearing and listening. The continued trajectory of post-Weber psychophysical research on touch—divorced from any explicitly articulated aesthetic concerns—suggests that each distinct tract of psychophysical investigation had its own idiosyncratic motivations. Alexandra Hui, *The Psychophysical Ear* (Cambridge, Mass.: MIT Press, 2013).

31   In Canales's excellent historiography, experiments on touch perception contributed prominently to the ongoing debate in the nineteenth century around "the personal equation"; researchers timed subjects' responses to tactual stimuli as a purported means of getting at the hard-coded biological limits of reaction time. Enclosed in the experimental system of reaction time studies, these investigations instrumentalized touch as a way of answering a broader set of questions about the relationship among sensation, perception, and time. Jimena Canales, *A Tenth of a Second* (Chicago: University of Chicago Press, 2009), 21.

32   Steven Shapin and Simon Schaffer, *Leviathan and the Air-Pump: Hobbes, Boyle, and the Experimental Life* (Princeton, N.J.: Princeton University Press, 1985).

33   Recently, media theorists (including Lori Emerson, Garnet Hertz, Marcel O'Gorman, Jussi Parikka, Isabel Pedersen, and Darren Wershler) have turned their attention to the lab, formulating it as a site crucial to the genesis of media both at the technical and cultural levels. See for example Jussi Parikka, "The Lab Imaginary: Speculative Practices In Situ," in *Across and Beyond: A Transmediale Reader on Post-Digital Practices, Concepts and Institutions,* ed. Ryan Bishop, Kristoffer Gansing, Jussi Parikka, and Elvi Wilk (Berlin: Sternberg Press, 2017), 78–91. The forthcoming book on media labs from Lori Emerson, Jussi Parikka, and Darren Wershler is described in "The Lab Book," http://whatisamedialab.com/.

34   Mark Paterson's move to bring these specialist discourses around touch into dialogue with cultural and philosophical treatments of the subject prompted him to embrace a broad definition of "haptic" as "relating to the sense of touch in all its forms." Paterson lists proprioception, the vestibular sense, kinesthesia, the cutaneous senses (pressure, pain, and pain), the tactile sense, and force feedback. Paterson, *Senses of Touch,* ix.

35   In *Touch: Sensuous Theory and Multisensory Media* (Minneapolis: University of Minnesota Press, 2002), Laura Marks uses this term to

designate a type of subjectivity activated by what she calls "haptic images"—images that cue or invite a felt embodied response, thus signaling the limits of the visual. Her formulation of the haptic subject depends on embracing a particular phenomenology of vision, one where an act of looking (or an active looking) can serve as a substitute for the act of touching and being touched. For Marks, "haptic" denotes a type of vision; in contrast, in my argument, "haptic" refers to a historically specific set of touching practices aimed at the production of technical knowledge about the tactile processes.

36   Crary, *Techniques of the Observer*, 16.

37   Ibid., 6.

38   The notion of touch as a multiplicity of senses not limited to physiology animates Paterson's thesis in *Senses of Touch*, as he tracks touch's heterogeneous manifestation across the fields of philosophy, computing, aesthetics, and phenomenology. In contrast, my approach, by focusing on the empirical subdivision of touch by psychophysical psychology, shows the process by which hegemonic scientific models of touch were inscribed in emerging touch media.

39   Frank Geldard, "The Perception of Mechanical Vibration," *Journal of General Psychology* 22 (1940): 261.

40   E. B. Titchener, H. C. Warren, and James Baldwin, "Laboratory," in *Dictionary of Philosophy and Psychology*, 613.

41   Horst Gundlach, "What Is a Psychological Instrument?" in *Psychology's Territories: Historical and Contemporary Perspectives from Different Disciplines*, ed. Mitchell Ash and Thomas Sturm (Mahwah, N.J.: Lawrence Erlbaum, 2007), 196.

42   Kittler, "Thinking Colours and/or Machine," 42.

43   I draw on Deborah Coon's "Standardizing the Subject: Experimental Psychologists, Introspection, and the Quest for a Technoscientific Ideal," *Technology and Culture* 34, no. 4 (1993): 757–83 for this argument, developed further in Interface 2.

44   Giorgio Agamben, *"What Is an Apparatus?" and Other Essays* (Stanford, Calif.: Stanford University Press, 2009), 15.

45   Ibid., 15. Conveniently for my argument here, Agamben includes in his exhaustive list of apparatuses the cellular phone, a device in whose operation haptic feedback and haptic alerts feature prominently.

46   Ibid., 17.

47 Immersion Corporation, *Harnessing Human Touch* (San Jose, Calif.: Immersion Corporation, 2009).

48 Gitelman claims that a supporting protocol both of science and media is the "eventual abnegation and invisibility of supporting protocols." Lisa Gitelman, *Always Already New: Media, History and the Data of Culture* (Cambridge, Mass.: MIT Press, 2006), 5–7.

49 Ibid., 7.

50 Alexander Galloway, *Protocol: How Control Exists after Decentralization* (Cambridge, Mass.: MIT Press, 2004), 12, 82.

51 Ibid., 87.

52 See Druckrey's foreword to Siegfried Zielinski's *Deep Time of the Media: Toward an Archaeology of Hearing and Seeing by Technical Means* (Cambridge, Mass.: MIT Press, 2006), vii.

53 Kittler, *Optical Media,* 119.

54 Erkki Huhtamo, "From Kaleidoscomaniac to Cybernerd: Notes toward an Archaeology of the Media," *Leonardo* 30, no. 3 (1997): 221.

55 Siegfried Zielinski, "Media Archaeology," *CTheory* (1996): ga11.

56 Before the articulation of sound studies as a coherent subfield, scholars interested in the history of sound reproduction and audio culture understood themselves to be in a similarly subordinate relationship to the visualists, and in response, successfully carved out dedicated spaces for conversations about sound media.

57 In other words, the visualist account of media—even the purportedly radical, nonlinear, deconstructivist account of visual media embraced by media archaeologists—provides a hegemonic master narrative of media history for those of us interested in touch media to push back against.

58 Eric Kluitenberg, "On the Archaeology of Imaginary Media," in *Media Archaeology: Approaches, Applications, and Implications,* ed. Errki Huhtamo and Jussi Parikka (Berkeley, Calif.: University of California Press, 2011), 51.

59 Fiona Candlin, "The Dubious Inheritance of Touch: Art History and Museum Access," *Journal of Visual Culture* 5, no. 2 (2006): 140.

60 Wolfgang Ernst, "Media Archaeography," in *Media Archaeology,* 253.

61 Thomas A. Edison, "Patent Caveat 110," October 8, 1888, Edison National Historical Site Archives, West Orange, N.J., http://edison .rutgers.edu/NamesSearch/glocpage.php?gloc=PT031AAA&.

62 "How Touching," *Economist Technology Quarterly,* March 8, 2007, http://www.economist.com/node/8766116.

63 This analog formulation appears often in both the historical and contemporary literature on tactile media. In *Brave New World,* Aldous Huxley famously described a multisensory cinema, dubbed "the Feelies," where tactual effects were as important as audio and visual ones (and the "the Feelies" were an analog to the then-new "Talkies" that Huxley deplored). A. Michael Noll, in his landmark dissertation "Tactile Man-Machine Communication" (PhD dissertation, Polytechnic Institute of Brooklyn, 1971), speculated that adding the tactile channel to the already extended senses of seeing and hearing would bring teleportation closer to reality. Informed by interviews with haptic interface designers, Howard Rheingold claimed that one day, cyberspace engineers would prioritize building realistic sensations as a design aim: "just as photorealism is the goal of one branch of computer graphics, 'tactile realism' has the potential to grow into a general goal for cybernetic clothing designers." Howard Rheingold, *Virtual Reality* (New York: Touchstone, 1991), 323. A more recent formulation similarly understands haptics as a furthering of the project initiated by computer graphics: "extending the frontier of visual computing and auditory display as the two dominant forms of man-machine interfaces, haptic interfacing has the potential to further increase the quality of human-computer interaction." Ming Lin and Miguel Otaduy, "Introduction," in *Haptic Rendering: Foundations, Algorithms, and Applications,* ed. Ming Lin and Miguel Otaduy (Wellesley, Mass.: A. K. Peters, 2008), 1. For further detail on the Feelies as a reaction to contemporaneous trends in cinema, see Laura Frost, "Huxley's Feelies: The Cinema of Sensation in Brave New World," *Twentieth-Century Literature* 52, no. 4 (2006): 443–73.

64 Throughout *Archaeologies of Touch,* I use the terms medialization, mediatization, and mediatic; each of which is being theorized in more precise detail as a way to explain the totalizing impacts of media on daily life. Mainly deployed in European accounts of media inspired by Marshall McLuhan and Friedrich Kittler, these are admittedly awkward words, but they are useful in my argument for the way that they encourage us to think about media as both technologies and processes. As processes, media are "gradually systematizing," increasingly organizing and ordering "the relatively unstructured realm of the everyday" (Friesen and Hug, 65) according to their individual logics. Specific to *Archaeologies of Touch,* I use medialization to draw attention to the way that touch has been intentionally and consciously remade as media by the engineers, psychologists, and programmers responsible for building haptic interfaces. Norm Friesen

and Theo Hug, "The Mediatic Turn: Exploring Concepts for Media Pedagogy," in *Mediatization: Concepts, Changes, Consequences,* ed. Knut Lundby (New York: Peter Lang, 2009), 65. For other representative works on the subject of mediatization, see Andreas Hepp, *Cultures of Mediatization,* trans. Keith Tribe (Malden, Mass.: Polity, 2013); Eliseo Verón, "Mediatization Theory: A Semio-anthropological Perspective," in *Mediatization of Communication,* ed. Knut Lundby (Berlin: De Gruyter, 2014).

65 For example, Antonio Bicchi et al., in their opening to a highly technical collection of essays on the current state of haptics research, cite a passage from Bertrand Russell's *The ABC of Relativity.* Antonio Bicchi, Martin Buss, Marc Ernst, and Angelika Peer, "Introduction," in *The Sense of Touch and Its Rendering,* eds. Antonio Bicchi et al. (Berlin: Springer, 2008), 1.

66 John Durham Peters, *Speaking into the Air: A History of the Idea of Communication* (Chicago: University of Chicago Press, 1999), 270. Similar claims are echoed throughout the discursive history of touch. These are taken up in greater detail in Interface 3; for a summary, see Frances Herring, "Touch: The Neglected Sense," *Journal of Aesthetics and Art Criticism* 7, no. 3 (1949).

67 Rheingold, *Virtual Reality.*

68 Declarations that haptics technology will transform society are often occasioned by some notable and well-publicized step forward in a corporate design lab. In 2012, for example, touchscreens that promised to allow their users to feel fine textures of onscreen object became a popular topic, owing in part to the company Senseg announcing that it had unlocked the key to finely stimulating the fingertips with the screen. Consequently, *Wired* named haptics one of 25 "big ideas for 2012." Tom Cheshire, "25 Big Ideas for 2012: The New Haptics," *Wired,* January 2012, http://www.wired.co.uk/magazine/archive/2012/01/features/the-new-haptics.

69 Jacques Derrida, *On Touching—Jean-Luc Nancy* (Stanford, Calif.: Stanford University Press, 2005), 300.

70 Martin Grunwald, ed., *Human Haptic Perception: Basics and Applications* (Boston: Birkhäuser, 2008).

71 Denis Diderot, *Diderot's Early Philosophical Works,* trans. Margaret Jourdain (Chicago: Open Court Publishing, 1916), 90.

72 K. Ludwig Pfeiffer, "From the Materiality of Communication to an Anthropology of Media," in *Materialities of Communication,* ed. Hans

Ulrich Gumbrecht and K. Ludwig Pfeiffer (Stanford, Calif.: Stanford University Press, 1994), 11.

73 Alois Riegl, *Late Roman Art Industry,* trans. Rolf Winkes (Rome: Giorgio Bretschneider, 1985).

74 Even for the initiated, Riegl's argument concerning the various phases of perception—the movement from the tactile to the optical-tactile and then to the optical stage—is fairly arcane. For productive discussions, see Candlin, "The Dubious Inheritance of Touch," 137–54; Antonia Lant, "Haptical Cinema," *October* 74 (Autumn 1995): 45–73; Marks, *Touch,* 4–7.

75 Alois Riegl, "Late Roman or Oriental," in *German Essays on Art History,* ed. Gert Schiff (New York, Continuum, 1988), 190n1.

76 Gilles Deleuze and Félix Guttari, *A Thousand Plateaus: Capitalism and Schizophrenia* (Minneapolis: University of Minnesota Press, 1987), 492. Laura Marks also cites Riegl's definition in *The Skin of the Film: Intercultural Cinema, Embodiment, and the Senses* (Durham, N.C.: Duke University Press, 2000), 162.

77 Marks, *Touch,* 7.

78 Riegl, "Late Roman or Oriental," 190n1.

79 I use this term loosely to include such thinkers as Walter Benjamin, Marshall McLuhan, Deleuze, Marks, and Mark Hansen, each of whom refers back to the aesthetic formulation of the term Riegl offered. For an example of Benjamin's use, see "On Some Motifs in Baudelaire," where he claims that with the coming of photography "a touch of the finger now sufficed to fix an event for an unlimited period of time. The camera gave the moment a posthumous shock. . . . *Haptic* experiences of this kind were joined by optic ones" (italics added). Walter Benjamin, "On Some Motifs in Baudelaire," in *Illuminations: Essays and Reflections* (New York: Random House, 1988), 175.

80 The "Texture Matters: The Optical and the Haptical in Media" project represents a positive move in this direction. Organized by Klemens Gruber and Antonia Lant and conducted from 2012 to 2014 at the University of Vienna, the project's participants included Bruna Petreca, a designer active in the field of haptics along with media archaeologists and art historians. For details, see http://texturematters.univie.ac.at/about-texture-matters/.

81 A project of designers Ali Israr and Ivan Poupyrev, Surround Haptics uses vibratory actuators distributed across a user's back as "tactile brushes" to paint sensations onto their skin. In their framing of Surround Haptics, Israr and Poupyrev differentiate the new technology

from previous generations of haptics, claiming "users do not feel 'buzzes' on their body that are common in current haptics technologies, instead they feel smooth tactile strokes moving across their body." Disney projects Surround Haptics to have uses both in videogaming and cinema. See Israr and Poupyrev, "Surround Haptics," https://www.disneyresearch.com/project/surround-haptics/.

82 New high-end wearable sex toys, such as the blueMotion NEX|1 from OhMiBod, allow the devices to be remotely controlled by dedicated apps running on networked smartphones. For one user's experience with the device, see Samantha Greene, "I Did It for Science: A Stranger Remotely Controlled My Vibrator," *Nerve,* February 9, 2015, http://www.nerve.com/science/i-did-it-for-science-a-stranger -remotely-controlled-my-vibrator. Jeffrey Bardzell and Shaowen Bardzell's excellent empirical study of the new user-centered design practices at work in luxury sex toys suggests that the current generation of designer, computer-controlled sex toys hails users as sociotechnical, sexual subjects. Jeffrey Bardzell and Shaowen Bardzell, "'Pleasure Is Your Birthright': Digitally Enabled Designer Sex Toys as a Case of Third-Wave HCI," Computer–Human Interaction 2011, May 7–12, 2011 Vancouver, B.C., http://dl.acm.org/citation .cfm?id=1978979.

83 Almost without comment, Microsoft added two additional motors to its Xbox One controller ("impulse triggers"), but game developers have been slow to code vibrations specifically for the extra motors, leaving the new resource untapped.

84 See, for example, Stanisa Raspopvic, Marco Capogrosso, Francesco Maria Petrini, and Marco Bonizzato, "Restoring Natural Sensory Feedback in Real-Time Bidirectional Hand Prostheses," *Science Translational Medicine* 6, no. 222 (Feb. 5, 2014): 222.

85 Barack Obama, "2015 State of the Union Address: Enhanced Version," White House Office of Digital Strategy, https://www .whitehouse.gov/blog/2015/01/20/watch-president-obamas -2015-state-union.

86 I briefly take up Immersion Corporation's many patent infringement lawsuits in Interface 5. In the summer of 2015, one patent infringement lawsuit brought the teledildonics development to a virtual standstill. See Joe Mullin, "'Teledildonics' Patent Used to Sue Six Nascent Cybersex Companies," *Ars Technica,* July 23, 2015, http://arstechnica.com/tech-policy/2015/07/teledildonics-patent-used -to-sue-six-nascent-cybersex-companies/.

87 Gitelman, *Always Already New,* 8–9.

1. The Electrotactile Machine

1   Siegfried Zielinski, *Deep Time of the Media: Toward an Archaeology of Seeing and Hearing by Technical Means* (Cambridge, Mass.: MIT Press, 2006), 271.

2   Friedrich Kittler, *Gramophone, Film, Typewriter* (Stanford, Calif.: Stanford University Press, 1999), 10.

3   The orienting subtitle to Zielinski's *Deep Time of the Media,* this framing distracts from the importance that a body irreducible to the organs of seeing and hearing played in the narrative he crafted.

4   Henry Cavendish and Michael Faraday, for example, were each praised for their ability to make finely graded distinctions among currents of different strengths and types. Faraday effortlessly moved between using instruments and using his body to measure the strength of electrical currents, displaying a highly refined and practiced electrocutaneous sensitivity. Michael Faraday, "Experimental Researches in Electricity—Ninth Series," *Philosophical Transactions of the Royal Society of London for the Year 1835, Part 1,* 41–56.

5   This characterization is admittedly reductive, as the torpedo fish had been previously used as a curative in ancient medicine. Stanley Finger and Marco Piccolino position the fish's "rediscovery" by European scientists as a point of connection between Egyptian medicine and Enlightenment philosophy. Stanley Finger and Marco Piccolino, *The Shocking History of Electric Fishes: From Ancient Epochs to the Birth of Modern Neurophysiology* (Oxford: Oxford University Press, 2011).

6   In spite of being featured in key media archaeological works, scholars have not theorized or understood these encounters between human bodies and electrical machines as primarily tactile. Zielinski, for example, devotes a chapter to early electrical experiments in his *Deep Time of the Media,* and he gives substantive attention to the German Romantic Johann Wilhelm Ritter, who engaged in extensive and self-destructive autoexperimention with electric batteries. However, Zielinski's account of these experiments comes in the broader context of his attempt to forge an "archaeology of hearing and seeing by technical means"; consequently, he passes by many opportunities to unpack the technics of touch at work in the treated historical material. Zielinski, *Deep Time of the Media,* 169–82.

7   Lorraine Daston and Peter Galison, *Objectivity* (New York: Zone Books, 2007), 19.

8   Michel Foucault, *The Order of Things: An Archaeology of the Human Sciences* (New York: Random House, 1970), 133.

9   Although the Enlightenment closely linked the advance of knowledge to a paradigm of visualist knowledge, the other modes of sensory experience were not simply relegated to obscurity. Jonathan Sterne's concept of an "ensoniment" suggests that sound, too, became the object of new valorization rhetorics; combined with new technologies and practices of listening, this ensoniment helped "render the world audible in new ways." Jonathan Sterne, *The Audible Past: The Cultural Origins of Sound Reproduction* (Durham, N.C.: Duke University Press), 2.

10   Simon Schaffer details the importance of "self evidence," describing the efforts experimenters made to have their bodies counted among those qualified to give and report evidence obtained through sensory observation. The production of electrical facts was linked inextricably to both the individual experimenter's body, as the phenomenal register of the experiment, and to the broader "collective body of experimental philosophy" the individual's findings were aimed at convincing. Simon Schaffer, "Self Evidence," *Critical Inquiry* 18, no. 2 (1992): 330.

11   James Delbourgo specifies the body's essential role in sparking interest in electricity, as the interface between bodies and electrical machines transformed electricity into a cultural commodity, configuring what Delbourgo terms an "American economy of wonder" operating in colonial and post-Revolutionary America. James Delbourgo, *A Most Amazing Scene of Wonders: Electricity and Enlightenment in Early America* (Cambridge, Mass.: Harvard University Press, 2006), 282.

12   Carolyn Thomas de la Peña's study frames the body as a primary object of electricity; in the United States during the second half of the nineteenth century, Americans adorned themselves with electric belts that pumped electricity into modern bodies whose energies were depleted by the incessant demands made on them by urban life. Carolyn Thomas de la Peña, *The Body Electric: How Strange Machines Built the Modern American* (New York: New York University Press, 2003).

13   Carolyn Marvin's study of "new" electronic media in the United States in the late nineteenth century describes the body as "a probe and a point of reference for making strange electrical phenomena familiar." Carolyn Marvin, *When Old Technologies Were New: Thinking about Electric Communication in the Late Nineteenth Century* (Oxford: Oxford University Press, 1990), 151.

14   Jessica Riskin articulates the centrality of embodied knowledge in apprehending and making sense of electricity. In her chapter on the

Leyden jar, she frames this as a debate between sentimental empiricism and mechanism, illustrating her overarching theme of the Enlightenment as a meeting place for an Age of Reason and an Age of Sensibility. The "sentimental empiricism" she suggests was operating in the debates around electricity, however much it allowed and depended on evidence gathered by the nonvisual senses, remained somewhat unified in its treatment of sensation as a general orientation toward the external world, informed by an interplay of sentiments and sensations. Jessica Riskin, *Science in the Age of Sensibility: The Sentimental Empiricists of the French Enlightenment* (Chicago: University of Chicago Press, 2002), 69–103.

15 Delbourgo, *A Most Amazing Scene of Wonders,* 89.

16 From the mid-eighteenth century to the first two decades of the twentieth, electricity was tasked with curing or treating a staggeringly long list of ills. After the discrediting of electrotherapy in the early twentieth century, that list substantially contracted. For a summary of the ailments it was asked to cure in the eighteenth century, see Hannah Sypher Locke and Stanley Finger, "*Gentleman's Magazine,* the Advent of Medical Electricity, and Disorders of the Nervous System," in *Brain, Mind and Medicine: Essays in Eighteenth Century Neuroscience,* ed. Harry Whitaker, C. U. M. Smith, and Stanley Finger (New York: Springer, 2007), 260–63.

17 My argument around electrotherapy in this chapter can be understood as a variant on those put forth by Laura Otis and de la Peña. For de la Peña, the "palpable force" of electricity entering the body helped the exhausted patients undergoing electrotherapeutic treatments see themselves as coequal to the electrical forces that acted on them; electricity became part of the body's "reserve force" that could be driven, by therapeutic process, directly into the muscles and nerves. De la Peña, *The Body Electric,* 9. For Otis, understanding the body as a network of nerves preceded the later invention of communication networks that electrically transmitted "thoughts without wires"; it was the body's nervous network that provided the foundational metaphor for the emergence of telegraphy. Laura Otis, *Networking: Communicating with Bodies and Machines in the Nineteenth Century* (Ann Arbor: University of Michigan Press, 2001), 6.

18 Georg Simmel and Walter Benjamin each framed shock as an embodied reaction to the chaos of modern urban sensory experience, rendering the human nervous system as an electrical battery whose energies were depleted through repeated and constant exertion. However, this cultural model is consistent with accounts of neurasthenia, which understood modern life as both cut and cure:

nervous energies spent navigating the modern metropolis could be recharged through the structured application of medical electricity to the human battery. The notion of shock as a collective cultural psychic response to the introduction of new technologies was further embraced by Lewis Mumford in *Technics and Civilization* (New York: Harcourt, Brace & World, 1963), and subsequently by Marshall McLuhan in *Understanding Media: The Extensions of Man* (Cambridge, Mass.: MIT Press, 1994) and Alvin Toffler in *Future Shock* (New York: Bantam Books, 1984). It reverberates in Douglas Rushkoff's recent *Present Shock: When Everything Happens Now* (New York: Bantam Books, 2013). Naomi Klein's *The Shock Doctrine: The Rise of Disaster Capitalism* (New York: Metropolitan Books, 2008) moves productively between an embodied notion of shock and its metaphorization, but it situates shock as a technique for clearing away, drawing on electroconvulsive therapy for her model of shock. The genealogy I construct here predates the emergence of electroconvulsive therapy in the 1930s by two centuries.

19   Tim Armstrong, "Two Types of Shock in Modernity," *Critical Quarterly* 42, no. 1 (2000): 60–61. Armstrong's delineation of these two types of shock emphasizes both their continuities and discontinuities. The first type, designating "collision, battle, thrill and speed," overlaps with the electrocutaneous shock discussed in this chapter, though Armstrong never explicitly discusses electric shock. The second, shock as trauma, stresses shock as a type of psychic wounding, akin to the formulation also taken up in Wolfgang Schivelbusch's genealogy.

20   Wolfgang Schivelbusch, *The Railway Journey: The Industrialization of Space and Time in the 19th Century* (Berkeley: University of California Press, 1986), 153.

21   Nikolay Pirogoff, quoted in Schivelbusch, 156.

22   Elaine Scarry suggests that pain is always bound up with power: linguistic expressions, abstractions, and transcriptions of pain. Elaine Scarry, *The Body in Pain: The Making and Unmaking of the World* (New York: Oxford University Press, 1985), 11–13.

23   John Durham Peters, *Speaking into the Air: A History of the Idea of Communication* (Chicago: University of Chicago Press, 1999), 7–8.

24   Francis Hauksbee, quoted in Simon Schaffer, "Experimenters' Techniques, Dyers' Hands, and the Electric Planetarium," *Isis* 88, no. 3 (1997): 461. Hauksbee, in his 1709 work *Physico-Mechanical Experiments on Various Subjects,* centrally located the tactile relationship to electricity—the book contains "an account of several surprising

phenomena touching light and electricity." Francis Hauksbee, *Physico-Mechanical Experiments on Various Subjects* (London, 1709), 51.

25 Gray's early experiments aimed at producing attraction between electrified objects, where proof of electricity's effects came from its ability to generate the visible spectacle of objects—such as silk threads—defying gravity, lifting away from the ground, and being drawn toward electrified bodies.

26 During the eighteenth century, this term was commonly used to refer to those who studied and manipulated electricity.

27 J. L. Heilbron, *Elements of Early Modern Physics* (Berkeley: University of California Press, 1982), 180. Also see Delbourgo, *A Most Amazing Scene of Wonders*, 96, for a further description of the kiss and its life on the other side of the Atlantic. Julie Wosk locates the kiss centrally in her examination of the relationship between women and electricity, describing the framing of women as "electric goddesses" both in public demonstrations and in print culture. Julie Wosk, *Women and the Machine: Representations from the Spinning Wheel to the Electronic Age* (Baltimore, Md.: Johns Hopkins University Press, 2001), 68–74.

28 This translation appears in J. L. Heilbron, *Electricity in the 17th and 18th Centuries: A Study of Early Modern Physics* (Berkeley: University of California Press, 1979), 267n19.

29 While historiographers of science frequently describe the kiss as a game, they do so without considering the full ramifications of this positioning. However, as I have argued elsewhere, locating the kiss as a game with formal structures of victory and defeat among defined competing parties places it at the base of a genealogy of electric shock games that includes early twentieth-century shock arcade games, such as *Spear the Dragon* and more recent art games like *PainStation* and *Tekken Torture*. For a comprehensive treatment, see David Parisi, "Shocking Grasps: An Archaeology of Electrotactile Games Mechanics," *Game Studies* 13, no. 2 (2013), http://gamestudies.org/1302/articles/parisi.

30 Heilbron, *Electricity in the 17th and 18th Centuries*, 267.

31 See Heilbron, *Electricity in the 17th and 18th Centuries*, 270. For an account of these experiments' popularity in the American colonies, see Delbourgo, *A Most Amazing Scene of Wonders*, 109–19.

32 Joseph Priestley, *The History and Present State of Electricity* (London: C. Bathurst and T. Lowndes, 1767), 83. The story of the jar's invention is a curious sort of accident—in the span of a year, it was discovered independently by Musschenbroek in the Dutch town of Leiden and by

the German cleric Ewald Georg von Kleist in present-day Poland. Heilbron, *Elements of Early Modern Physics,* 183.

33   Priestley, *The History and Present State of Electricity,* 108.

34   Tubervill Needham, "Extract of a Letter from Mr. Tubervill Needham to Martin Folkes, Esq; Pr. R. S. Concerning Some New Electrical Experiments Lately Made at Paris," *Philosophical Transactions of the Royal Society* 44 (January 1, 1746): 254, doi:10.2307/104816.

35   Ibid.

36   Ibid.

37   Priestley, *The History and Present State of Electricity,* 106.

38   Tom Standage, *The Victorian Internet: The Remarkable Story of the Telegraph and the Nineteenth Century's On-line Pioneers* (New York: Walker and Company, 1998), 1–2.

39   Arthur Elsenaar and Remko Scha, "Electric Body Manipulation as Performance Art," *Leonardo Music Journal* 12 (2002): 19.

40   Henry Wiegand, "Applications of Electricity to Medicine," *Medical and Surgical Reporter* 73, no. 6 (August 10, 1895): 157–64.

41   Nollet was not the first to connect bodies in this way, and not all such networks allowed their participants to feel the shock. Heilbron, *Elements of Early Modern Physics,* 186.

42   In the concluding experiment, researchers passed a charge through a wire 12,276 feet long. The observer who held the ends felt the charge instantly on its introduction into the circuit, leaving them satisfied that "the velocity of the electric matter was instantaneous." Priestley, *The History and Present State of Electricity,* 101–9.

43   Gilles Deleuze, *Foucault* (Minneapolis: University of Minnesota Press, 1988), 40.

44   Dutch and French researchers noticed the similarities between the fish and the jar earlier than the British did, but I have opted to use the British reception to illustrate touch's role in apprehending the link between the two experiences.

45   The secondary literature on the electric fish produced by historians of science contains rich and exhaustive detail, which I have flattened in the interests of specifying touch's centrality to research on the fish. For a nuanced historical analysis of the electric fish's significance to medicine, see Finger and Piccolino, *The Shocking History of Electric*

Fishes. Delbourgo's chapter on the fish experiments emphasizes the way that colonial relations shaped attempts to study and commodify the animals. Delbourgo, *A Most Amazing Scene of Wonders,* 165–99. Marco Piccolino's *The Taming of the Ray: Electric Fish Research in the Enlightenment from John Walsh to Alessandro Volta* (Florence: LS Olschki, 2003) provides a comprehensive account both of John Walsh's experiments and of their influence on Volta's invention of the voltaic pile.

46   Umberto Rossi, "The History of Electrical Stimulation of the Nervous System for the Control of Pain," in *Electrical Stimulation and the Relief of Pain,* ed. Brian Simpson (New York: Elsevier, 2003), 6.

47   The torpedo fish was so named by the Romans for its ability to produce numbness, or sleep (*torpor*), in the affected area. The Greeks referred to it as *narke,* or numbness-producing, from which the term "narcosis" was derived. Ibid., 6.

48   Edward Bancroft, *An Essay on the Natural History of Guiana, in South America* (London, 1769), 196.

49   Delbourgo, *A Most Amazing Scene of Wonders,* 196.

50   Hans-Jörg Rheinberger, *Toward a History of Epistemic Things: Synthesizing Proteins in the Test Tube* (Stanford, Calif.: Stanford University Press, 1997), 28. I modify Rheinberger's notion of an epistemic thing here to refer to the way the fish was used as an instrument that induced electrical shock.

51   A 1774 advertisement in the *South Carolina Gazette* made precisely this comparison and offered spectators the chance to view the fish for "the small expense of one dollar." Quoted in Delbourgo, *A Most Amazing Scene of Wonders,* 165.

52   Heilbron, *Elements of Early Modern Physics,* 83.

53   John Walsh, "Of the Electric Property of the Torpedo," *Philosophical Transactions of the Royal Society* 63 (1773): 465.

54   The term "battery" was borrowed from military lexicon to refer to a series of linked jars, akin to a cluster of cannons that fired simultaneously.

55   Ibid., 463–64.

56   Delbourgo, *A Most Amazing Scene of Wonders,* 186–87.

57   Elaine Scarry's description of the power torturers had over the bodies of prisoners proves instructive here again: by configuring the

conditions under which prisoners experienced their bodies as sources of pain, the torturer had the capacity to construct the mental worlds of the tortured. Scarry, *The Body in Pain,* 9–13.

58　In one public demonstration, Walsh connected eight participants to the animal, with each of them receiving a shock when Walsh closed the human–fish circuit. Walsh, "Of the Electric Property of the Torpedo," 466.

59　Ibid., 471.

60　In spite of this demand, the challenges of keeping the fish healthy made exploiting the fish as a spectacular commodity difficult and frequently unprofitable. See Delbourgo, *A Most Amazing Scene of Wonders,* 190–93.

61　Finger and Piccolino discuss the difficulty capturing the fish as reported by Alexander von Humboldt during his 1800 visit to Spanish Guyana (present-day Venezuela). Finger and Piccolino, *The Shocking History of Electric Fishes,* 6.

62　Walsh, "Of the Electric Property of the Torpedo," 468.

63　Alessandro Volta, "On the Electricity Excited by the Mere Contact of Conducting Substances of Different Kinds," *Philosophical Magazine* 7, September 1800, 291. The original letter, written in French, was read for the Royal Society and published in *Philosophical Transactions of the Royal Society of London* 90 (1800): 403–31, doi: 10.1098/rstl.1800.0018. For further discussion of Volta's letter and a reprint of the English translation, see Bern Dibner, *Alessandro Volta and the Electric Battery* (New York: Franklin Watts, 1964).

64　Ibid., 303.

65　Joost Mertens, "Shocks and Sparks: The Voltaic Pile as a Demonstration Device," *Isis* 89 (1998): 301.

66　Marcello Pera, quoted in ibid.

67　The initial trials where Volta claimed to have disproven the theory of animal electricity took place in 1797, three years before his letter detailing the chain of cups and columnar apparatus. Ibid., 301.

68　Volta, quoted in ibid., 305.

69　Volta, "On the Electricity," 302.

70　Ibid., 291.

71   Ibid., 293.

72   Ibid., 295.

73   Ibid., 302–3.

74   "I shall therefore forbear myself from describing a great number of these experiments . . . and I shall relate only a few which are no less instructive than amusing"; ibid., 279.

75   The immodest Volta was not unaware of his experiments' ramifications, claiming not only to have uncovered "sufficiently numerous" facts that would be of interest to anatomists and physiologists but also to have paved the way, via his electrical apparatus, for the discovery of new facts through the multiplication and variation of the experiments he described. Volta, "On the Electricity," 130. I take up Müller's research into the doctrine of specific sense energies and its ramifications for the nascent psychophysics of sensation at the opening of Interface 2. While Jonathan Crary explains the significance of Müller's doctrine as it affected and shaped the nineteenth-century observer, his reading obscures the extent to which both Volta and Müller were concerned with the whole range of human senses rather than just vision alone. See Jonathan Crary, *Techniques of the Observer: On Vision and Modernity in the Nineteenth Century* (Cambridge, Mass.: MIT Press, 1990), 88–94.

76   Applied inside the nose, Volta found that electricity "does not excite any sensation of smell . . . only a pricking more or less painful, and commotions more or less extensive, according as the said current is weaker or stronger." Volta, "On the Electricity," 309.

77   I take up this notion of a common (or shared) sensibility in Interface 2; although Volta did not use it in his work, it became central to Ernst Heinrich Weber's scientific account of tactility later that century.

78   Volta, "On the Electricity," 305.

79   Ibid., 305.

80   Ibid., 307.

81   Ibid.

82   Ibid.

83   Ibid., 308.

84   Ibid.

85   Ibid., 304.

86   Although "electrification" and "electricization" (and the variant
      spellings "electrisation" and "electrisization") similarly refer to
      charging a body with electrical current, the former term has since
      become associated with infrastructure, while the latter has become
      somewhat of an anachronism. I have opted to use "electricization"
      in this chapter because it was the more common term during the
      historical period under investigation here.

87   Park Benjamin, *A History of Electricity* (New York: Wiley and Sons,
      1898), 500.

88   Cited in ibid., 501.

89   From 1750 to 1789, for example, one French journal published 26
      papers that addressed medical uses of electricity. See Timothy
      Kneeland and Carol Warren, *Pushbutton Psychiatry: A History of
      Electroshock in America* (Westport, Conn.: Praeger, 2002), 7.

90   De la Peña, *The Body Electric,* 99.

91   De la Peña attributes electrotherapy's decline to a confluence of factors,
      including the rise of immunizations that served to stave off many
      diseases the practice had been used to treat. According to her
      argument, it was not the triumph of any conclusive scientific evidence
      that diminished medical electricity's popularity; its persistent
      discrediting as quackery did little to diminish its widespread
      acceptance prior to the middle decades of the twentieth century.
      Ibid., 214.

92   Otis, *Networking,* 2–3, 129–31. I layer an additional level on Otis's
      argument by showing how touch and tactility were specified in
      electrotherapeutic discourse, feeding into changing conceptions about
      the nervous system and its structures.

93   "Electricity is life" was often repeated throughout the era, eventually
      becoming a branding slogan used on electrotherapeutic apparatuses
      and electricity-dispensing arcade machines. The phrase's implications
      shifted with the first use of the electric chair, in the execution of
      William Kemmler at an Auburn, New York, prison in 1890.
      Kemmler's death demonstrated electricity's power not just to give life
      but to take it. See T. H. Metzger, *Blood and Volts: Edison, Tesla & the
      Electric Chair* (New York: Autonomedia, 1996), 173. The electric chair
      was seen as an object of scientific and medical wonder until it failed so
      miserably in its first use. Kemmler's execution required the
      administration of two long volleys of current, the second of which

caused him to catch fire from the strength of the current. Prior to Kemmler's gruesome death, the chair's proponents (among them, Thomas Edison) hoped it would provide a more humane, painless, and effective means of executing the condemned than the common method of hanging. In spite of its inability to live up to this promise, the chair was quickly adopted, and by 1905, more than 100 people had been put to death by using the new machine.

94 De la Peña, *The Body Electric,* 111.

95 George Adams, *An Essay on Electricity* (London, 1785), 312.

96 Ibid., 324, 319.

97 Ibid., 324, xxxiv.

98 Celia Haynes, *Elementary Principles of Electro-Therapeutics* (Chicago: McIntosh Galvanic & Faradic Battery Company, 1884), 214–15.

99 Lowndes, cited in Kneeland and Warren, *Pushbutton Psychiatry,* 6. Originally in Francis Lowndes, *Observations on Medical Electricity* (London: D. Stuart, 1787).

100 Ernst Onimus, *A Practical Introduction to Medical Electricity,* trans. Armand de Watteville (London: H. K. Lewis, 1878), 103–4.

101 Ibid.

102 Adams, *An Essay on Electricity,* 323.

103 Haynes, *Elementary Principles of Electro-Therapeutics,* 276.

104 Onimus used this often-repeated phrase, but found it unsatisfactory, and sought to further define and quantify the resistances provided by the individual elements of the body. "The bones," Onimus found, "are the worst conductors." Onimus, *A Practical Introduction to Medical Electricity,* 30.

105 Ibid., 104. A rheophore is a wire attached to an electrical battery.

106 Ibid.

107 Haynes, *Elementary Principles of Electro-Therapeutics,* 276.

108 William Erb's *Handbook of Electro-Therapeutics* (New York: Wood, 1883), for example, contains many formulas for calculating the amount of current that ought to be applied to a given body.

109 Eugene Taylor, "The Electrified Hand—Psychotherapeutic Implications," *Medical Instrumentation* 17, no. 4 (July–August 1983): 282.

110 Haynes, *Elementary Principles of Electro-Therapeutics,* 228.

111 Electrotherapeutic discourse had wider-ranging ramifications than can be accounted for here. It was informed by shifting notions of sexuality, it reshaped ideas about the mutuality of human and artificial energies, and it promoted a cyborgian imaginary of human body. The use of electricity as a treatment for neurasthenia, as formulated by George Beard and A. D. Rockwell, was tied to the intensification of media flows in late-nineteenth-century U.S. culture.

112 Otis advances this argument throughout *Networking,* describing the process by which organic nerves came to be understood as structures analogous to electrical communication systems.

113 Alfred C. Garratt, *Medical Electricity: Embracing Electro-Physiology and Electricity as a Therapeutic* (Philadelphia: J. B. Lippincott, 1866), 33–34.

114 Nicholas Costa, *Automatic Pleasures: The History of the Coin Machine* (London: Bath Press, 1988), 144.

115 Haynes, *Elementary Principles of Electro-Therapeutics,* 417.

116 Costa, *Automatic Pleasures,* 147.

117 See, for example, Erkki Huhtamo, "Slots of Fun, Slots of Trouble: An Archaeology of Arcade Gaming," in *Handbook of Computer Game Studies,* ed. Joost Raessens and Jeffrey Goldstein (Cambridge, Mass.: MIT Press, 2005), 3–21.

118 Marvin, *When Old Technologies Were New,* 3.

119 For summaries of the telegraph's pre-Morse history, see J. J. Fahie, *A History of Electric Telegraphy, to the Year 1837* (New York, 1884); and Anton Huurdeman, *The Worldwide History of Telecommunications* (New York: Wiley, 2003), 48–83.

120 I appropriate Hansen's phrase here to call attention to the way that the body's electrocutaneous sensitivity ordered and structured electrical experiments. For Hansen, tactility unceasingly provides a structuring, "infraempirical" frame for human experience and thus lies at the seat of a technics of the body. Mark B. N. Hansen, *Bodies in Code: Interfaces with Digital Media* (New York: Routledge, 2006), 71.

121 Jacques Derrida, *On Touching—Jean-Luc Nancy* (Stanford, Calif.: Stanford University Press, 2005), 300.

122 Gina Perry, *Behind the Shock Machine: The Untold Story of the Notorious Milgram Psychology Experiments* (Brunswick, Australia: Scribe Publications, 2012).

123 Darius Rejali, "Electricity: The Global History of a Torture
    Technology," *Connect: art.politics.theory.practice,* June 2001, 101–9.

124 According to a recent *New York Times* estimate, roughly 100,000
    people each year receive implantable cardiac defibrillators. Paula Span,
    "A Heart Quandary," *New York Times,* June 7, 2012, http://newoldage
    .blogs.nytimes.com/2012/06/07/a-heart-quandary/.

125 Stern's project involves connecting players of the videogame *Tekken 3*
    to a custom-designed feedback mechanism that issues painful electric
    shocks to the player based on the damage done to their onscreen
    avatar. The shocks are described as "bracing but nonlethal." However,
    Stern's claim to map visual damage to electrical intensity masks the
    conversion process already engaged in by the game's software—as
    PlayStation uses a mechanical feedback technology known as rumble,
    it already projects tactile information onto the player's body, even
    before Stern's intervention. Instead, Stern allows the player to experience
    the electrical signal sent from the machine without the intervening
    apparatus that would ordinarily translate electrical signals into
    mechanical rumble sensations. Eddo Stern and Mark Allen, "Torture
    Tekken: Game Performance/Custom Hardware," C-level, May 5,
    2001, http://www.c-level.org/tekken1.html. See also Pau Waelder
    Laso, "Games of Pain: Pain as Haptic Simulation in Computer-
    Game–Based Media Art," *Leonardo* 40, no. 3 (June 2007): 238–42.

126 Elsenaar and Scha, "Electric Body Manipulation," 17. The authors
    describe Elsenaar's own performances, including *The Varieties of Human
    Facial Expressions* (1997), in which "a computer program enumerates
    all facial expressions that can be realized with a particular electrode
    configuration" on Elsenaar's face. Like Stelarc's pieces, Elsenaar's
    performance involves ceding control over the body to computer-
    modulated electrodes that contract and relax the muscles. By establishing
    "electric body manipulation" as a discrete field of performance art,
    the authors give form and structure to a range of previously disparate
    practices. Proceeding from "an information-theoretical, cybernetic
    standpoint," they "view the human body as a kinematic system whose
    motions can be steered by means of electrical control signals."
    Eighteenth-century demonstrations like Galvani's electrified frog legs
    and the *Venus electrificata* implied "the possibility of employing the human
    body as a display device for algorithms that run on digital computers."

127 The use of electricity to directly activate nerve fibers in the skin has
    shown promise, but in fits and spurts—researchers working out of
    Susumu Tachi's lab at the University of Tokyo put forward a model of
    "tactile primary colors," where targeted bursts of electricity selectively
    act on three different layers of tactile receptors, allowing for the

creation of complex tactile sensations in the fingertip. See Hiroyuki Kajimoto, Naoki Kawakami, and Susumu Tachi, "Electro-Tactile Display with Tactile Primary Color Approach," in *Proceedings of the International Conference on Intelligent Robots and Systems (IROS)* (Tokyo, 2004), http://www.tachilab.org/content/files/publication /ic/sato200708ROMAN.pdf. More recently, full-body haptic suits for virtual reality have employed electrical stimulation as a means of synthesizing tactile experience. The Teslasuit, developed as a prototype but not released for commercial sale, provides one example. Using 46 electrodes distributed throughout the suit, the Teslasuit can draw complex haptic patterns on the wearer's body. See "Teslasuit: virtual reality reinvented," https://teslasuit.io/.

128 Electricity-powered vibrators provide the most effective and earliest illustration of this principle. For a detailed examination of their history, see Rachel Maines, *The Technology of Orgasm: "Hysteria, the Vibrator, and Women's Sexual Satisfaction"* (Baltimore, Md.: Johns Hopkins University Press, 2001).

129 Friedrich Kittler, *Optical Media* (Malden, Mass.: Polity, 2010), 38.

130 Johannes Müller, *Elements of Physiology, Volume 2,* trans. William Baly (London, 1842), 1073–74.

2. The Haptic

1   While Weber did experiment broadly on the senses, his most systematic investigations centered on touch and the range of senses he associated with it, such as pain and the "common" senses, discussed later in this chapter.

2   Chris Green, "Institutions of Early Experimental Psychology: Laboratories, Courses, Journals, and Associations," in *Classics in the History of Psychology: An Electronic Resource Developed by Christopher D. Green* (Toronto: York University, 2000), http://psychclassics.yorku.ca /Special/Institutions/labsintro.htm.

3   Friedrich Kittler, "Thinking Colours and/or Machines," *Theory, Culture & Society* 23 (2006): 42.

4   Ibid.

5   Notable exceptions to this general omission are Alexandra Hui, *The Psychophysical Ear* (Cambridge, Mass.: MIT Press, 2013); and Jimena Canales, *A Tenth of a Second* (Chicago: University of Chicago Press, 2009).

6   The Scottish metaphysician Thomas Reid anticipated Weber in
     suggesting touch be subdivided into its component parts. Writing in
     1764, Reid (1710–1796) argued that touch should be treated as a set of
     distinct systems for constructing impressions of the external world.
     The error that prior philosophers made, according to Reid, was in
     collapsing tactile sensations onto our experience of them, failing to
     attend to the processes by which our sensations are assembled into
     coherent and structured ideas about the external world. Reid asserted
     that we have no way of distinguishing between sensation and
     perception. The process by which a subject feels a stone to be hard, the
     muscles at play in registering the existence of the stone, and the nerves
     that compress against it in the process of touching exist only as what
     Reid termed "fugitive sensations": learning to assemble these various
     sensations occurs so instinctively that distinguishing their component
     parts—the sensation of sensations—proves almost impossible. Thomas
     Reid and Dugald Stewart, *The Works of Thomas Reid; With an Account
     of His Life and Writings* (New York: Duyckinck, Collins, and Hannay,
     1822), 120.

7   Such a positioning is emblematic of the creation story told around
     haptic interfaces. But their chapter, and the other historical essays on
     touch contained in the volume, also provides a valuable map of the
     variety and scope of research devoted to touch prior to 1950. Martin
     Grunwald and Matthias John, "German Pioneers of Research into
     Human Haptic Perception," in *Human Haptic Perception: Basics and
     Applications,* ed. Martin Grunwald (Boston: Birkhäuser-Verlag,
     2008), 15.

8   What Sterne terms "perceptual technics" refers to a process of
     mobilizing perceptual research in the service of economizing the
     transmission of sensory signals in media technologies. In short, signals
     that will generate unnoticeable (or imperceptible) sensations can be
     eliminated, as they are functionally unnecessary for the conveyance of
     a given message or aesthetic experience. Jonathan Sterne, *MP3: The
     Meaning of a Format* (Durham, N.C.: Duke University Press, 2012), 19.

9   Max Dessoir, "Über den Hautsinn," *Archiv für Anatomie und Physiologie*
     (1892), 242.

10   The term "haptics" had previously been employed by the English
     mathematician Isaac Barrow, in his 1735 work *The Usefulness of
     Mathematical Learning Explained,* to refer to "the science of touches,"
     but neither Dessoir nor Titchener (responsible for introducing the
     word to English-speaking audiences) referenced Barrow's work, in
     spite of the similarity between their definitions. Isaac Barrow, *The

Usefulness of Mathematical Learning Explained and Demonstrated, trans. John Kirby (London, 1735), 24.

11   Grunwald and John, "German Pioneers," 21.

12   Edward Bradford Titchener, "Haptics," in *Dictionary of Philosophy and Psychology, Volume 1,* ed. James Baldwin (New York: Macmillan, 1901), 441. During his time at Cornell, Titchener engaged in a far-ranging and exhaustive program of experimental research into the touch senses, collaborating with investigators such as G. Stanley Hall to establish what the historian of psychology Rand Evans describes as a "familial line of most psychologists researching in the field of touch in the United States." Rand Evans, "Haptics in the United States before 1940," in *Human Haptic Perception,* 72.

13   Weber's experiments predated and foregrounded the later founding of institutionally recognized experimental psychology labs. The structure of the lab that Wilhelm Wundt established at the University of Leipzig in 1879 had a particularly strong influence on the field, as researchers from Europe and the United States trained there and cloned its structure on returning from their visits. See Wolfgang Bringmann and Gustav Ungerer, "The Foundation of the Institute for Experimental Psychology at Leipzig University," *Psychological Research* 42, no. 1–2 (1980): 5–18.

14   Max Weber, *From Max Weber: Essays in Sociology* (New York: Oxford University Press, 1946), 155, 144.

15   David Harvey, *The Condition of Postmodernity: An Enquiry into the Origins of Cultural Change* (New York: Wiley-Blackwell, 1991), 12–13.

16   Robert Jütte, *A History of the Senses: From Antiquity to Cyberspace* (Malden, Mass.: Polity, 2005), 180. Echoing Wolfgang Schivelbusch's argument in *The Railway Journey: The Industrialization of Time and Space in the Nineteenth Century* (Berkeley: University of California Press, 1986), Jütte claims that nineteenth-century technology altered human spatiotemporal consciousness. New technologies like the railroad circulated bodies with increasing rapidity. Perceptual "disorders" signified the challenges posed to the sensorium, as sense experience had to be managed in response to changing technologies.

17   Claudia Benthien, *Skin: On the Cultural Border between Self and World* (New York: Columbia University Press, 2002), 185–234. See also Mark Paterson, *Senses of Touch: Haptics, Affects and Technologies* (New York: Berg Press, 2007); Erin Manning, *The Politics of Touch* (Minneapolis: University of Minnesota Press, 2007); Constance

Classen, ed., *The Book of Touch* (New York: Berg Press, 2005); Elizabeth Harvey, ed., *Sensible Flesh: On Touch in Early Modern Culture* (Philadelphia: University of Pennsylvania Press, 2003).

18 Michel Foucault, *Discipline and Punish: The Birth of the Prison* (New York: Vintage Books, 1995), 296. Foucault himself downplayed Weber's importance in comparison to the contemporaneous development of the carceral subject in the penal colony, suggesting in a dismissive reference to scientific psychology that "Weber manipulating his little compass for the measurement of sensations" was of considerably less significance than the mode of disciplining that emerged out of the nineteenth-century prisons (Foucault, *Discipline and Punish*, 295).

19 Jonathan Crary, *Techniques of the Observer: On Vision and Modernity in the Nineteenth Century* (Cambridge, Mass.: MIT Press, 1990), 128.

20 Ibid., 62.

21 Kittler, "Thinking Colours," 42. Like Crary and other media historians, Kittler focuses on the construction of vision and hearing in the early psychophysical trials, excluding any discussion of Weber's experiments on touch. He locates Fechner as the original figure in this discourse, despite Fechner's designation of Weber as "the father of psychophysics." Gustav Theodor Fechner, *Elements of Psychophysics* (New York: Holt, 1966), xxviii.

22 Martin Heidegger, *The Question Concerning Technology and Other Essays* (New York: Harper, 1977), 4.

23 These titles are shortened to *De Tactu* and *Der Tastsinn* throughout this chapter. Translations appear in Ernst Heinrich Weber, *E. H. Weber on the Tactile Senses,* trans. Helen Ross and David Murray (Hove, UK: Erlbaum, 1996).

24 G. Stanley Hall, "The New Psychology," *Harper's Monthly Magazine,* November 1901, 727.

25 Vladislav Krurta, "Ernst Heinrich Weber," in *Dictionary of Scientific Biography* (New York: Scribner's, 1974), 200.

26 According to Susan Lederman, monastic scribes copying medical texts were prohibited from transferring any knowledge relating to the skin. René Verry, "Don't Take Touch for Granted: An Interview with Susan Lederman," *Teaching of Psychology* 25, no. 1 (1998): 64.

27 As Robert Mitchell suggests, the relationship between vitalism and experimentalism was a complex one and was frequently synergistic

rather than antagonistic. I do not want to intervene in this debate, other than to note the role speculative nature philosophy played in orienting Weber toward meticulous experimentation. Weber possessed an absolute distaste for speculations that were not grounded in or could not be confirmed through experimental trials. Each new hypothesis he offered was followed immediately by a series of experiments designed to illuminate his intuitions. In a passage where he theorized on the physical operations responsible for the neurological production of sensation, for example, Weber cut himself off, noting that "these speculations go beyond the bonds of experience, making it impossible to prove them by observation or experiment; I shall therefore dwell no longer on them, and will prove nothing from it." Weber, *E. H. Weber on the Tactile Senses,* 163. For the relationship between vitalism and experimentation, see Robert Mitchell, *Experimental Life: Vitalism in Romantic Science & Literature* (Baltimore, Md.: Johns Hopkins University Press, 2013).

28 For a summary of these positions, see Benthien, *Skin: On the Cultural Border between Self and the World,* 195–200. A consideration of Enlightenment philosophies of touch is beyond the scope of this book; both Jacques Derrida, *On Touching—Jean-Luc Nancy* (Stanford, Calif.: Stanford University Press, 2005) and Paterson's *Senses of Touch* provide valuable resources on this topic.

29 Helen Ross and David Murray, "Introduction," in *E.H. Weber on the Tactile Senses,* 7–8.

30 Charles Bell, "Charles Bell on Spinal Nerve Roots," in *A Source Book in the History of Psychology,* ed. Richard J. Herrnstein and Edwin Boring (Cambridge, Mass.: Harvard University Press, 1965), 17–19.

31 Francois Magendie, "Francois Magendie on Spinal Nerve Roots," trans. Mollie D. Boring, in Herrnstein and Boring, *Source Book,* 19–22.

32 Ibid., 20.

33 Ibid., 21.

34 This segmentation of motion and sensation would become crucial more than 100 years later in the design of remote manipulation systems; see Interface 4 for discussion.

35 Aldini, the nephew of animal electricity proponent Luigi Galvani, traveled Europe in the early nineteenth century showing how electricity could be used to animate the corpses of animals. Human corpses were eventually folded into these experiments, further illustrating electricity's power to control and manipulate the human

body. For a summary of Aldini's 1803 experiments on the body of convicted murderer George Foster, see Helen MacDonald, *Human Remains: Dissection and Its Histories* (New Haven, Conn.: Yale University Press, 2006), 14–17.

36 Müller published his doctrine in the 1838 *Handbuch der Physiologie des Menchen.* Though Bell's formulation came earlier, Müller's is generally taken to be the more influential, especially given its widespread acceptance in Germany and the significance of his students, among them Hermann von Helmholtz and Wilhelm Wundt. For a comprehensive history of Müller's influence on nineteenth-century scientific networks, see Laura Otis, *Müller's Lab* (New York: Oxford University Press, 2007). Otis traces the significant differences in the way that Müller's students understood and deployed his work, arguing that each, in their differing appropriations of his research, played crucial roles in shaping his legacy.

37 Charles Bell, "Bell on the Specificity of Sensory Nerves," in Herrnstein and Boring, *Source Book,* 23.

38 Ibid., 25.

39 Emphasis added. Johannes Peter Müller, *Elements of Physiology,* trans. William Baly (Philadelphia: Lea and Blanchard, 1843), 707.

40 Ibid., 712.

41 Ibid., 714.

42 Crary, *Techniques of the Observer,* 92.

43 Ulf Norsell, Stanley Finger, and Clara Lajonchere, "Cutaneous Sensory Spots and the 'Law of Specific Nerve Energies': History and Development of Ideas," *Brain Research Bulletin* 48, no. 5 (1999): 457–65.

44 See, for example, Jutta Schickore, "The 'Philosophical Grasp of the Appearances' and Experimental Microscopy: Johannes Müller's Microscopical Research, 1824–1832," *Studies in History and Philosophy of Science Part C: Studies in History and Philosophy of Biological and Biomedical Sciences* 34, no. 4 (December 2003): 569–92.

45 Crary, *Techniques of the Observer,* 62.

46 In tracing the impact of this new science of perception on vision, Crary's narrative disentangled touch from sight. With the development of the stereoscope, he contends, the tactile was folded back into the optical. But the new tactility was one that could only be invoked by deploying the eye as a mediating agent, or "a tangibility that has been

transformed into a purely visual experience." Crary, *Techniques of the Observer,* 124.

47   This shift away from philosophical psychology toward scientific, experimentally informed psychology was a significant movement in the history of thought. A host of thinkers writing in the early to mid-1800s were emblematic of this move, among them Weber, Johannes Peter Müller (1801–1858), and later, Hermann von Helmholtz (1821–1894), Gustav Theodor Fechner, and Wilhelm Wundt.

48   Though Weber insisted early on in *De Tactu* that "the experiments should be repeated and compared for many subjects" (p. 30), he did not identify or account for differences among subjects, particularly those of gender. He often referred to his subjects simply as his "colleagues" (his brother Eduard frequently assisted him in both designing and executing experiments). In one trial, where he attempted to judge the passive subject's ability to discriminate between different weights, Weber identified several different subjects, among them "a woman, a girl, a merchant, a student of literature and a mathematician" (p. 64). But Weber was often content to draw conclusions based solely on autoexperimentation.

49   "We must discover what may be perceived by touch if we do not move the touch-organs" (Weber, *E.H. Weber on the Tactile Senses,* 29).

50   To quote this passage at greater length: "The senses are fatigued and tired by continuous work, so that we can no longer perceive differences which were formerly noticed very easily and clearly" (ibid., 28). This was no mere speculation for Weber; like most of his ideas, it originated in his experiments, in this case ones designed to measure the subject's ability to distinguish repeatedly differences between stimuli. Later in the nineteenth century, this idea of sensory fatigue became a motive in the development of electrotherapy, where mental reserves depleted by the constant sensory stimulation of the modern urban environment were recharged by batteries connected to the body. For a discussion on the scientific and cultural origins of fatigue, see Anson Rabinbach, *The Human Motor: Energy, Fatigue, and the Origins of Modernity* (Chicago: University of Chicago Press, 1992), 149.

51   Weber, *E.H. Weber on the Tactile Senses,* 30.

52   Ibid., 181.

53   Ibid.

54   Weber compared the perception of shape in the fingertip and tongue to the ability to perceive shape when an object was pressed against the abdomen and back. Weber, *E.H. Weber on the Tactile Senses,* 193.

55 Friedrich Nietzsche, *The Will to Power* (New York: Vintage, 1968), 317.

56 Crary also cites this passage from Nietzsche, arguing that it summarizes and embodies the function of the camera obscura. According to Crary, the camera obscura's status as a representational apparatus took priority over the individual's flawed, inadequate, and unaided sense of vision. Crary, *Techniques of the Observer*, 40.

57 Weber, *E.H. Weber on the Tactile Senses*, 92–93, 213.

58 Daniel Heller-Roazen, *The Inner Touch: Archaeology of a Sensation* (New York: Zone, 2007), 237–51.

59 For an expanded analysis of nineteenth-century physiological theories of pain, see Andrew Hodgkiss, *From Lesion to Metaphor: Chronic Pain in British, French and German Medical Writings, 1800–1914* (Atlanta, Ga.: Rodopi Bv Editions, 2000), 75–88.

60 Weber, *E.H. Weber on the Tactile Senses*, 148.

61 Weber's choice of the term "observer" here is curious, as it indicates his ability to detach himself from his own sensory experience—a type of autoexperimentation common in early experimental science in general and experimental psychology in particular. Further, it designated a particular type of nonvisual observer and observation, where observing is not tied to any one particular sense organ, designating instead a careful perceptual attentiveness even to disagreeable sensations. In contrast, other models of the observer, such as Crary's, emphasize the visual nature of observation (see ibid., 156). In the history of physiological psychology, being a "good observer" meant being one who trained in the techniques and methods of observing one's own sensations in the context of the psychological laboratory. By the beginning of the twentieth century, such training in observation was accomplished through the institutionalization of experimental psychology and was standardized through the use of common instruments, training manuals, and university labwork. See Deborah Coon, "Standardizing the Subject: Experimental Psychologists, Introspection, and the Quest for a Technoscientific Ideal," *Technology and Culture* 34, no. 4: 776.

62 Weber, *E.H. Weber on the Tactile Senses*, 156.

63 Ibid.

64 Weber was not alone in subjecting his body to uncomfortable trials for the sake of research on the senses. Johann Wilhelm Ritter, whose experiments Weber read with great interest, ruined his health in his trials with electricity; see Siegfried Zielinski, *Deep Time of the Media:*

Toward an Archaeology of Hearing and Seeing by Technical Means (Cambridge, Mass.: MIT Press, 2006), 175–78. Fechner's repeated investigations into the nature of retinal afterimages, which involved long periods of staring directly at the sun, left him in an extended state of blindness and nervous exhaustion. With research on touch in particular, this sort of painful autoexperimentation frequently provided the most expedient means of gathering data; Grunwald and John describe Dessoir's extensive experiments on his body, executed, like Weber's, with the goal of furthering knowledge about the structure of the temperature sense. See Grunwald and John, "German Pioneers," 22.

65 Weber, *E.H. Weber on the Tactile Senses,* 153.

66 Ibid., 217.

67 Elsewhere in *Der Tastsinn,* Weber reviewed the experiments Volta and others carried out to test electricity's effects on the senses. Although he did not systematically repeat these experiments, his brother Eduard attempted to replicate Volta's experiments on hearing. Filling the entrances to both ears with water (to ensure that the current would be strongly conducted), Eduard passed a strong current through his head. Rather than hearing the detailed crackling that Volta described, Eduard reported perceiving "a light appearing to jump slant-wise over the head, but heard no tone, in fact not even a sound" (Weber, 166).

68 Ibid., 153.

69 Ibid.

70 Paul Ilie, *The Age of Minerva, Volume 2: Cognitive Discontinuities in Eighteenth-Century Thought* (Philadelphia: University of Pennsylvania Press, 1995), 340. Ilie attempts, with varying degrees of success, to tease a counternarrative out of Enlightenment thought, where a valorization of touch and tactile knowledge haunts vision's claims to absolute knowledge.

71 Frances Herring, in outlining the reasons for touch's "neglect" in aesthetics, grouped Schopenhauer's position together with a body of arguments against touch designated as "psychological." As opposed to vision and hearing, with touch it is difficult (according to the psychological perspective) to suspend the biological imperative for self-preservation. Frances Herring, "Touch: The Neglected Sense," *Journal of Aesthetics and Art Criticism* 7, no. 3 (1949): 200.

72 The common sensibility can be seen as akin to the deodorizing project Alain Corbin describes in *The Foul and the Fragrant,* where disagreeable

odors came to be the target of a programmatic "olfactory vigilance" perpetuated by doctors and clinicians. As a consequence of this program, a new set of associations between positive and negative scents—a new ordering of odors—emerged, bound up with French social modernization more generally. Alain Corbin, *The Foul and the Fragrant: Odor and the French Social Imagination* (Cambridge, Mass.: Harvard University Press, 1986), 29.

73 For an index of industrial labor's brutal impact on workers' bodies, see Jamie Bronstein, *Caught in the Machine: Workplace Accidents and Injured Workers in Nineteenth Century Britain* (Stanford, Calif.: Stanford University Press, 2008).

74 Constance Classen details the transformations to touch as a result of urbanization and industrialization, suggesting that these developments, along with shifts in the material composition of domestic life, constituted a drastic reorganization of touch that expressed the values of mechanization more generally. See Constance Classen, *The Deepest Sense: A Cultural History of Touch* (Champaign: University of Illinois Press, 2012), 178–91.

75 Weber's experimental investigations into touch perception ignored sexual pleasure, and the sexual organs, a particularly curious exclusion in light of his extensive work throughout the rest of his career on both the male and female reproductive anatomy.

76 For one account, see Classen, *The Deepest Sense,* 60–70.

77 Later in the nineteenth century, the investigation of pain became a dedicated area of sensory research, complete with its own measurement instruments and techniques. The Austrian physiologist Max von Frey, who was connected to Helmholtz's research network, was responsible for the isolation of "pain spots" (*Schmerzpunkte*) in the skin. Using a mix of instruments, including electrical stimulators, von Frey participated in a project of mapping separate spots on the skin responsible for producing sensations of pain, touch, coolness, and warmth. These investigations, however, did little to settle the debate about pain's status as a distinct sensory system, to the extent that Norsell, Finger, and Lajonchere suggested in a 1999 review that pain, in spite of widespread agreement on the existence of specialized "nociceptors" in the skin, remains "the most puzzling of the skin senses" (464).

78 By the late 1800s, the "new psychology" was used almost interchangeably with "experimental psychology" and "sensory psychology." G. Stanley Hall's "The New Psychology" in *Harper's Monthly Magazine,* November 1901, 727–32, presents a valorization of the emergent field,

declaring its superiority over prior modes of investigating the human mind.

79   Edward Bradford Titchener, H. C. Warren, and James Baldwin, "Laboratory," in Baldwin, *Dictionary of Philosophy and Psychology, Vol. 1,* 611–13. This entry grouped lab equipment together based on the sense modality it is used to study; of the five senses, the haptical and visual modes had the most affiliated instruments.

80   Ibid., 613.

81   Merriley Borell discusses the rise of medical instrument companies that occurred toward the end of the nineteenth century, differentiating such companies from the "medical and surgical supply houses" that they came to complement. These apparatuses made use of both mechanical and electrical means to act on and quantify the senses. Merriley Borell, "Training the Senses, Training the Mind," in *Medicine and the Five Senses,* ed. William Bynum and Roy Porter (Cambridge: Cambridge University Press, 1993), 256.

82   Joseph Stevens and Barry Green, "History of Research on Touch," in *Pain and Touch,* ed. Lawrence Kruger (San Diego, Calif.: Academic Press, 1996), 2. As far back as Aristotle, Stevens and Green note, it was suspected that touch had several different attributes. But Weber succeeded in providing proof of their operation grounded in an appeal to structured, empirical verification.

83   Ibid.

84   Though Dessoir was the first to suggest this division, it did not see its fullest expression until David Katz's 1925 monograph *Der Aufbau der Tastwelt* (The World of Touch). Katz valorized and celebrated the role of the active, moving hand in forming tactile perceptions, while also arguing for the existence of vibration as a psychophysiologically distinct sensory system. David Katz, *The World of Touch,* trans. Lawrence Krueger (Hillsdale, N.J.: L. Erlbaum, 1989).

85   For a summary, see Edwin Boring, *Sensation and Perception in the History of Experimental Psychology* (New York: D. Appleton-Century, 1942), 463–522. Also see Norsell et al., "Cutaneous Sensory Spots."

86   The crafting of these categories did not occur without controversy. See E. G. Jones, "The Development of the 'Muscular Sense' Concept during the Nineteenth Century and the Work of H. Charlton Bastian," *Journal of the History of Medicine and Allied Sciences* 27, no. 3 (July 1972): 298–311.

87   Titchener, "A Psychological Laboratory," *Mind* 7 (1898): 315.

88 For an analysis of the psychological instruments used to isolate just-noticed difference thresholds, see Rand Evans, "The Just Noticeable Difference: Psychophysical Instrumentation and the Determination of Sensory Thresholds," in *Proceedings of the Eleventh International Scientific Instrument Symposium, Bologna, Italy, 9–14 September 1991,* 137–41. Stephen Ward uses Latour's concept of "nonhuman actants" to describe the psychological instruments as they migrated out of the experimental psychologist's lab into broader social and institutional contexts. Stephen Ward, *Modernizing the Mind: Psychological Knowledge and the Remaking of Society* (Westport, Conn.: Praeger, 2002), 111–25.

89 Michel Foucault, *The Order of Things: An Archaeology of the Human Sciences* (New York: Random House, 1970), 360.

90 E. B. Angell, "Hypesthesia and Hypalgesia and Their Significance in Functional and Nervous Disturbances," *Journal of Nervous & Mental Disease* 33, no. 5 (1906): 324.

91 Ben Singer, "Modernity, Hyperstimulus, and the Rise of Popular Sensationalism," in *Cinema and the Invention of Modern Life,* ed. Leo Charney and Vanessa Shwartz (Berkeley: University of California Press, 1995). Brigid Doherty, "See: 'We Are All Neurasthenics!' or, the Trauma of Dada Montage," *Critical Inquiry* 24, no. 1 (1997): 82–131.

92 See Angell, "Hypesthesia and Hypalgesia," 324.

93 Foucault refers to mathesis as "the science of calculable order." Foucault, *The Order of Things,* 73.

94 See Boring, *Sensation and Perception,* 480–81, and Rabinbach, *The Human Motor,* 149.

95 Guy Montrose Whipple, *Manual of Mental and Physical Tests* (New York: Arno Press, 1910), 207.

96 Charles Sherrington, cited in Borell, "Training the Senses," 248.

97 Maria Montessori, *The Montessori Method: Scientific Pedagogy as Applied to Child Education in "the Children's Houses"* (New York: Stokes, 1912), 4–5.

98 Ibid., 168.

99 Ibid., 185.

100 Montessori used this term to refer to the sense of weight and its associated discriminatory capacities. Ibid., 187.

101 Ibid., 188.

102 F. T. Marinetti, "Tactilism," in *Marinetti: Selected Writings,* ed.
     R. W. Flint (New York: Farrar, Strauss and Giroux, 1971), 111. The
     self-aggrandizing rhetoric in Marinetti's manifesto obscures the extent
     to which he borrowed the concept of tactile education from
     pedagogues like Montessori. The Hungarian artist Lazlo Moholy-
     Nagy, while teaching at the Bauhaus school, also stressed the
     importance of tactile education, designing materials intended to help
     his students cultivate their tactile sensibilities. See Leah Dickerman,
     "Bauhaus Fundaments," in *Bauhaus 1919–1933: Workshops for
     Modernity,* ed. Barry Bergdoll and Leah Dickerman (New York:
     Metropolitan Museum of Art, 2009), 32–33.

103 Coon, in "Standardizing the Subject," points to the pragmatic
     arguments advanced by Edward Scripture and Hall (both of whom
     investigated touch extensively) in suggesting that technoscientific
     ideals shaped the discipline in the United States. Scripture pressed the
     case that experimentation had been responsible for much of the
     technological progress in modern society, suggesting that its embrace
     by psychologists could yield similarly tangible benefits: "It is to the
     introduction of experiment that we owe our electric cars and lights,
     our bridges and tall buildings, our steam-power and factories, in fact,
     every particle of our modern civilization that depends on material
     goods. It is to the lack of experiment that we must attribute the
     medieval condition of the mental sciences." Edward Scripture,
     Thinking, Feeling, Doing (Meadville, Penn.: 1895), 24–25. For Coon,
     the standardization occurring in experimental psychology labs, both of
     subject and of instrument, mapped neatly onto the standardization of
     manufacture in the factory.

104 Dessoir, "Über den Hautsinn," 242.

105 Titchener, "Haptics," 441.

106 William Krohn, "An Experimental Study of Simultaneous
     Stimulations of the Sense of Touch," *Journal of Nervous and Mental
     Diseases* 18, no. 3 (1893): 172.

107 G. Stanley Hall, "The New Psychology," *Andover Review* 3 (1885): 126.

108 O. T. Mason, "Notes," *American Journal of Psychology* 1, no. 3 (1888): 552.

109 Frank Geldard, "Adventures in Tactile Literacy," *American Psychologist*
     12, no. 3 (1956): 115. Susan Lederman, one of the most prominent
     researchers on the psychology of touch, echoed Geldard's sentiments
     on touch's neglect (Verry, "Don't Take Touch for Granted," 64).

110 See Norsell et al., "Cutaneous Sensory Spots," for a summary of the nineteenth-century work on sensory spots. For a summary of work on the muscular senses, see E. G. Jones, "The Development of the 'Muscular Sense' Concept during the Nineteenth Century and the Work of H. Charlton Bastian," *Journal of the History of Medicine and Allied Sciences* 27, no. 3 (July 1972): 298–311.

111 I do not want to overstate the consensus that emerged around the science of touch, as a host of controversies and debates surrounded the publication of each new experimental finding. Rather, I want to highlight the myriad exertions psychophysicists, psychologists, and physiologists made in their attempts to build a science of touch during this period.

112 Mandayam Srinivasan and Cagatay Basdogan, "Haptics in Virtual Environments: Taxonomy, Research Status, and Challenges," *Computers & Graphics* 21, no. 4: 393.

113 Alfred Goldscheider, a German physician who carried out extensive research on tactile sensitivity, developed methods for meticulously plotting the skin's sensitivity to electrical currents, described in *Diagnostik der Krankheiten des Nervensystems* (Berlin, 1897). See Andreas Killen, *Berlin Electropolis: Shock, Nerves, and German Modernity* (Berkeley: University of California Press, 2006), 98–102.

114 Sander Gilman, *Inscribing the Other* (Lincoln: University of Nebraska Press, 1991), 32.

3. The Tongue of the Skin

1   Robert Gault, "Recent Developments in Vibro-Tactile Research," *Journal of the Franklin Institute* 221, no. 6 (June 1936): 705.

2   Ibid.

3   Ibid.

4   "New Instrument Enables the Deaf to Hear with Their Hands," *Popular Science Monthly,* March 1926, 60.

5   Robert Gault and George Crane, "Tactual Patterns from Certain Vowel Qualities Instrumentally Communicated from a Speaker to a Subject's Fingers," *Journal of General Psychology* 1, no. 2 (1928): 353. Gault repeated the phrase "grafting an ear upon the skin" frequently in his publications on the Teletactor, but this was the only passage where he distinguished the mechanical ear from its organic variant.

6   Though it informed and structured research on touch communication
     systems from the 1920s on, the formulation was made most explicitly
     in Carl Sherrick's "The Art of Tactile Communication," where he
     provided a detailed overview of the successes, failures, and lessons
     learned from the preceding decades of labor dedicated to engineering
     touch communication systems. Carl Sherrick, "The Art of Tactile
     Communication," *American Psychologist* 30, no. 3 (March 1975):
     353–60.

7   Gault understood the implications of his experiments on tactile
     sensitivity to be so radical that they were positioned at "the frontier of
     a pioneer state." Robert Gault, "'Hearing' through the Sense Organs
     of Touch and Vibration," *Journal of the Franklin Institute* 204, no. 3
     (September 1927): 357.

8   This is not to suggest that they were uninterested in the philosophical
     ramifications of their work, but instead to note experimental
     psychology's mechanistic orientation to touch. Many of the
     nineteenth- and early twentieth-century debates centered around
     which experimental protocols would best isolate touch's submodalities
     from one another rather than emphasizing the broader ramifications
     implied by the newly recut touch.

9   Frank Geldard, "Adventures in Tactile Literacy," *American Psychologist*
     12, no. 3 (1956): 115.

10   Sherrick, "The Art of Tactile Communication," 353.

11   Though many researchers contributed to this genealogy, Geldard
     articulated it most coherently across his many publications, particularly
     in his landmark essay "Tactile Communication," where he reviewed
     the pre-twentieth-century forerunners to electromechanical tactile
     communication systems. Frank Geldard, "Tactile Communication," in
     How Animals Communicate, ed. Thomas A. Sebeok (Bloomington:
     Indiana University Press, 1977).

12   Gault and Crane, "Tactual Patterns," 357. Louis Goodfellow, who
     attempted to further refine some of Gault's research into the tactual
     perception of vibrations, later defined the tactogram as "a record of a
     person's sensitivity to vibratory stimuli at various frequency levels,"
     and he directly established the analogy with the audiogram. Louis
     Goodfellow, "The Sensitivity of Various Areas of the Body to
     Vibratory Stimuli," *Journal of General Psychology* 11, no. 2 (1934): 436n4.

13   Frank Geldard, "The Perception of Mechanical Vibration," *Journal of
     General Psychology* 22, no. 2 (1940): 261.

14   In spite of leveling this critique, Katz himself employed experimental methods and instruments in search of ever more specific knowledge about the properties of the skin senses. For example, his theory that the vibration sense is distinct from the pressure sense relied on observations that grew out of repeated experiments applying psychophysical instruments to different parts of the body, each intended to isolate sensations of pressure from those of vibration.

15   It is worth recalling here Kittler's observation that the success of psychophysical machines hinged precisely on their capacity to produce sensations that had no referent outside of the lab, allowing "the subconscious mechanisms responsible for the construction of psychophysical reality" to be isolated from "the cultural—that is, language-dependent—functions responsible for concept formation." Friedrich Kittler, "Thinking Colours and/or Machines," *Theory, Culture & Society* 23, no. 7–8 (2006): 42. In the case of sensory substitution systems, the experimental disassembly of touch aimed precisely at a reassembly that would allow language-dependent concepts to flow through a tactility now understood as a pathway for the transmission of intelligence.

16   Geldard, "Tactile Communication," 222.

17   Frances Herring, "Touch: The Neglected Sense," *Journal of Aesthetics and Art Criticism* 7, no. 3 (1949): 200–201.

18   Geldard, "Tactile Communication," 215. Geldard used this phrase to designate the eighteenth-century pre-Braille systems, most of which involved some form of raised script.

19   Ibid., 89.

20   Ibid., 90.

21   Ibid.

22   Ibid.

23   Shortly after Diderot's "Letter," Jean-Jacques Rousseau's 1762 *Émile* advanced the possibility of learning to hear through the fingers. By using a highly cultivated sense of touch to interpret vibrations as sounds, Rousseau thought that "one could easily speak to the deaf by means of music." Arguing that "tone and measure are no less capable of regular combination than voice and articulation," he suggested that these criteria "might be used as elements of speech." Jean-Jacques Rousseau, *Émile* (1911; repr., Mineola, N.Y.: Dover, 2013), 122–23. Geldard cited this passage in his genealogy of sensory substitution systems.

24   For a deeper engagement with Diderot and Saunderson that positions
     them in a broader tradition of attempting to translate the active
     movement of the body into visual impressions, see Mark Paterson's
     Seeing with the Hands: Blindness, Vision, and Touch after Descartes
     (Edinburgh: Edinburgh University Press, 2016). Paterson's suggestion
     that Braille provides a language of touch, demonstrated through a
     richly detailed account of its emergence from earlier systems of
     finger-reading, adds needed layers of depth to the genealogy I present
     in this chapter.

25   Yvette Hartwel and Edouard Gentaz, "Early Psychological Studies on
     Touch in France," in *Human Haptic Perception—Basics and Applications,*
     ed. Martin Grunwald (Boston: Basel, 2008), 56–57.

26   Barbier's interest in night writing grew out of his experience as a
     captain in Napoleon's army, where he had seen a gun platoon
     bombarded after they gave away their position by lighting a lamp to
     read written orders. These military origins of night writing position it
     in a continuum with later investments by the U.S. Office of Naval
     Research into touch communication. Such efforts attempted to pass
     messages through the tactile channel as part of an overall combat
     strategy that tacitly recognized the importance of rapid message
     transmission to battlefield success.

27   As discussed in Interface 2, Weber determined the two-point threshold
     for the fingertip to be one Paris line, or 2.25 mm. Weber, *E. H. Weber
     on the Tactile Senses,* trans. Helen Ross and David Murray (Hove:
     Erlbaum [UK], 1996), 43.

28   Although Geldard would later credit Braille for its widespread
     adoption and utility, he ultimately found it to be inadequate,
     criticizing it both for its "arbitrary coding" and for "being developed
     on logical rather than psychological lines." Geldard, "Tactile
     Communication," 215.

29   Jan Eric Olsén, "Vicariates of the Eye: Blindness, Sense Substitution,
     and Writing Devices in the Nineteenth Century," *Mosaic: A Journal for
     the Interdisciplinary Study of Literature* 46, no. 3 (September 2013): 76.

30   Vanessa Warne presents an alternative argument, suggesting that the
     nineteenth-century reevaluation of the relationship between vision
     and sight taking place in Europe was prompted primarily by the wave
     of new finger-reading technologies rather than the nascent
     physiological studies on the dependency of the two sense modalities.
     Vanessa Warne, " 'So That the Sense of Touch May Supply the Want
     of Sight': Blind Reading and Nineteenth-Century Print Culture," in
     Media, Technology and Literature in the Nineteenth Century: Image, Sound,

Touch, ed. Collette Colligan and Margaret Linley (Burlington, Vt.: Ashgate Press, 2011), 46.

31 Olsén, "Vicariates of the Eye," 85.

32 For further detail, see Yvonne Eriksson, *Tactile Pictures: Pictorial Representations for the Blind, 1784–1940* (Göteborg, Sweden: Acta Universitatis Gothoburgensis, 1998). Eriksson suggests that, in spite of enjoying some successes, these models were often unhelpful, precisely because they relied on "sighted people's frames of reference in the choice of content and design" (ibid., 244). In this argument, the translation of image to touch accomplished by tactile pictures failed to escape ocularcentric modeling, as tactile media unconsciously imported the structural logic of vision.

33 John Fahie, *A History of Electric Telegraphy, to the Year 1837* (New York, 1884), 5, 19–20.

34 Eric Kluitenberg, "On the Archaeology of Imaginary Media," in *Media Archaeology: Approaches, Applications, and Implication,* ed. Erkki Huhtamo and Jussi Parikka (Berkeley: University of California Press, 2011), 48.

35 Ibid., 19.

36 Ibid., 20. Fahie's descriptions are aggregated from a variety of sources. This particular description of the sympathetic flesh telegraph likely originated in Joseph Glanvill's 1661 work *Scepsis Scientifica, or the Vanity of Dogmatizing* (London, 1885). In Glanvill's account, this system is referred to as "sympathetic hands." Glanvill, *Scepsis Scientifica,* 177.

37 Many of these sympathetic needle telegraphs were well known at the time to be stage tricks for popular amusement. Simon During, *Modern Enchantments: The Cultural Practice of Secular Magic* (Cambridge, Mass.: Harvard University Press, 2002), 91.

38 This system of skin writing recalls the one imagined in Franz Kafka's "In the Penal Colony." Kakfa's electromechanical apparatus writes the sentencing of the condemned on the body, not just for purposes of spectacular display but also to inform the condemned of the violation for which they are being executed. Through a complex series of motions by carefully controlled vibrating needles, the sentence is conveyed to the condemned in a grotesque act of skin writing. Communication through the skin here acts as a replacement for oral communication. As an officer in the tale explains to the traveler: "There would be no point in telling him. He'll learn it on his body." Franz Kafka, "In the Penal Colony," in *The Complete Stories* (New York: Schocken Books, 1971), 170. Unlike the skin graft telegraph, Kafka's machine wrote in script, rather

than through point-based signaling. But because this script had been precisely calibrated to the particular resistances of the human body, it was "difficult to decipher the script with the eyes" (ibid., 175). The script's structure embodied a technique for engraving the sentence over a protracted period of 12 hours, during which it would penetrate progressively deeper into the skin in successive drafts. In the story, "enlightenment"—that moment when the repeated drafts become intelligible not as the noise of pain but as the signal of language—"comes to the most dull-witted" around the sixth hour of inscription. Though the eyes fail at reading its message, the condemned "deciphers it with his wounds" (ibid.). During their extended time in the disciplinary inscription machine, the machine's victims are gradually able to bring their own perceptual apparatus into harmony with sensations the machine impresses on it, such that reading through touch becomes possible through a compulsory sensorial training.

39   The "talking gloves" proposed by William Terry in the early twentieth century, where letters of the alphabet were written on the gloves and then pressed as if they were typewriter keys, are also a close analog to the signaling mechanism employed by the sympathetic flesh telegraph. For a description, see Harold Clark, *Talking Gloves for the Deaf and Blind: Their Value to Men Injured in the Present War* (Cleveland, Ohio: Harold Clark, 1917). Mara Mills provides a historical contextualization that locates talking gloves as part of a broader effort to route speech through the skin. Mara Mills, "On Disability and Cybernetics," *Differences* 22, no. 2–3 (2011): 74–111.

40   Edward Knight, *Knight's American Mechanical Dictionary, Volume 3* (New York: J. B. Ford and Co., 1884), 2504. In his typology, Knight grouped telegraphs into three categories (optical, audible, and tangible) based on the sense modality used to receive their transmissions. Knight's division between mechanical and electrical signal reception practices anticipated Geldard's later division of skin-based communication systems into the categories of electrical, mechanical, and electromechanical, depending on the mechanism used to stimulate the skin.

41   Jonathan Sterne, *The Audible Past: The Cultural Origins of Sound Reproduction* (Durham, N.C.: Duke University Press), 141.

42   William Plum, *The Military Telegraph during the Civil War in the United States, Volume 1* (Chicago: Jansen, McClurg & Co., 1882), 190–91.

43   Sterne, *The Audible Past,* 141. Italics original. Sterne borrows this phrase from Rick Altman.

44 While my primary focus here concerns touch's use as a message
reception sense, touch was also recognized as crucial for sending
messages; building on Kittler's claims about the typewriter, Ivan
Raykoff describes felt similarities in the tactile materiality of piano,
typewriter, and telegraph keys. Ivan Raykoff, "Piano, Telegraph,
Typewriter: Listening to the Language of Touch," in *Media, Technology,
and Literature in the Nineteenth Century*, 180–81.

45 Building on methodological points made by Marvin, Gitelman, and
Sterne concerning the concretization of previously unstable and
historically contingent media forms, Colligan and Linley suggest that
the retroactive construction of rigid categories to account for the
chaotic mix of technologies that developed during the nineteenth
century obscures the complexity of the transformation these categories
seek to explain. Collette Colligan and Margaret Linley, "Introduction:
The Nineteenth-Century Invention of Media," in *Media, Technology,
and Literature in the Nineteenth Century*, 1–4.

46 Kluitenberg, "Imaginary Media," 49.

47 In this version, sound captured by a pair of microphones panned from
one hand to the other, permitting the spatial localization of sound
sources. Gault, "Recent Developments in Vibro-Tactile Research,"
709–11.

48 In her fascinating exploration of the "hearing glove" project, Mills
takes the MIT research, associated mostly notably with Weiner, as her
starting point and works backward to position the MIT project in a
longer tradition of hearing gloves. Focusing on the intersection
between hearing through the skin and cybernetics-era concerns about
information processing, Mills details the project's life after it migrated
from Gault's lab to MIT, devoting close attention to the role Helen
Keller played in driving forward the glove's development. However,
because I am interested in isolating a conversation that developed
around touch's capacity as a language reception sense, I prioritize
Gault's protocybernetic articulation of touch as the container sense for
hearing, which set the stage for Geldard's subsequent investigations
into the possibilities of coding language for tactile transmission. See
Mills, "On Disability and Cybernetics," 96.

49 For a recent example, see Charles Spence, "The Skin as a Medium
for Sensory Substitution," *Multisensory Research* 27, no. 5–6
(2014): 298.

50 Gault, "'Hearing' through the Sense Organs of Touch and Vibration,"
330.

51 Warren Jones patented the multiunit version of the device in 1928, and immediately transferred ownership of the patent to Bell Labs. See Warren Jones, "US Patent #US 1,733,605—Tactual Interpretation of Vibrations," October 29, 1929.

52 Sherrick specifically identified the instruments used in Gault and Knudsen's research as part of this parasitic tradition. Sherrick's own invention, a compact, Velcro-mounted motor dubbed "the Sherrick vibrator," presents a notable exception to the overall trend he identified. Intended specifically as a means of facilitating tactile communication, the vibrator had the advantage of not simply being a clumsy modification of a device (typically a speaker) originally intended for the transmission of sounds. Sherrick, "The Art of Tactile Communication," 353.

53 Gault's phrasing here anticipates McLuhan's famous framing of media as extensions of man. Gault, "'Hearing' through the Sense Organs of Touch and Vibration," 329.

54 By 1935, the Teletactor had been adopted at the Illinois School for the Deaf, where large groups of students gathered together in the classroom received speech from the teacher through networked Teletactors. Individual students were also asked to take turns speaking into the Teletactor microphone, "so that the entire class will feel his message." See Robert Gault, "New Instrument Brings Back Girl's Lost Senses," *Spartanburg Herald,* April 12, 1935.

55 Gault, "New Instrument Brings Back Girl's Lost Senses."

56 The formulation of touch as the base on which the other senses are constructed is not entirely unique to Weber and Darwin, but in the discursive context Gault worked in, these were the two primary figures associated with the claim. See in particular David Katz, *The World of Touch,* trans. Lawrence Krueger (Hillsdale, N.J.: L. Erlbaum, 1989), 196–97.

57 This link between hearing and touching is embedded in the design of contemporary cinematic sound reproduction systems: what are known as "low-frequency effects" cross over into a subaudible frequency range (also known as sub-bass), intended to act both on the sense of touch and of hearing.

58 Gault, "'Hearing' through the Sense Organs of Touch and Vibration," 330–31.

59 According to Rand Evans, just in the 10-year period from 1924 to 1934, Gault published 27 articles on routing sound through the fingers.

Rand Evans, "Haptics in the United States before 1940," in *Human Haptic Perception,* 80.

60   Katz, who influenced both the philosophical and practical dimensions of Gault's research, devoted a great deal of energy to rejecting the previous analogy between touch and vision. Gault embraced Katz's argument without specifically engaging it. See Katz, *The World of Touch,* 39–44.

61   Gault, "'Hearing' through the Sense Organs of Touch and Vibration," 358.

62   Punctuation worked strategically throughout Gault's many publications on the subject, as the quotation marks frequently placed around "hearing" reflected its repositioning as a subset of the tactile senses.

63   As is the case throughout *Archaeologies of Touch,* my use of the term here is filtered through Agamben's reading of Foucault; see Giorgio Agamben, *"What Is an Apparatus?" and Other Essays* (Stanford, Calif.: Stanford University Press, 2009), 24.

64   Vern Knudsen, "'Hearing' with the Sense of Touch," *Journal of General Psychology* 1, no. 2 (January 1928): 322–23.

65   Gault, "'Hearing' through the Sense Organs of Touch and Vibration," 331.

66   Ibid., 330.

67   Ibid., 358.

68   Ibid., 341.

69   Ibid., 345.

70   Ibid., 354.

71   Geldard, "Adventures in Tactile Literacy," 118.

72   Ibid., 117.

73   Geldard, "Tactile Communication," 217.

74   Geldard, "Adventures in Tactile Literacy," 115.

75   Ibid. The eyes and ears dominated "in business and industry, on the highway, in sports and recreational activities, in the scientific laboratory, in the operation of military equipment, indeed, in practically all our comings and goings."

76 Ibid.

77 Ibid.

78 Ibid.

79 Ibid., 116.

80 The research that informed the design of Vibratese characters had been conducted at Geldard's University of Virginia laboratory with the express intent of incorporating this knowledge about touch's psychophysical parameters into a communication system. The location of the vibrators, the intensity of the vibrations they produced, and the division of vibrations into three stages of duration were each arrived at through independent experiments, conducted by different investigators, all with the same overarching goal.

81 Ibid., 122. Though he failed to mention it in the 1956 address, this figure was a theoretical maximum; as he later noted, actual speeds recorded during the trials only reached 38 words per minute. Geldard found the discrepancy between the theoretical and the actual relatively inconsequential, given the advantages Vibratese enjoyed over International Morse: "closing the gap between 38 and 67 words per minute we leave to those interested in establishing world's records." Geldard, "Some Neglected Possibilities of Communication," *Science* 131, no. 3413 (May 1960): 1568.

82 During World War II, Geldard served for four years in the Army Air Corps, with much of his work involving the training of personnel for air crews. Throughout and beyond his military career, Geldard maintained an interest in showcasing the utility of psychological research. As a major, he was chief of the Psychology Section of the Flying Training Command, and after the war, he traveled to the Philippines to help the nation launch a psychology unit for the nascent Philippine Air Force. His investigations into cutaneous communication overlapped with his chairmanship of the NATO Advisory Group on Human Factors, which he held from 1959 until 1965. For further biographical detail, see Frank Finger and Carl Sherrick, "Frank Geldard: 1904–1984," *American Journal of Psychology* 99, no. 1 (Spring 1986): 135–42.

83 As a result of the Great Depression, funding for psychology research had decreased precipitously. With the discipline in a state of existential crisis, Geldard argued successfully for the immediate and practical utility of psychologists' research. Frank Geldard, "Military Psychology: Science or Technology?" *American Journal of Psychology* 66, no. 3 (July 1953): 341.

84   Geldard, "Adventures in Tactile Literacy," 122.

85   Ibid.

86   N. Katherine Hayles, *How We Think: Digital Media and Contemporary Technogenesis* (Chicago: University of Chicago Press, 2012), 130.

87   Geldard, "Some Neglected Possibilities," 1588.

88   Paul Virilio, *War and Cinema: The Logistics of Perception* (New York: Verso, 1989), 10.

89   The James quote, according to Geldard, is: "Common-sense says, we lose our fortune, are sorry and weep; we meet a bear, are frightened and run . . . the more rational statement is that we feel sorry because we cry, angry because we strike . . . etc." ("Adventures in Tactile Literacy," 123). For the full passage, see William James, "What Is an Emotion?" *Mind* 9 (1885): 188–205.

90   Geldard, "Some Neglected Possibilities," 1585.

91   William Howell, whose master's research Geldard supervised at the University of Virginia, mapped out five different possibilities for a Vibratese language, including a pair of systems that employed six vibrators rather than the five used in the demonstration. William Howell, "Training on a Vibratory Communication System" (master's thesis, University of Virginia, 1956), 48–51.

92   Geldard, "Some Neglected Possibilities," 1587. I take this discussion of electrocutaneous language transmission systems up in greater detail later on in this section.

93   Geldard used this military metaphor frequently, with a curious double meaning: on one hand, "attack" referred to a strategy for approaching the problem and on the other, "attack" designated the resultant experimental apparatus that would attempt to "force a receptor system to perform" according to the parameters of the experiment. Geldard, "Some Neglected Possibilities," 1584.

94   The many projects pursued at the lab, along with their associated researchers and funding streams, are cataloged and archived at the organization's Web site. See "The Cutaneous Communication Lab at Princeton: Publications of Research," http://www.tactileresearch.org/pucclabs/pagePublicn.html. Roger Cholewiak, who assumed the directorship of the lab in 1991, also headed the Tactile Research Laboratory at the Naval Aerospace Medical Research Laboratory from 1995 to 2005.

95   Glenn Hawkes, ed., *Symposium on Cutaneous Sensitivity* (Fort Knox, Ky.: U.S. Army Medical Research and Development Command, 1960). For full proceedings, see http://www.dtic.mil/dtic/tr/fulltext/u2/249541.pdf.

96   Carl Sherrick, "Current Prospects for Cutaneous Communication," in *Conference on Cutaneous Systems and Devices* (Austin, Tex.: Psychonomic Society, 1974), 106.

97   Ibid.

98   Geldard, "Some Neglected Possibilities," 1585.

99   Hall himself had a keen interest in studies of touch, having been involved, while at Harvard, with H. H. Donaldson's research into the perception of space and temperature in the skin. Hall and Donaldson each designed apparatuses to investigate the skin; Hall's Kinesimeter and Donaldson's Temperature Stimulator helped provide the map that Geldard would later project Vibratese signals onto. G. Stanley Hall and H. H. Donaldson, "Motor Sensations in the Skin," *Mind* 10 (1885): 557–72; H. H. Donaldson, "On the Temperature Sense," *Mind* 10 (1885): 399–416. For a summary, see Rand Evans, "Haptics in the United States before 1940," in *Human Haptic Perception,* 68–71.

100 Geldard, "Some Neglected Possibilities," 1584.

101 Beverly von Haller Gilmer, "Possibilities of Cutaneous Electro-Pulse Communication," *Symposium on Cutaneous Sensitivity* (Fort Knox, Ky.: U.S. Army Medical Research and Development Command, 1960), 79–80.

102 Glenn Hawkes, *Tactile Communication* (Oklahoma City, Okla.: Civil Aeromedical Research Institute, 1962), 6.

103 Von Haller Gilmer, "Possibilities of Cutaneous Electro-Pulse Communication," 77.

104 Von Haller Gilmer's full list of fourteen potential applications for cutaneous communication anticipates some of the present-day uses of vibration as a signaling system in smartphones and wearables. Notably, he suggested that vibration could be used "as an aid to spatial orientation," "to quickly alert or warn," and "where environmental conditions handicap both visual and auditory presentation." Ibid., 78.

105 Again, the present state of vibratory alerts from smartphones offers an illustration of this problem: though these vibrations are intended to be read by touch, the vibrations frequently produce sound as a by-product.

106 John Hennessy, "Cutaneous Sensitivity Communications," *Human Factors* 8, no. 5 (October 1966): 466.

107 These researchers acknowledged the immense research required to raise tactile communication up to a point where it could leave the lab. Noting the dearth of dedicated knowledge on tactile sensitivity, von Haller Gilmer explained that, before installing a "practical communication system . . . in a military, or other situation, we will have to have measures of 'skin deafness' as well as knowledge about channel loads, the effects of distraction and error ranges." Von Haller Gilmer, "Possibilities of Cutaneous Electro-Pulse Communication," 79.

108 Gibson's apparatus for mapping the skin's capacity to distinguish between electrical stimuli was capable of providing graded steps of current to 48 spots on the body simultaneously. Like Geldard's Vibratese machine, Gibson's system both recalls the Apparatus for Simultaneous Touches and pushes beyond its limits: the precise control afforded by the computer helped the experimenter specify touch and divide it from pain with a degree of accuracy impossible with prior touch machines. Robert Gibson, "Electrical Stimulation of Pain and Touch," in *The Skin Senses,* ed. Dan Kenshalo (Springfield, Ill.: Thomas, 1968), 232–35.

109 Discussions of gender difference were virtually absent from these conversations and experiments. One paper presented at the 1966 International Symposium on the Skin Senses attempted to update Weber's research on two-point threshold discriminations to account for possible differences between the sexes in tactual sensitivity, but especially in military research, both experimental subjects and the intended users of the devices they yielded were assumed to be male. See Sidney Weinstein, "Intensive and Extensive Differences in Tactile Sensitivity as a Function of Body Part, Sex, and Laterality," in *The Skin Senses,* 195–222.

110 Gibson, "Electrical Stimulation of Pain and Touch," 229.

111 Von Haller Gilmer, "Possibilities of Electro-Pulse Communication," 82.

112 In their 1970 summary of their findings, Bach-y-Rita et al. identified Geldard's address as the inspiration for their investigations. Paul Bach-y-Rita, Benjamin White, Frank Saunders, Lawrence Scadden, and Carter Collins, "Seeing with the Skin," *Perception & Psychophysics* 7, no. 1 (1970): 23.

113 This phrase was used frequently in the literature on TVSSs.

114 Paul Bach-y-Rita, *Brain Mechanisms in Sensory Substitution* (New York: Academic Press, 1972), 1.

115 Geldard, "Tactile Communication," 221.

116 The OPTACON itself has a curious lineage, deriving its name from the Optophone, a device built in the early 1910s to translate light into tones that could be assigned to individual letters. The OPTACON, along with Geldard's Optohapt, drew conceptual inspiration from the device, but both substituted the Optophone's acoustic coding of light with a tactual one, allowing for the transmission of words through dynamic, refreshable vibrations targeted at the fingertips. For a first-hand account of the Optophone's early development and efficacy as a reading device, see Mary Jameson, "The Optophone: Its Beginning and Development," *Bulletin of Prosthetics Research* (Spring 1966): 25–28. Mara Mills has productively established the links between the Optophone and tactile reading systems. See Mara Mills, "Optophones and Musical Print," *Sounding Out,* January 1, 2015, http://soundstudiesblog.com/2015/01/05/ optophones-and-musical-print/.

117 Bach-y-Rita, *Brain Mechanisms in Sensory Substitution,* ix.

118 Bach-y-Rita et al., "Seeing with the Skin," 26.

119 Bach-y-Rita, "Sensory Plasticity: Applications to a Vision Substitution System," *Acta Neurology Scandinavia* 43 (1967): 417.

120 G. A. Miller, "The Magical Number Seven, Plus or Minus Two: Some Limits on Our Capacity for Processing Information," *Psychological Review* 63 (1956): 81–97.

121 These findings are from Gerhard Werner and Vern Mountcastle, "Quantitative Relations between Mechanical Stimuli to the Skin and Neural Responses Evoked by Them," in *The Skin Senses,* 112–37.

122 For a comprehensive summary, including a review of the various methods of investigating tactual perception in the wake of Information Theory, see David Sinclair, *Cutaneous Sensations* (New York: Oxford University Press, 1967), 19–32.

123 Bach-y-Rita's pilot study used only a three-by-three array to map the televisual image onto the skin. Paul Bach-y-Rita, *Seeing with the Skin: Development of a Tactile Television System* (San Francisco: Smith-Kettlewell Institute of Visual Sciences, 1970), 3.

124 Bach-y-Rita, *Seeing with the Skin,* vi.

125 Bach-y-Rita acknowledged that Tactile Television would only receive its true baptism of fire once it left the lab. For example, a blind person walking down a busy street would be confronted by changing light conditions, rapidly moving objects, and the inconsistent pressure of the vibrators against their own moving skin. Bach-y-Rita, *Brain Mechanisms in Sensory Substitutions,* 152.

126 Ibid.

127 Bach-y-Rita, *Seeing with the Skin,* vi.

128 Bach-y-Rita, *Brain Mechanisms in Sensory Substitutions,* 153.

129 The story of Tactile Television remains unfinished. Subsequent iterations of the apparatus increased its resolution, miniaturized it, substituted vibration with electricity, and moved the stimulator array to different points on the skin. Still, it was slow to migrate from the lab. In a 1995 essay published in an anthology on human–computer interfaces for virtual environments, Bach-y-Rita lamented the turn away from TVSS research in favor of work on force feedback computing displays, suggesting that the new machines had diverted intellectual resources away from efforts at passing images through the skin. In spite of the comparative lack of attention TVSSs received, he continued work on the project undeterred. By the early 2000s, owing in part to the miniaturization of computers, Bach-y-Rita had developed a more compact version of Tactile Television, dubbed the BrainPort, that employed a small head-mounted camera that could relay tactual images not to the back or chest, but instead to the far more sensitive space on the tongue. Weber's experiments had demonstrated that the tip of the tongue was generally accepted as the bodily location with the greatest capacity for localizing stimuli. Projecting images onto it was intended to facilitate the perception of a higher resolution image while also making it less cumbersome for the user. The improved system met with some well-publicized success. The story of Eric Weihenmayer, who used the BrainPort's 611 electrode display to guide him in mountain-climbing expeditions, spread quickly in the popular scientific press. In June 2015, after years of clinical trials, the U.S. Food and Drug Administration approved an updated version of the BrainPort for sale in the United States. Some side effects of using the BrainPort v100 identified during these trials include "burning, stinging or metallic taste associated with the intra-oral device." See Kurt Kaczmarek and Paul Bach-y-Rita, "Tactile Displays," in *Virtual Environments and Advanced Interface Design* (Oxford: Oxford University Press, 1995), 350. For popular press coverage of Weihenmayer's adventures, see Buddy Levy, "The Blind Climber Who 'Sees' with His

Tongue," *Discover,* June 23, 2008, http://discovermagazine.com/2008
/jul/23-the-blind-climber-who-sees-through-his-tongue. For the Food
and Drug Administration's press release on the BrainPort v100, see "FDA
Allows Marketing of New Device to Help the Blind Process Visual
Signals via Their Tongues," http://www.fda.gov/NewsEvents
/Newsroom/PressAnnouncements/ucm451779.htm.

130 Marshall McLuhan, *Understanding Media: The Extensions of Man*
(Cambridge, Mass.: MIT Press, 1994), 313. There is also a kinship in
the language of "sensory extension" deployed by both authors.

131 Following Riegl's *Late Roman Art Industry,* the notion of tactile vision
that McLuhan embraced had become fashionable with Bauhaus artists,
such as Paul Klee and László Moholy-Nagy. Alois Riegl, *Late Roman
Art Industry,* trans. Rolf Winkes (Rome: Giorgio Bretschneider, 1985).
For an examination of the link between Riegl, Moholy-Nagy, and
McLuhan, see Klemens Gruber, "Medien des Taktilen: Theorien aus
der Vorgeschichte," *Maske und Kothurn* 62, no. 2–3 (2016): 207–34.

132 McLuhan, *Understanding Media,* 332.

133 Ibid., 334.

134 Sherrick, "The Art of Tactile Communication," 353.

135 This argument, which weaves together strands of biological and
technological determinism, has been advanced in some form or
another throughout many of the discussions on touch communication
systems. Specific to sensory substitution systems, Spence provides a
detailed and clear case for the limits of the tactile channel as an
alternative pathway for aural and visual sensations. Spence, "The Skin
as a Medium for Sensory Substitution."

136 Sensory prostheses like the Teletactor and Tactile Television can be
understood biopolitically, as attempts to help bodies labeled as deficient
"approximate the historically specific expectations of normalcy." See
David Mitchell and Sharon Snyder, *Biopolitics of Disability: Neoliberalism,
Ablenationalism, and Peripheral Embodiment* (Ann Arbor: University of
Michigan Press, 2015), 2.

137 Geldard, "Tactile Communication," 212.

138 Diderot, "Letter on the Blind," 89.

139 Ibid.

140 See Lux Research, *Getting Back in Touch with Electronics: Finding
Opportunity in Emerging Haptics* (New York: Lux Research, 2013). The
report predicts that the market for haptics technologies, in particular

the electrical components used to produce tactile feedback in smartphones, will grow from its 2012 value of $842 million to $13.8 billion by 2025.

141 Sherrick, "The Art of Tactile Communication," 359.

142 Marvin Minsky, "Telepresence," *OMNI Magazine,* June 1980, 52. Emphasis original. The device that can "translate print into feel" was a direct reference to J. C. Bliss and J. G. Linville's OPTACON.

4. Human–Machine Tactile Communication

1 Ivan Sutherland, "The Ultimate Display," *Proceedings of the IFIP Congress* 2 (1965): 506–8. Haptics engineers hoping to realize Sutherland's vision cited the essay frequently, but notably, some demilitarized his imaginary machine by excluding Sutherland's wish that the display have the capacity to kill its user. See, for example, Margaret Minsky, "Computational Haptics: The Sandpaper System for Synthesizing Texture for a Force-Feedback Display" (PhD dissertation, MIT, 1995), 22.

2 Sutherland, "The Ultimate Display," 506.

3 Ibid., 508.

4 Ibid.

5 Ibid.

6 In "The Ultimate Display," Sutherland made explicit reference to GE's well-publicized Handyman remote manipulation system, discussed in the second section of this chapter.

7 Batter and Brooks coined this phrase to describe their GROPE-1 interface. J. J. Batter and Frederick Brooks, Jr., "GROPE-1: A Computer Display to the Sense of Feel," *Proceedings of the IFIP Congress* 1 (1971): 759–63. In a recent interview, occasioned by the renewed interest in virtual reality, Brooks noted that he had been inspired by Sutherland's 1965 address to pursue the project of making virtual worlds move, feel, and interact in a realistic way. See Kent Bye, "Fred Brooks on Ivan Sutherland's 1965 'Ultimate Display' Speech," *Road to VR* (May 10, 2016), https://www.roadtovr.com/fred-brooks-ivan -sutherlands-1965-ultimate-display-speech/.

8 The uppercase titling of the device throughout the publications devoted to it suggests that "GROPE" is an acronym, but curiously,

none of the many publications on the various GROPE systems define
the acronym.

9   Brooks and Batter borrowed the GROPE-1 from MIT professor
     Thomas Sheridan. Ibid., 763.

10   William Aktinson and Guy Tribble were later part of the development
     team that produced the Apple Macintosh computer. Kent Wilson, a
     chemistry professor at University of California–San Diego, headed a
     research group he referred to as the "Senses Bureau," which was
     devoted to developing innovative techniques for the multisensory
     representation of scientific data. "Touchy Feely" and "Touchy Twisty"
     can be understood as part of a larger, lifelong project for Wilson that
     involved searching for creative ways to engage with and communicate
     scientific knowledge. For a summary, see http://www.cireeh.org
     /pmwiki.php/Main/SensesBureau.

11   William Atkinson, Karen Bond, Guy Tribble III, and Kent Wilson,
     "Computing with Feeling," *Computers & Graphics* 2, no. 2 (1977): 97–103.

12   Noll's use of the gendered phrase "man–machine tactile
     communication" throughout his work expresses dominant assumptions
     about the gender of computer users. In appropriating Noll's phrase
     for the title of this chapter, I have altered it to the gender-neutral
     "human–machine tactile communication." A similar gendering of the
     user appears in the engineering literature on remote manipulators
     discussed in the next section of this chapter, and I use the original
     phraseology when dealing with historical sources there as well.
     However, the use of such gendered language in the fields of human–
     machine and human–computer interaction is not an artifact of a
     historical period when scientists and engineers could claim to be
     blissfully unaware of language's marginalizing potential, as the use
     of gendered terminology persists in technical publications and
     patent documents. A. Michael Noll, "Man–Machine Tactile
     Communication," *Journal for the Society of Information Display* 1, no. 2
     (1972).

13   Mandayam Srinivasan and Cagatay Basdogan, "Haptics in Virtual
     Environments: Taxonomy, Research Status, and Challenges,"
     Computers & Graphics 21, no. 4 (1997): 393.

14   Hiroo Iwata, "History of Haptic Interface," in *Human Haptic Perception:
     Basics and Applications,* ed. Martin Grunwald (Boston: Birkhäuser,
     2008), 356.

15   Lev Manovich, *Software Takes Command* (New York: Bloomsbury,
     2013), 5.

16   The tactile pixel/taxel has a long history worthy of further
      exploration. Like haptic interfaces more generally, the taxel originated
      in robotics, used in that context to describe a mechanism for giving
      robots the sense of touch. Akin to the mechanoreceptors found in the
      human skin, an array of taxels could provide an image of the material
      properties of an object, the detail of which corresponded to the density
      of taxels in the array. For an early summary of robotics research and
      applications, see Joseph Engelberer, *Robots in Service* (Cambridge,
      Mass.: MIT Press, 1989), 75–81.

17   Katherine Kuchenbecker, "Haptography: Capturing the Feel of Real
      Objects to Enable Authentic Haptic Rendering," in Proceedings of
      Haptic in Ambient Systems (HAS) Workshop, in conjunction with
      the First International Conference on Ambient Media and Systems,
      February 2008, http://www.seas.upenn.edu/~kuchenbe/pub/pdf
      /Kuchenbecker08-HAS-Haptography.pdf.

18   Though many designers disagree on what constitutes this Holy Grail,
      they each describe questing after it. In a recent article about ultrasound
      haptics, which allows the generation of haptic sensations without
      physically touching an interface device, one designer positioned this
      touchless haptics as "probably the Holy Grail" of the research
      paradigm. Geoff Merrett, quoted in Stephen Harris, "Researchers
      Create Screen You Can Feel without Touching It," *The Engineer,*
      October 8, 2013, http://www.theengineer.co.uk/electronics/news/
      researchers-create-screen-you-can-feel-without-touching-it/1017259
      .article.

19   Bernard Stiegler, *Technics and Time, 1: The Fault of Epimetheus*
      (Stanford, Calif.: Stanford University Press, 1998), 35.

20   Blake Hannaford, "Feeling Is Believing: History of Telerobotics
      Technology," in *The Robot in the Garden: Telerobotics and Telepistemology
      in the Age of the Internet,* ed. Ken Goldberg (Cambridge, Mass.: MIT
      Press, 2001), 246–75.

21   Ibid., 254.

22   Raymond Goertz and W. M. Thompson, "Electronically Controlled
      Manipulator," *Nucleonics* 12, no. 11 (1954): 46–47.

23   The device was also referred to informally in the lab as the "Waldo," a
      name derived from a similar mechanism described by Robert Heinlein
      in his 1920 novel *Waldo.* In a 1967 lecture, Heinlein reflected on how
      he came to the idea after he learned of their "invention" in Goertz's
      lab. He drew inspiration from a *Popular Mechanics* article about a "poor
      fellow afflicted with myasthenia gravis." The disease left the man's

muscular control system intact but rendered his muscles so weak that basic tasks were impossible. In response to his condition, he "devised complicated lever arrangements to enable him to use what little strength he had and he became an inventor and industrial engineer, specializing in how to get maximum result for least effort. He turned his affliction into an asset." Thus dexterous remote manipulation systems originated from an intermingling of physiology, science fiction, and engineering. See Heinlein, "Science Fiction: Its Nature, Faults and Virtues," in *The Science Fiction Novel* (Chicago: Advent Publishers, 1959), http://sciencefiction.loa.org/biographies/heinlein _science.php.

24   Ralph Mosher, "From Handyman to Hardiman," in *Automotive Engineering Congress 1967* (New York: Society of Automotive Engineers, 1967).

25   General Electric Specialty Materials Handling Products Operation, "Cybernetic Anthropomorphous Machine Systems" (Schenectady, N.Y.: General Electric, 1968), 7.

26   Ibid., 2, 6.

27   Ibid., 2–3.

28   Ibid., 7. This framing of machines as extensions, later echoed by McLuhan in his definition of media as "extensions of man," had also been suggested by J. D. North in an essay that proved influential for the design of early computer interfaces. See Thierry Bardini, *Bootstrapping: Douglas Engelbart, Coevolution, and the Origins of Personal Computing* (Stanford, Calif.: Stanford University Press, 2000), 20.

29   Ibid.

30   The company was fond of using its iconic light bulbs in demonstrations of its CAMs—in a promotional video for its "quadruped" walking truck, GE showed the one of the machine's legs gingerly pressing a bulb into a pillow. "GE's Walking Truck—ca. 1965," YouTube video, posted by Schenectady InventTech, February 21, 2012, https://www .youtube.com/watch?v=coNO9FpDb6E.

31   General Electric Specialty Materials Handling Products Operation, "Cybernetic Anthropomorphous Machine Systems," 7.

32   In a 1969 *New York Times* article on the new wave of "cyborgs," the author noted that GE's Handyman possessed "only 10 degrees of freedom," compared to the 22 in a normal human hand. Walter Troy Spencer, "They're Not Robots, They're Cyborgs," *New York Times,* December 14, 1969.

33 Billy Crawford, "Measures of Remote Manipulator Feedback: Differential Sensitivity for Weight," WADD Technical Report 60–591 (I), 1961, http://contrails.iit.edu/DigitalCollection/1960/WADDTR60 –591volume02.pdf.

34 Thomas Sheridan and William Verplank, "Human and Computer Control of Undersea Teleoperators," Technical Report to the Office of Naval Research, July 14 (Arlington, Va.: Office of Naval Research, 1978), 7–2.

35 Capowski's research is summarized in Paul Jerome Kilpatrick, "The Use of Kinesthetic Supplement in an Interactive System" (PhD dissertation, University of North Carolina at Chapel Hill, 1976), 9–12.

36 Ming Ouh-Young, "Force Display in Molecular Docking" (PhD dissertation, University of North Carolina at Chapel Hill, 1990), 14.

37 As Thierry Bardini explains, concerns over tactile feedback were not entirely absent in the design of the mouse/keyboard interface. Among Douglas Engelbart's early collaborators when he worked on the Chord Keyset was James Bliss (referenced in Interface 3), who had developed one of the first systems for dynamically rendering images as tactile sensations. Engelbart suggested using electric impulse stimuli to feed information back from machine to user. See Bardini, *Bootstrapping*, 60–62.

38 Martin Jay provides a comprehensive summary of these positions in his narrative. Very generally, optical media became synonymous with a hegemonic visuality, where using these image-making media entailed the submission to a positivist, instrumental rationality. In contrast, touch became idealized as a sense with a set of essential characteristics that allowed it to stand in opposition to the visual; where vision and modernity were understood as coterminous, touch was equated with a premodern or countermodern mode of knowing and relating to the world. See Martin Jay, *Downcast Eyes: The Denigration of Vision in Twentieth-Century French Thought* (Berkeley: University of California Press, 1993), 435–91.

39 Marshall McLuhan, *Understanding Media: The Extensions of Man* (Cambridge, Mass.: MIT Press, 1994), 333.

40 Laura Marks, *Touch: Sensuous Theory and Multisensory Media* (Minneapolis: University of Minnesota Press), 2002, 7.

41 Michael Noll, "Tactile Man–Machine Communication" (PhD dissertation, Polytechnic Institute of Brooklyn, 1971), iv.

42   See Interface 3 for a rough chronicling of this research.

43   Noll, "Tactile Man–Machine Communication," repr. Society for Information Display (1972), 6, http://noll.uscannenberg.org/PDF papers/SID%20Tactile.pdf.

44   Noll, "Tactile Man–Machine Communication" (1971), 29.

45   Noll and other early researchers consistently placed the word "feel" in quotation marks, as if to differentiate this new virtual touch from its earlier counterpart.

46   Ibid., 26.

47   Because the tactile communication facilitated by Noll's device only involved a single point of interaction between the user and the computer-generated object, he argued that it was less demanding than visual displays, which had to render "the complete surface in all its fine and global details." Noll, "Tactile Man–Machine Communication" (1972), 10.

48   Ibid., 11.

49   Noll, "Tactile Man–Machine Communication" (1971), 71.

50   Noll, "Tactile Man–Machine Communication" (1972), 10.

51   Batter and Brooks, "GROPE-1," 759.

52   Frederick Brooks, Jr., "The Computer 'Scientist' as Toolsmith," *IFIP Conference Proceedings* (1977): 629, http://cs.unc.edu/xcms/wpfiles/toolsmith/The_Computer_Scientist_as_Toolsmith.pdf.

53   Batter and Brooks, "GROPE-1," 759.

54   Atkinson et al., "Computing with Feeling," 102–3.

55   Noll dubbed two of these changes "humanistic." Noll, "Tactile Man–Machine Communication" (1972), 10. The first and most significant was touch computing's ability to aid the blind; properly deployed, it would allow them "to feel the shape of graphs and other curves and surfaces and even objects" (ibid.). The second concerned intersubjective tactile communication, where the tactile systems of two remote users could be interfaced, allowing for the experience of a bidirectional distanced touching. On a more pragmatic level, Noll's immediate imagined usage for the joystick was as a mechanism for the virtual prototyping of telephone handsets, which would serve as a way of reducing Bell's development costs. Noll, "Tactile Man–Machine

Communication" (1971), 67. Atkinson et al. suggested an even wider range of potential applications. Taking note of a recent symposium on tactile aesthetics, they speculated that touch computing could pair synergistically with that emerging field to fuel an increased role for touch in aesthetic judgments. In the area of medicine, touch computing could be used to help surgeons learn the feel of organs by touching and handling [computer-generated] anatomical models. By enabling particle forces to be felt, by scaling "the speed of light closer to the hand," touch computing could also push forward new understandings of the physical world. Atkinson et al., "Computing with Feeling," 101.

56   Ibid., 97.

57   Ibid., 101.

58   Noll, "Tactile Man–Machine Communication" (1972), 11.

59   James Carey, "Technology and Ideology: The Case of the Telegraph," in *Communication as Culture: Essays on Media and Society* (New York: Routledge, 2009), 156.

60   Atkinson et al., "Computing with Feeling," 103.

61   Batter and Brooks, "GROPE-1," 760.

62   Atkinson et al., "Computing with Feeling," 103.

63   Larissa Hjorth, "Domesticating New Media: A Discussion on Locating Mobile Media," in *The New Media and Technocultures Reader*, ed. Seth Giddings and Martin Lister (New York: Routledge, 2011), 445.

64   Brooks, "The Computer 'Scientist' as Toolsmith," 631.

65   Ibid.

66   Ken Hillis, *Digital Sensations: Space, Identity, and Embodiment in Virtual Reality* (Minneapolis: University of Minnesota Press, 1999), xxii. Hillis's claims were informed by his experience using the PHANToM (a three-dimensional force-feedback interface, described in detail later in this chapter) and his reading of Margaret Minsky's research.

67   Ibid.

68   Caroline Jones, "The Mediated Sensorium," in *Sensorium: Embodied Experience, Technology, and Contemporary Art,* ed. Caroline Jones (Cambridge, Mass.: MIT Press, 2006), 8.

69   Examining later instantiations of these microtelepresence
     technologies, Sacha Loeve understands the visual and haptic
     representation of microscale objects as implicating these imaging
     practices in political and ethical debates. Sacha Loeve, "Sensible
     Atoms: A Techno-Aesthetic Approach to Representation," *Nanoethics*
     5 (2011): 203–22.

70   Bruno Latour, "Visualization and Cognition: Thinking with Eyes and
     Hands," in *Knowledge and Society: Studies in the Sociology of Culture Past
     and Present,* ed. Robert Alun Jones and Henrika Kuklick, vol. 6
     (Baltimore, Md.: Jai Press, 1986), 10.

71   Paul Jerome Kilpatrick's dissertation, carried out under Brooks's
     supervision, presented the most comprehensive review of the
     psychophysical literature on touch of any experimental era publications
     on force displays. Kilpatrick, "The Use of a Kinesthetic Supplement in
     an Interactive Graphics System," 6–8.

72   It is difficult to pinpoint the exact moment when "haptic" migrated
     to interface design, but it appears to have spiderwebbed out from the
     increasing points of contacts between the cognitive psychologists (who
     used the term frequently) and the computer scientists (who were
     reading their work with increasing interest). Given her direct
     connections to Frederick Brooks, Margaret Minsky, and other leading
     researchers in touch feedback computing, the shift is perhaps most
     attributable to Lederman. Brooks et al. first used the term "haptic
     display" in a 1990 publication updating earlier research on the GROPE
     system. Among the piece's four coauthors was Margaret Minsky,
     whose doctoral thesis at MIT was supervised by Lederman and
     Nicholas Negroponte. Under the heading of haptic displays, they
     grouped "all displays to the haptic senses," differentiating "haptic"
     displays from "tactile" ones by the former's appeal to "the
     propriopositional senses" and the "pressure sense," and the latter's
     appeal to "the sense of touch proper." Frederick Brooks, Ming
     Ouh-Young, James J. Batter, and Paul Jerome Kilpatrick, "Project
     GROPE—Haptic Displays for Scientific Visualization," *Computer
     Graphics* 24, no. 4 (1990): 177–85. Crucially, the term "haptic
     interface" was first proposed by Srinivasan in 1992 and was shared in
     several conference presentations and publications. See, for example, the
     sections on haptic interfaces in Kenneth Salisbury and Mandayam
     Srinivasan, "Virtual Environment Technology for Training," *BBN
     Report 7661* (Cambridge, Mass.: BBN Systems and Technologies,
     1992); Gary Bishop and Henry Fuchs, "Research Directions in Virtual
     Environments: Report of an NSF Invitational Workshop," *Computer
     Graphics* 26, no. 3 (1992): 153–77.

73 See, for example, Ruzena Bajcsy, Susan Lederman, and Roberta Klatzky, "Object Exploration in One and Two Fingered Robots," *Proceedings of the 1987 IEEE International Conference on Robotics & Automation* 3 (Raleigh, NC: IEEE, 1987), 1806–10; Roger Browse and Susan Lederman, "Feature-Based Robotic Tactile Perception," *Proceedings of Compint85, IEEE Conference on Computer Aided Technologies* (Montreal, Queb.: IEEE, 1985), 455–58.

74 Susan Lederman, Roberta Klatzky, and Dianne Pawluk, "Lessons from the Study of Biological Touch for Robotic Haptic Sensing," in *Advanced Tactile Sensing for Robots,* ed. H. R. Nichols (Singapore: World Scientific, 1992), 193.

75 Ibid., 158.

76 Ibid., 151.

77 For example, William Schiff and Emerson Foulke, eds., *Tactual Perception: A Sourcebook* (Cambridge: Cambridge University Press, 1982). This volume gathered together research on tactile and haptic perception. Lederman contributed a chapter to the volume.

78 See the detailed discussion of haptics as the science of touch in Interface 2.

79 I use each of these terms here, and throughout the chapter, as a shorthand for several related fields: in the case of cognitive psychology, these fields include neuroscience, biology, physiology, and experimental psychology; in the case of computer science, these fields include robotics, interface design, information display, and information processing.

80 Iwata, "History of Haptic Interface," 356. Grunwald's volume evidences what has become an inextricable link between the psychophysical study of touch and the design of haptic interfaces. An impressive sourcebook, it positions studies devoted to understanding touch's physiology alongside applications of that knowledge to the design of haptic human–computer systems. Grunwald, *Human Haptic Perception: Basics and Applications.*

81 In the editor's preface to *Tactual Perception,* Schiff and Foulke carefully spelled out the constituencies who would benefit from the research contained in their volume. My claim here is not that, prior to the emergence of haptic interfaces, research into haptic perception lacked utility, but rather that those who studied haptics constantly attempted to legitimate their work by framing it in instrumental terms. See

William Schiff and Emerson Foulke, "Editorial Introduction," in *Tactual Perception,* xi.

82 For one example, see Thomas Sheridan, "Human and Machine Haptics in Historical Perspective," paper presented at HMH: Workshop on Human-Machine Haptics, Pacific Grove, Calif., December 1997, http://www.universelle-automation.de/1961_Boston .pdf.

83 Ibid., 2.

84 In the process of disseminating knowledge about this technology to the nonspecialist, designers, researchers, and marketers gradually got in the habit of shortening the phrase "haptic interfacing" to simply "haptics." Consequently, the two are frequently conflated in popular and even scientific discourses on the subject, as the manufacture of computer-generated haptic sensations becomes coterminous with and inseparable from the scientific study of touch. The Haptics entry on Whatis.com, for example, defines the term as "the science of applying touch (tactile) sensation and control to interaction with computer applications." Margaret Rouse, "What Is Haptics?," http://whatis .techtarget.com/definition/haptics.

85 Srinivasan and Basdogan, "Haptics in Virtual Environments," 393. At the time that the term was adopted by virtual reality engineers and roboticists, psychologists defined haptics as "active touch," and located it as a relatively underexplored area of touch perception in comparison to the more bountiful research on the skin senses. For a summary and speculative explanation for why active touch had gone understudied relative to passive touch, see Joseph Stevens and Barry Green, "History of Research on Touch," in *Pain and Touch,* ed. Lawrence Kruger (San Diego, Calif.: Academic, 1996), 17–18.

86 Srinivasan and Basdogan, "Haptics in Virtual Environments," 393. Publications on haptic interfaces during the 1990s concisely defined the term as a way of introducing it to their audiences, a practice that remains curiously common and suggests that haptics retains its exotic connotation.

87 Howard Rheingold, *Virtual Reality* (New York: Simon & Schuster, 1991), 14. The first chapter of *Virtual Reality* focuses extensively on Brooks and the work carried out at his Chapel Hill laboratory, which Rheingold visited when researching the book. Rheingold also interviewed Margaret Minsky, Japanese roboticist Susumo Tachi, and MIT Touch Lab founder Mandayam Srinivasan; these conversations were positioned alongside Rheingold's colorful and often

philosophical accounts of his own hands–on experiences using the prototypes these researchers were in the process of developing.

88 Rheingold cited those who put forth this formulation and embraced it himself. Rheingold, *Virtual Reality,* 321, 325–26.

89 Though it is worth exploring further, the capacity of robots to feel and touch untethered from human operators is not a subject I take up in this book. By the time Srinivasan cleaved touch research into human, machine, and human–machine haptics in the 1990s, a well-developed body of literature had been devoted to the subject. Many of the engineers who later worked on human–machine haptics started off researching robotic touch, as illustrated by J. Kenneth Salisbury's narration of his pathway "from robot fingers to haptic interface." See J. Kenneth Salisbury, "Making Graphics Physically Tangible," *Communications of the ACM* 42, no. 8 (1999): 75–81.

90 The Touch Lab's homepage clearly broke up publications and projects into the three groupings. See "Touch Lab Research," http://touchlab .mit.edu/oldresearch/index.html.

91 Srinivasan and Basdogan, "Haptics in Virtual Environments," 393, 396. Emphasis original.

92 Ibid.

93 Ibid.

94 Ibid.

95 Sheridan, echoing Geldard's earlier claims about touch's evolutionary primacy among the sense organs, noted the flipped timelines of the biological evolution of the senses and their evolution by media technology: "Haptic capability is primitive on the evolutionary time scale. Early creatures had force sensitive skin or discrete structures such as hairs or antennae. . . . Exteroceptor organs for vision and hearing generally came later in evolution. Curiously, exactly the reverse occurred in human development of sensing technology. Edison is credited with inventing the telephone and gramophone (record player) early in the 1900s. RCA scientists and others are credited with the invention of television in the 1930s. Haptics, though of scientific interest to sensory psychologists and physiologists, did not have an equivalent technological reproduction until very recently." Sheridan, "Human and Machine Haptics in Historical Perspective," 1.

96 Grigore Burdea, *Force and Touch Feedback for Virtual Reality* (New York: Wiley Interscience, 1996), 14–21.

97 Martin Heidegger, *The Question Concerning Technology and Other Essays* (New York: Garland, 1977), 18.

98 M. Cenk Cavusoglu, Frank Tendick, and S. Shankar Sastry, "Haptic Interfaces to Real and Virtual Surgical Environments," in *Touch and Virtual Environments,* ed. Margaret L. McLaughlin, Gaurav Sukhatme, and Jooao Hespanha (Upper Saddle River, N.J.: Prentice Hall, 2001), 219.

99 Iwata, "History of Haptic Interface," 355–61.

100 For a summary of applications across these fields, see Margaret McLaughlin, João Hespanha, and Gaurav Sukhatme, "Introduction to Haptics," in *Touch in Virtual Environments: Haptics and the Design of Interactive Systems* (Upper Saddle River, N.J.: Prentice Hall, 2002), 2–11.

101 Margaret Minsky worked on Atari's *Hard Drivin'* (1989) arcade racing simulator. It was one of the first videogames to incorporate force feedback, and her experience designing the game inspired the Sandpaper project she pursued in her dissertation research. In the late 1990s, force feedback started to become a standard feature in videogame controllers for both consoles and the personal computer. For an overview of these early devices and their manufacturers, see Chuck Walters, "Cop a Feel . . . with Haptic Peripherals," *Gamasutra,* December 19, 1997, http://www.gamasutra.com/view/feature/131653/cop_a_feelwith_haptic_.php.

102 See Interface 5 for an expanded discussion of teledildonics applications.

103 Stiegler, *Technics and Time,* 35.

104 McLaughlin, Hespanha, and Sukhatme surveyed these "issues in haptic rendering," explaining that latency and "haptic data compression" provided two of the most significant limits on the ability to produce realistic, convincing, and engaging haptic simulations. McLaughlin, Hespanha, and Sukhatme, "Introduction to Haptics," 11–18.

105 These include the Messner corpuscles (responsible for sensing heavy pressure), Pacinian corpuscles (vibration), Merkel discs (light touch), and Ruffini endings (sensations of skin stretching). For a detailed summary, particularly concerning the distribution of these receptors in the hand, see Lynette Jones and Susan Lederman, *Human Hand Function* (Oxford: Oxford University Press, 2006), 25–31.

106 Derrida, in a brief examination of contemporary developments in the engineering of touch feedback computing, cited such research as evidence against what he termed the "haptocentric intuitionalism" that leads to the belief in touch's countermediatic qualities. Jacques

Derrida, *On Touching* (Stanford, Calif.: Stanford University Press, 2005), 300.

107 The difference in both cost and accuracy between Novint's Falcon and the PHANToM illustrates this tension—where the PHANToM ranged in cost from $2,400 to $79,500, the Falcon retailed for roughly $200.

108 Iwata's survey catalogs the full range of haptic interfaces built during the epoch, which he divides into three general categories, each developed beginning in the 1990s: 1) exoskeleton systems (for example, the CyberGrasp discussed later in this section), 2) tool-handling-type haptic interfaces (or point-contact systems; see the PHANToM, discussed later in this section), and 3) object-oriented-type haptic interfaces (such as Susumo Tachi's HapticSpace). Iwata, "History of Haptic Interface," 357–61.

109 Frustrated by the constructivist reduction of the body to a function of discourse, Hansen urges a reading of mixed reality that understands it not as an expression of "the autonomy of the technical" (as it appears in Kittler's work), but rather as a demonstration of the way technics can "stimulate or provoke the power of the body open to the world." While I do not contest his understanding of the body as a site of continual negotiation between the transcendental/biological and technics, the history traced throughout *Archaeologies of Touch* shows how touch was gradually reshaped through its contacts with specific institutional and economic imperatives. As a result, it eventually gave rise to haptics as a new paradigm that enabled the positive deployment of technical knowledge about touch. Mark Hansen, *Bodies in Code: Interfaces with Digital Media* (New York: Routledge, 2006), 10.

110 Srinivasan and Basdogan, "Haptics in Virtual Environments," 396.

111 Though Weber separated pain from touch, locating the former in the nebulous category of the "common sensibility," later investigators repositioned it as part of the tactile system. See the summary provided by Joseph Stevens and Barry Green, "History of Research on Touch," in *Pain and Touch,* ed. Lawrence Kruger (San Diego, Calif.: Academic Press, 1996), 1–19.

112 Linking the category up to social norms of performativity, Crowley defines comfort as "the self-conscious satisfaction with the relationship between one's body and its immediate physical environment." John Crowley, *The Invention of Comfort: Sensibilities & Design in Early Modern Britain & Early America* (Baltimore, Md.: Johns Hopkins University Press, 2001), ix.

113 The PainStation, a videogame art installation developed by the artist collective //////////fur////, similarly used a heating element brought into direct contact with its user. However, as suggested by the title of the installation, the PainStation's heating pad aimed at a distracting infliction of pain. See //////////fur////, "PainStation: From Art to Arcade" (2001), http://www.painstation.de/history.html.

114 As some physiological accounts of touch cleave the tactile senses, including the temperature sense, off from haptic perception, locating the VR Thermal Kit as a haptic interface might seem problematic. In spite of the contested boundaries of these submodalities, devices like the VR Thermal Kit were routinely dubbed "haptic interfaces" by their creators, and research findings were frequently published in journals and presented at conferences under the banner of haptics. In the misleadingly titled "Virtual Hell: A Trip through the Flames," Dionisio specifically stated he would, "for now," "avoid the extremes of heaven and hell." Jose Dionisio, "Virtual Hell," *IEEE Computer Graphics and Applications* 7, no. 3 (1997): 11.

115 Gabriel Robles-De-La Torre, "Principles of Haptic Perception in Virtual Environments," in *Human Haptic Perception: Basics and Applications,* 368.

116 Massie and Salisbury, "The PHANToM Haptic Interface: A Device for Probing Virtual Objects," *Proceedings of the ASME Winter Annual Meeting, Symposium on Haptic Interfaces for Virtual and Teleoperator Systems* (New York: American Society of Mechanical Engineers, 1994), 2, https://alliance.seas.upenn.edu/~medesign/wiki/uploads/Courses /Massie94-DSC-Phantom.pdf.

117 Ibid., 1.

118 Ibid.

119 Ibid., 2. A newton is a unit of force that produces an acceleration of one meter per second^2 on a mass of one kilogram.

120 The later addition of a PHANToM-style robot arm dubbed the "CyberForce" to the CyberGrasp gave the device the ability to simulate weight but added cost and bulk to the system.

121 Zhihua Zhou, Huagen Yan, and Shuming Gao, "A Realistic Force Rendering Algorithm for CyberGrasp," *Ninth International Conference on Computer Aided Design and Computer Graphics,* December 7–10, 2005, doi: 10.1109/CAD-CG.2005.13.

122 Robert Lindeman, Robert Page, Yasuyuki Yanagida, and John Sibert, "Towards Full-Body Haptic Feedback: The Design and Deployment of

a Spatialized Vibrotactile Feedback System," in *Proceedings of the ACM Virtual Reality Software and Technology 2004* (New York: Association for Computing Machinery), 146–49.

123 Grigore Burdea, "Keynote Address: The Challenges of Large-Volume Haptics," Virtual Reality International Conferences 2000, Laval, France, May 18–21, 2000.

124 Iwata, "History of Haptic Interface," 355.

5. The Cultural Construction of Technologized Touch

1   Chief technology officer at Immersion Corporation, quoted in an interview with Baratunde Thurston, "Pop Sci's Future of Communication: Haptics," *Popular Science,* http://science.discovery .com/tv-shows/pop-scis-future-of/videos/popscis-future-of-haptics .htm.

2   The trope of dematerialization is common in discussions of transitioning from buttons and keys to the touchscreen. The term, however, is misleading: while the touchscreen may feel like it lacks the materiality of the buttons and keys it displaces, concealed beneath its flat glass surface is a dense series of layers, each of which contains the sensors required to accurately register the user's contact with the screen. The inaccessibility of this materiality to touch does not negate its existence.

3   Maurizio Lazzarato, *Signs and Machines: Capitalism and the Production of Subjectivity* (Los Angeles: Semiotext(e), 2014), 34.

4   Friedrich Kittler, *Gramophone, Film, Typewriter* (Stanford, Calif.: Stanford University Press, 1999), 3.

5   This trial, carried out using two Internet-connected PHANToMs, allowed remote users to collaboratively manipulate a virtual object in a shared workspace. See Jung Kim, Hyun Kim, Boon Tay, Manivannan Muniyandi, and Mandayam Srinivasan, "Transatlantic Touch: A Study of Haptic Collaboration over Long Distance," *Presence* 13, no. 3 (June 2004): 328–37. For further discussion, see Mark Paterson, *The Senses of Touch: Haptics, Affects, and Technologies* (New York: Berg Press, 2007), 140–43.

6   For a comprehensive list of devices available in the late 1990s along with an explanation of their accompanying rendering software, see Chuck Walters, "Cop a Feel . . . with Haptic Peripherals," *Gamasutra,*

December 19, 1997, http://www.gamasutra.com/view/feature/3245
/cop_a_feelwith_haptic_.php?print=1.

7   Starting in the early 2000s, edited volumes, special issues, and
     published conference proceedings that gathered together recent
     research in haptics became increasingly common. Monographs
     intended to introduce aspiring haptic interface designers to the
     fundamentals of the field also provided discursive sites for the
     formation and concretization of standardized design practices. During
     the ten-year period preceding the publication of the inaugural issue of
     the *IEEE Transactions on Haptics* in 2008, there were more than forty
     special journal issues devoted to the subject. For further detail, see
     J. Edward Colgate, "Editorial," *IEEE Transactions on Haptics* 1, no. 1 (2008):
     2. Examples of edited volumes include Margaret McLaughlin, João
     Hesphana, and Gaurav Sukhatme, *Touch in Virtual Environments: Haptics
     and the Design of Interactive Systems* (Upper Saddle River, N.J.: Prentice
     Hall, 2002); Ming Lin and Miguel Otaduy, eds., *Haptic Rendering:
     Foundations, Algorithms, and Applications* (Wellesley, Mass.: A. K. Peters,
     2008); Antonio Bicchi, Martin Buss, Marc Ernst, and Angelika Peer,
     eds., *The Sense of Touch and Its Rendering: Progress in Haptics Research*
     (Berlin: Springer-Verlag, 2008).

8   This software is still available, though many of the haptic mice the
     software controls are incompatible with current generation operating
     systems. For a comprehensive list of devices, see http://www.ifeelpixel
     .com/download/mice.htm.

9   For one overview of possibilities for integrating robust haptic
     interfaces in museum displays, see David Prytherch and Marie
     Jefsioutine, "Touching Ghosts: Haptic Technologies in Museums," in
     The Power of Touch: Handling Objects in Museum and Heritage Contexts,
     ed. Elizabeth Pye (Walnut Creek, Calif.: Left Coast Books, 2007),
     223–40. Fiona Candlin suggests that an aesthetics activated by direct
     tactile contact with art works can counteract the hegemony of vision.
     Fiona Candlin, *Art, Museums, and Touch* (Manchester, UK: Manchester
     University Press, 2010).

10   For a critical engagement with the medicalization of phantom
     vibrations, see David Parisi, "Banishing Phantoms from the Skin:
     'Vibranxiety' and the Pathologization of Interfacing," *Flow* 17, no. 5
     (2013), http://www.flowjournal.org/2013/01/
     banishing-phantoms-from-the-skin/.

11   Executives from both Sony and Microsoft participated in rumble's
     denigration, with Sony's Phil Harrison declaring rumble to be a
     "last-generation" feature in 2007, and Microsoft's Kudo Tsunoda

describing rumble as a "rudimentary form of haptic feedback."
See Ludwig Kietzman, "Sony's Phil Harrison: No Pressure to Drop
PS3 Price," *Joystiq,* February 26, 2007, http://www.joystiq.
com/2007/02/26/sonys-phil-harrison-no-pressure-to-drop-ps3-price;
Edge Staff, "Kudo Tsunoda: Rumble Is Rudimentary," *Edge Online,*
http://www.edge-online.com/features/kudo-tsunoda-rumble
-rudimentary/.

12   Robert Jütte, *A History of the Senses: From Antiquity to Cyberspace*
(Malden, Mass.: Polity, 2005), 238. Jütte took the possibility of a
"haptic age" from an unnamed German writer for *Stuttgarter
Nachrichten.* Jütte later revisited this question in his contribution to
Human Haptic Perception, where his historical account of touch was
juxtaposed with chapters that chronicled recent developments in haptic
interfacing and the neurophysiology of haptic perception. Robert
Jütte, "Haptic Perception: A Historical Approach," in *Human Haptic
Perception: Basics and Applications,* ed. Martin Grunwald (Boston:
Birkhäuser, 2008).

13   Michael Arlen, *Thirty Seconds* (New York: Macmillan, 1980). As
Claude Fischer explains, the formulation of telephonic communication
as touch had a long history in the discourse around telephony that
predates the AT&T campaign. Claude Fischer, " 'Touch Someone':
The Telephone Industry Discovers Sociability," *Technology and Culture*
29, no. 1 (1988): 32–61.

14   The figuration of touch as a form of affective contact has a complex
history, owing in part to etymological linkages in the German
language between feeling (*Gefühl*) and emotion (*Empfindung*). For
further discussion, see Claudia Benthien, *Skin: On the Cultural Border
Between Self and World* (New York: Columbia University Press, 2004),
185–88.

15   Nadia Seremetakis, "The Memory of the Senses: Historical
Perception, Commensal Exchange, and Modernity," *Visual
Anthropology Review* 9, no. 2 (1993): 2.

16   The narrative in these ads is reminiscent of Norbert Elias's argument
about the civilizing process, which Elizabeth Harvey, through a
haptocentric reading of Elias, applies specifically to touch. Civilizing
becomes a constant process of tactile disengagement from the
materiality of the world. This withdrawal of the body participates in
the relocation of the physical qualities of touch from the body into the
category of affect. Elizabeth Harvey, "Introduction: The Sense of All
Senses," in *Sensible Flesh,* ed. Elizabeth Harvey (Philadelphia:
University of Pennsylvania Press, 2003), 9, 12–13.

17   "Touching Is Good," print ad for the Nintendo DS portable gaming system, http://Touchingisgood.com.

18   Nadia Seremetakis, "The Memory of the Senses," in *In The Senses Still: Perception and Memory as Material Culture in Modernity,* ed. Nadia Seremetakis (Chicago: University of Chicago Press, 1994), 3.

19   Paul Virilio, *War and Cinema: The Logistics of Perception* (New York: Verso, 1989), 24.

20   Ibid.

21   Charting the range of articulated responses to the campaign, Heidi Campbell and Antonio La Lastina link the debate around the image to questions concerning the cultural relationship between technology and the divine more broadly. Heidi Campbell and Antonio La Lastina, "How the iPhone Became Divine: New Media, Religion and the Intertextual Circulation of Meaning," *New Media & Society* 12, no. 7 (2010): 1191–207. Brett Robinson describes consistent mobilization of religious themes in Apple's design and marketing practices, situating the Touching Is Believing campaign as part of a spiritual ethos and aesthetics that animated the company's history. Brett Robinson, *Appletopia: Media Technology and the Religious Imagination of Steve Jobs* (Waco, Tex.: Baylor University Press, 2013).

22   In contrast, the Finnish mobile phone company Nokia had long featured hands that were reaching out to touch in its corporate iconography (frequently accompanied by the "Connecting People" slogan the company adopted in 1992), building on AT&T's metaphor of telephonic touch as a means of affective contact.

23   Nanna Verhoeff's compelling analysis of mobile touchscreen interfacing resists making the history of touch subservient to the history of vision, but it also proceeds from a theoretical framework informed primarily by visual and screen studies. Verhoeff embraces a theory of tactile seeing that weaves a new logic of touching into the act of screenic vision and by doing so, invites those in screen studies to consider the reconfigured relationship between finger and eye prompted by the embrace of touchscreens. See Nanna Verhoeff, *Mobile Screens: The Visual Regime of Navigation* (Amsterdam: Amsterdam University Press, 2012), 20.

24   For example, Lev Manovich's foundational *Language of New Media* (Cambridge, Mass.: MIT Press, 2001) reduced the history of new media to an operation of cinema's history, and consequently it emphasized the reconfiguration of the screen over and above the more general rearranging of the body accomplished by computing.

25 Nanna Verhoeff and Heidi Rae Cooley, "The Navigational Gesture: Traces and Tracing at the Mobile Touchscreen Interface," *NECSUS: European Journal of Media Studies* (Spring 2014), http://www.necsus-ejms.org/navigational-gesture-traces-tracings-mobile-touchscreen-interface/.

26 See, for example, Hanna Rosin, "The Touch-Screen Generation," *Atlantic Monthly,* April 2013.

27 Heidi Rae Cooley, "It's All About the *Fit*: The Hand, the Mobile Screenic Device and Tactile Vision," *Journal of Visual Culture* 3, no. 2 (2004): 137; Verhoeff, *Mobile Screens*; Nanna Verhoeff, "Theoretical Consoles: Concepts for Gadget Analysis," *Journal of Visual Culture* 8, no. 3 (2009): 279–98.

28 David Parisi, "A Counterrevolution in the Hands," *Journal of Games Criticism* 2, no. 1, http://gamescriticism.org/articles/parisi-2–1/.

29 Owing to the popularity of smartphones and tablets, straightforward technical accounts of the touchscreen abound. For a critical media archaeological treatment, see Wanda Strauven, "The Archaeology of the Touchscreen," *Maske und Kothurn* 58, no. 4 (2012): 69–79.

30 "The social and cultural construction of the personal computer user," as Bardini argues, "has led to an overwhelming hegemony of the visual sense" that the touchscreen intensifies rather than interrupts. Thierry Bardini, *Bootstrapping: Douglas Engelbart, Coevolution, and the Origins of Personal Computing* (Stanford, Calif.: Stanford University Press, 2000), 231.

31 I am not suggesting here that the touchscreen has erased the importance of the keyboard–mouse interface. Rather, the move to touchscreens has created an awareness of the physicality of interfaces, prompting new demand for high-end, consumer-grade mechanical keyboards differentiated by the tactile feedback of keystrokes, as well as expensive computer mice whose selling point is their feel in the user's hand.

32 Zoe Sofia, "Container Technologies," *Hypatia* 15, no. 2 (2000): 181–201.

33 Immersion Corporation Chief Technology Officer Christophe Ramstein, quoted in Thurston, "Pop Sci's Future of Communication: Haptics."

34 For background on the use of haptic feedback in mobile phones, see Immersion Corporation, *The Value of Haptics* (San Jose, Calif.: Immersion Corporation, 2010).

35 A market research report conducted in 2013 projects the market for haptics to swell from its 2012 value of $842 million to $13.8 billion by 2025, driven primarily by the widespread adoption of piezoelectric actuators used to produce haptic feedback in mobile communication devices. Jonathan Melnick and Michael Holman, "Getting Back in Touch with Electronics: Finding Opportunity in Emerging Haptics" (New York: Lux Research, 2013), 2.

36 Robyn Schwartz, "IBM 5 in 5 2012: Touch," December 17, 2012, http://ibmresearchnews.blogspot.com/2012/12/ibm-5-in-5–2012 -touch.html.

37 Immersion's partnership with Microsoft during the mid-1990s to develop DirectX-compatible force feedback joysticks and steering wheels was almost solely responsible for the incorporation of haptic feedback in PC games. DirectX protocols, a collection of application program interfaces designed by Microsoft primarily for processing videogame-related multimedia content on its Windows operating system, helped ensure that the game could successfully communicate force sensations from the virtual environment to the motors in the joystick that communicated force sensations to the player's hands. The development of standard protocols for sending touch information from software to hardware was crucial to the spread of early haptic interface applications, as these protocols served to inscribe the specificity of both sensory feedback channels and corporate partnerships in the materiality of code. Microsoft Corporation, "Microsoft and Immersion Continue Joint Efforts to Advance Future Development of Force Feedback Technology," Microsoft News Center, February 3, 1998, http://news.microsoft.com/1998/02/03 /microsoft-and-immersion-continue-joint-efforts-to-advance-future -development-of-force-feedback-technology/.

38 See Immersion Corporation, "Immersion. We Are Haptics," http:// www.immersion.com/about/.

39 The most notable and protracted of Immersion's many patent infringement lawsuits involved the rumble feedback mechanism used in the controllers for Microsoft's Xbox and Sony's PlayStation consoles, where Immersion used favorable court rulings to pressure the companies into licensing agreements. The 2003 settlement between Immersion and Microsoft involved a $26 million payout from the latter to the former, in addition to Microsoft acquiring a 10 percent stake in Immersion and licensing rights for its patents. In contrast, the suit with Sony dragged out over a five-year period, with Immersion winning court battles at several levels of appeal, including an injunction that would have prevented Sony from selling its PlayStation 2 consoles in

the United States. Sony eventually settled in 2007 after initially opting
to leave rumble out of its third-generation PlayStation controller.
Faced with growing pressure from gamers disappointed with the lack
of rumble, and embroiled in a public battle with Immersion CEO
Victor Viegas about the technical feasibility of combining rumble with
motion sensing, Sony grudgingly entered into a partnership with
Immersion that allowed them to release a new, rumble-enabled version
of the PlayStation 3 controller. Robert Guth, "Microsoft to Settle
Immersion Lawsuit," *Wall Street Journal,* July 29, 2003, http://www
.wsj.com/articles/SB10594431058937200; Chase Murdey, "Ready to
Rumble? Immersion's Victor Viegas on PlayStation 3's Lack of
Vibration," *Gamasutra*, May 17, 2007, http://www.gamasutra.com
/view/feature/2693/ready_to_rumble_immersions_.php; Immersion
Corporation, "Immersion and Sony Computer Entertainment
Conclude Litigation and Enter into Business Agreement," March 1,
2007, http://immr.client.shareholder.com/ReleaseDetail.cfm
?ReleaseID=232152.

40 For a full list, see Immersion Corporation, "US Patents as of
September 1, 2015," http://www.immersion.com/haptics-technology
/patents/.

41 In addition to Sony and Microsoft, Immersion also filed patent
infringement lawsuits against Motorola, HTC, and Apple. The list of
companies Immersion has entered into licensing agreements with is a
who's who of major technology firms, including LG Electronics,
Samsung, Volkswagen, Microsoft, Sony, Nokia, and Logitech. For a
detailed discussion of Immersion's aggressive patent protection
strategy, see Kristen Osenga, "Formerly Manufacturing Entities:
Piercing the 'Patent Troll' Rhetoric," *Connecticut Law Review* 47, no. 2
(December 2014): 460–62.

42 Immersion Corporation, *Haptics in Touchscreen Handheld Devices*
(San Jose, Calif.: Immersion Corporation, 2012), 9.

43 Figures are taken from VGChatz, "Platform Totals" (2015), http://
www.vgchartz.com/analysis/platform_totals/.

44 Immersion Corporation, "What Is Haptics?" (2002), www.immersion
.com/corporate/press_room/what_is_haptics.php.

45 Emphasis added. Immersion Corporation, "Haptics Glossary" (2008),
http://web.archive.org/web/20090101133958/http://www.immersion
.com/corporate/haptics/glossary.php.

46 As Rachel Plotnick suggests in her examination of the electric
pushbutton's early history, this notion of a moral panic around the

point of material and tactile contact between humans and machines was not a new one. See Rachel Plotnick, "At the Interface: The Case of the Electric Push Button, 1880–1923," *Technology and Culture* 43, no. 4 (2012): 815–45.

47   Immersion Corporation, "Harnessing Human Touch" (San Jose, Calif.: Immersion Corporation, 2009), 3.

48   Emphasis original. Immersion Corporation, "Harnessing Human Touch."

49   Ibid., 3. As with electrotherapy in the late nineteenth century and Geldard's failed Vibratese system from the 1950s, machinic touch became the way to restore vital energies depleted by the overtaxing burdens that modern life had placed on the sensorium.

50   Ibid., 2.

51   Ibid., 5.

52   The technical system is in an age of "perpetual transformation and structural instability," with the enmeshing of the human in this system prompting a reconsideration of the empirical relationship between humans and technics. Bernard Stiegler, *Technics and Time, 1: The Fault of Epimetheus* (Stanford, Calif.: Stanford University Press, 1998), 70, 43.

53   T. Jackson Lears, "From Salvation to Self-Realization: Advertising and the Therapeutic Roots of Consumer Culture, 1880–1930," in *The Culture of Consumption: Critical Essays in American History, 1880–1980* (New York: Pantheon Books, 1983), 11.

54   This phrase was not, of course, original to Immersion. By the time it first used the phrase in "Harnessing Human Touch," it was already becoming a buzzword in marketing literature, serving as a way to describe a new focus on the lived experiences of consumption. The concept also intersects with the growing focus among advertisers in the early part of this century on sensory (or multisensory) marketing, which took aim at passing advertising messages through a broader range of senses than ads conventionally targeted at the audiovisual senses. For further details, see David Howes, "Hyperesthsia, or The Sensual Logic of Late Capitalism," in *Empire of the Senses,* ed. David Howes (Oxford: Berg Press, 2004), 281–303; David Howes, "The Race to Embrace the Senses in Marketing: An Ethnographic Perspective," lecture delivered at the Ethnography Praxis in Industry Conference (London, 2013), http://onlinelibrary.wiley.com/doi /10.1111/j.1559–8918.2013.00002.x/pdf.

55   See, for example, Langdon Winner, *The Whale and the Reactor: A Search for Limits in an Age of High Technology* (Chicago: University of Chicago

Press, 1986); Fred Turner, *From Counterculture to Cyberculture: Stewart Brand, the Whole Earth Network, and the Rise of Digital Utopianism* (Chicago: University of Chicago Press, 2006).

56 For example, Immersion designed its TouchSense 3000 software to control a single actuator or motor, limiting the range of sensations the device could provide. Subsequently, the TS 4000 could control up to sixteen motors or actuators simultaneously, exponentially expanding the possibilities for coding and generating haptic effects.

57 Howes, "Hyperaesthsia," 287.

58 Immersion Corporation, "Immersion and LeTV to Launch Tactile Video Experience," press release, February 17, 2015, http://ir .immersion.com/releasedetail.cfm?releaseid=896583.

59 The term is commonly attributed to Nelson's sprawling manifesto *Computer Lib/Dream Machines,* but while Nelson used "dildonics," he did not affix "tele" to it in his original work. The single, densely packed page of text Nelson devoted to "body electronics" shows that the designs of early remote manipulation systems were central to informing his imagination of the technology. He also drew inspiration from a patent by San Francisco hacker How Wachspress that proposed the Auditac—a device for transposing sound into touch, unintentionally similar in operation to Gault's Teletactor from fifty years earlier. However, the Auditac functioned as what Wachspress described at one point as a "Radio Dildo"; it depended on translating sounds into tactile stimulation, rather than transmitting feel through feel. See Ted Nelson, *Computer Lib/Dream Machines* (self-published 1974). The popular adoption of the term shot up drastically after Rheingold began using it in the 1990s.

60 Howard Rheingold, *Mondo 2000,* 2 (Summer 1990): 52–54. This article was soon reprinted in Howard Rheingold, "Teledildonics and Beyond," in *Virtual Reality* (New York: Simon & Schuster, 1991). Later the article appeared in *Net.seXXX: Readings on Sex, Pornography, and the Internet,* ed. Dennis Waskul (New York: Peter Lang, 2004), 319–22.

61 Rheingold, *Virtual Reality,* 348.

62 Several excellent Web resources engage in precisely this sort of cataloging. Since 2005, Kyle Machulis has written about teledildonics and cybersex on his Slashdong Web site (recently renamed Metafetish) at https://www.metafetish.com/about/. Machulis positions electronically controlled sex toys as part of maker culture more generally and is himself active as a hacker, modder, and advocate for the technology. While his engagements with teledildonics oscillate

between playful and sober, they are always technically adept and cognizant of the gender politics invoked by the machines.

63   For cyberutopians, teledildonics—like other virtual reality technologies—promised to bring about a host of liberatory ends. By freeing users from the constraints of immediacy, these touch machines would allow distant bodies to be brought into pleasurable contact. Performed behind the shielding provided by the teledildonic interface, sex work would become safer, more humane, and dignified. Particularly given the growing concerns over the spread of HIV when Rheingold published his *Mondo 2000* essay, fully embodied sexual encounters free from the risk of disease transmission offered an appealing path toward safer sex, with the interface serving as a sort of virtual prophylactic—allowing pleasurable sensations to pass through the interface, while shielding the user from the contamination threatened by bodily contact. But where the utopians saw promise, cyberdystopians saw peril. The ability to remotely manipulate another's sexual organs threatened to further the dehumanization of social relations perpetuated by technologies of mediated communication. Rheingold identified these ethical concerns in *Virtual Reality,* and they persist unresolved today. For one echo of this longstanding concern over the threat of computer-mediated sexual touch to replace and alienate users from their lived bodies, see the philosopher Richard Kearney's claim that "digital eros may be removing us further from the flesh." Kearney, "Losing Our Touch," *New York Times,* August 30, 2014, http://opinionator.blogs.nytimes .com/2014/08/30/losing-our-touch/?_r=0. In spite of their differences, both sides embraced a deterministic perspective on teledildonics that assumed its ascent would inevitably accompany the proliferation of networked computing. Popular culture representations of teledildonics only served to cement the deterministic imagination. For example, the 1992 film *The Lawnmower Man* depicted a full-body, neoprene cybersuit that allowed its wearer to be fully embodied both visually and haptically in virtual space. In one of the more famous episodes in the history of teledildonics, the adult film company Vivid Entertainment built and attempted to commercialize a similar suit in 1999, only to abandon the project after it failed to pass Federal Communications Commission review.

64   As it was in early VR systems, synchronizing the streams of data for the eyes, ears, and skin was key to constructing a convincingly immersive simulation. Initially, the RealTouch used two distinct streams—one for the conventional audiovideo file and another with haptic data—but network latency issues caused the haptic stream to lag behind the audiovisual one. Their solution was to combine the data

together in one stream at the source, using a proprietary streaming system to ensure a seamless coupling of sensations. See Daniel Cooper, "Adult Themes: The Story behind the Rise and Fall of America's First Digital Brothel," *Engadget,* February 26, 2015, http://www.engadget .com/2015/02/26/adult-themes-digital-brothel/.

65   "Haptic Technology Product Delivers Piracy Proof Experience," *Business Wire,* February 12, 2010, http://www.businesswire.com /news/home/20100212006285/en/AEBN-CEO-Scott-Coffman -Groundbreaking-RealTouch-Product.

66   Cooper, "Adult Themes."

67   See *Sex/Now,* directed by Chris Moukarbel (HBO: 2014).

68   Cooper, "Adult Themes."

69   "RealTouch, How It Works," http://web.archive.org/ web/20120419150349/http://www.realtouch.com/device. The kinship between this definition and the one provided in Immersion's promotional materials is likely attributable to Ramon Alarcon, whose seven-year career with Immersion began in 1994, shortly after the company's founding. During his time as director of Immersion's gaming and entertainment business unit, he was responsible for bringing over twenty gaming-related peripherals to market, and oversaw the establishment of licensing agreements with Immersion's many corporate partners.

70   RealTouch Homepage, http://web.archive.org/web/20130719075726/ http://www.realtouch.com/device.

71   Moukarbel, *Sex/Now.*

72   The positioning of sex work as a form of affective labor is taken up by Melissa Ditmore, who builds on Maurizio Lazzarato to suggest that the concept of affective labor provides a means of capturing the value of work not traditionally considered as such. The fundamental question Ditmore grapples with concerns the potential of tactical organization to empower female sex workers, as they struggle for legitimacy and agency in their labor. Melissa Ditmore, "In Calcutta, Sex Workers Organize," in *The Affective Turn: Theorizing the Social,* ed. Patricia Clough and Jean Halley (Durham, N.C.: Duke University Press, 2007).

73   Cited in Cooper, "Adult Themes."

74   In Zabet Patterson's early analysis of pornography's transformation by its migration online, she suggests that Internet-based cybersexual labor will exist in a constant state of transformation as new technologies of

interfacing change the relationship between remote subjects. Those studying online pornography thus need to attend to the "material specificity and the embodied cultural history of this particular interface." See Zabet Patterson, "Going On-Line: Consuming Pornography in the Digital Era," in *Porn Studies,* ed. Linda Williams (Durham, N.C.: Duke University Press, 2004), 119.

75   The material interface between these devices and the male body varied greatly, but they shared the common purpose of restoring potency to the depleted male organ. See Carolyn Thomas de la Pena, *The Body Electric: How Strange Machines Built the Modern American* (New York: New York University Press, 2003), 139–56.

76   See for instance, EJ Dickson, "The Future of Sex Toys Is Tied Up in Patent Hell," *Daily Dot,* August 27, 2014, http://www.dailydot.com /technology/sex-toy-teledildonics-patent/. (Please note that the name EJ Dickson, the author of this piece, refers to the writer and editor, whereas the EJ referred to throughout the rest of this chapter is the pseudonym that the former RealTouch product manager uses in press interviews and other public venues.) Machulis also chronicles these issues extensively on the Metafetish Web site.

77   Allison Okamura directs the CHARM (Collaborative Haptics and Robotics in Medicine) Lab at Stanford and teaches a wide range of courses in haptic interface design.

78   Rachel Maines, *The Technology of the Orgasm: "Hysteria," the Vibrator, and Women's Sexual Satisfaction* (Baltimore, Md.: Johns Hopkins University Press, 1999), 66.

79   Jackie Cohen, "What Ever Happed to Cyber Dildos?" *AVN Online* 111 (February 2003): 118–22.

80   Jeffrey Bardzell and Shaowen Bardzell, " 'Pleasure Is Your Birthright': Digitally Enabled Designer Sex Toys as a Case of Third-Wave HCI," Computer–Human Interaction 2011, May 7–12, 2011, Vancouver, BC, 3, http://dl.acm.org/citation.cfm?id=1978979.

81   According to Internet lore, the Trance Vibrator peripheral for the PlayStation 2 game *Rez* was quickly repurposed as a tool for sexual stimulation. In another semifamous incident, enterprising modders routed signals that control the two rumble motors in an Xbox so that they would instead modulate the motor in a sex toy. Machulis even provided instructions for this "stupidly easy" mod at his Web site. See "SeXBox—Using Force Feedback Signals for Sex Toys," February 24, 2005, https://www.metafetish.com/2005/02/24/sexbox -using-force-feedback-signals-for-sex-toys/.

82   Bardzell and Bardzell, " 'Pleasure Is Your Birthright,' " 4.

83   John Durham Peters, *Speaking into the Air: A History of the Idea of Communication* (Chicago: University of Chicago Press, 1999), 269–70.

84   Some of the circuit-bending practices involved in repurposing the Falcon interface share a kinship with those used to reanimate "zombie media." But ultimately, Garnet Hertz and Jussi Parikka encourage us to refuse the metaphor, suggesting instead that "media never dies: it decays, rots, reforms, remixes, and gets historicized, reinterpreted, and collected." See Garnet Hertz and Jussi Parikka, "Zombie Media: Circuit Bending Media Archaeology into Art Practice," *Leonardo* 45, no. 5 (2012): 430.

85   For an expanded analysis of the Falcon's rise and fall, see David Parisi, "Reach In and Feel Something: On the Strategic Reconstruction of Touch in Virtual Space," *Animation: An Interdisciplinary Journal* 9, no. 2 (2014).

86   In its promotional video, FriXion stresses the inclusivity of the new platform, claiming that its "advanced peripherals can be configured for any device or orientation," and promising "an unprecedented opportunity for sexual empowerment, freedom, security, and access." This focus on openness is also reflected in the network's compatibility with a wide range of control and manipulator devices. See "The FriXion Revolution," https://www.youtube.com/watch?list=SPIAYjV uUOsYLHKThbB06oLduazoOa_PZG&v=haBM4GFu9Bs.

87   Although no longer in production, the Falcon can be purchased on the secondhand market for less than $200; the Force Dimension haptics devices the Falcon derived from—versions of which are used by NASA for telepresent surgery—sell for more than $20,000.

88   Caroline Jones, "The Mediated Sensorium," in *Sensorium: Embodied Experience, Technology, and Contemporary Art,* ed. Caroline Jones (Cambridge, Mass.: MIT Press, 2006), 11.

89   Peters, *Speaking into the Air,* 271.

90   Apple visually represents the Taptic Engine in print ads for the Apple Watch and in a video of the iPhone 6s. At one point, an animated dissembling of the device (narrated by Apple Chief Design Officer Jony Ive) shows where the Engine is housed in the machine and visualizes its operation. See "Introducing the iPhone and iPhone 6 Plus with 3D Touch," https://www.youtube.com/ watch?v=cSTEB8cdQwo.

91 "iPhone 6s—3D Touch," https://www.youtube.com/watch?v =bdg7iEiXQAg.

Coda

1 In its development documentation, Oculus recommends only minimal use of the Touch's haptic feedback system, cautioning that extensive activation of the haptic feedback motors can cloud the device's position-sensing capability: "Prolonged high levels of vibration may reduce positional tracking quality. Right now, we recommend turning on vibration only for short periods of time." See "Haptic Feedback," https://developer.oculus.com/documentation/pcsdk/latest/concepts /dg-input-touch-haptic/.

2 Though he did not reference Sutherland's address to the International Federation for Information Processing, Abrash's keynote followed its template nicely, walking through each of the five major senses and describing the potential of extant computing technologies to act on them, just as Sutherland had done fifty years prior.

3 Michael Abrash, "Oculus Connect 2 Keynote," October 3, 2015, https://www.youtube.com/watch?v=tYwKZDpsjgg.

4 William Atkinson, Karen Bond, Guy Tribble III, and Kent Wilson, "Computing with Feeling," *Computers & Graphics* 2, no. 2 (1977): 101.

5 The idea of touch as irreducible is central to the account of communication provided by John Durham Peters, where he claims that touch provides a guarantee of sincerity that is impossible to fake. John Durham Peters, *Speaking into the Air: A History of the Idea of Communication* (Chicago: University of Chicago Press, 1999), 270–71. The theme has also reemerged in Sherry Turkle's recent findings based on her research into media and teen sociability. Sherry Turkle, *Alone Together: Why We Expect More from Technology and Less from Each Other* (New York: Basic Books, 2011).

6 This is already happening in the area of networked sex toys. At the 2016 hacker conference Def Con, the hackers goldfisk and follower showed how to access user data from the We Vibe Plus, an Internet-connected, app-controlled vibrator developed by We Vibe as a way to help enhance feelings of intimacy between romantic partners who were physically distant. See Alex Hern, "Someone Made a Smart Vibrator, So of Course It Got Hacked," *The Guardian,* August 10, 2016, https://www.theguardian.com/technology/2016/aug/10 /vibrator-phone-app-we-vibe-4-plus-bluetooth-hack.

7   Howard Rheingold, *Virtual Reality* (New York: Simon & Schuster, 1991), 352.

8   Chris Salter, "Haptic Field" (2016), http://chrissalter.com/projects/haptic-field-2016/.

9   Benjamin Peters, "And Lead Us Not into Thinking the New Is New: A Bibliographic Case for New Media History," *New Media & Society* 11, nos. 1, 2 (2009): 13–30.

10   Even today, with the maturation of computer haptics into an institutionalized discipline, this feeling persists among many who work on touch-communication systems. This shared sense of marginalization and isolation helps provide the discipline with a definitional habitus, as the refrain of touch's neglect is echoed throughout introductory essays to dedicated anthologies and journal issues.

11   Particularly in the area of game studies, touch has attracted considerable attention over the past decade, with a substantial upward spike following Nintendo's release of the Wii in 2007. See especially Martti Lahti, "As We Become Machines: Corporealized Pleasures in Video Games," in *The Video Game Theory Reader,* ed. Mark Wolf and Bernard Perron (New York: Routledge, 2003); Graeme Kirkpatrick, "Controller, Hand, Screen: Aesthetic Form in the Computer Game," *Games and Culture* 4, no. 2 (2009): 127–43; Bryan Behrenshausen, "Toward a (Kin)Aesthetic of Video Gaming: The Case of *Dance Dance Revolution,*" *Games and Culture* 2, no. 4 (2007): 335–54; Tom Apperley, *Gaming Rhythms: Play and Counterplay from the Situated to the Global* (Amsterdam: Institute of Network Cultures, 2010); Brendan Keogh, "Across Worlds and Bodies: Criticism in the Age of Video Games," *Journal of Games Criticism* 1, no. 1 (2014); Peter McDonald, "On Couches and Controllers: Identification in the Video Game Apparatus," in *Ctrl-alt-play: Essays on Control in Video Gaming,* ed. Matthew Wysocki (Jefferson, N.C.: McFarland, 2013); Nadav Lipkin, "Controller Controls: Haptics, Ergon, Teloi and the Production of Affect in the Video Game Text," in Wysocki, *Ctrl-alt-play.*

12   The terminological problem W. J. T. Mitchell raises around the terms "image media" and "sound media" proves particularly vexing with the designation "haptic media." For Mitchell, defining a medium by its targeting of a single sense modality is as much ideological and normative as it is descriptive, performing a sort of "sensory hygiene" that forecloses consideration of the way media invoke a mixing of sensory modalities. Mitchell suggests that what we really mean when we designate a medium by reference to a unitary modality is

predominance rather than purity. As "all media are, from the standpoint of sensory modality 'mixed media,'" Mitchell prompts us to consider the McLuhanesque *sense ratios* involved in using a given medium. Media always invoke a blending—or what Mitchell terms a "braiding" or layering—of the senses. I keep the term, in spite of these concerns, as a way of calling attention to the hierarchies it implies. W. J. T. Mitchell, "There Are No Visual Media," *Journal of Visual Culture* 4, no. 2 (2005): 257–66. Quotes are from 259, 257.

Index

Page numbers in italics refer to illustrations.

David Parisi is associate professor of emerging media in the Department of Communication at the College of Charleston. He is coeditor of the *New Media & Society* special issue on Haptic Media Studies.